Classic Problems
of Probability

Classic Problems of Probability

Prakash Gorroochurn
Department of Biostatistics, Columbia University, New York, NY

WILEY

A JOHN WILEY & SONS, INC., PUBLICATION

Library of Congress Cataloging-in-Publication Data:

Gorroochurn, Prakash, 1971-
    Classic problems of probability / Prakash Gorroochurn.
        p. cm.
    Includes bibliographical references and index.
    ISBN 978-1-118-06325-5 (pbk.)
1.   Probabilities–Famous problems. 2.   Probabilities–History. I. Title.
    QA273.A4G67 2012
    519.2–dc23
                                                2011051089
Printed in the United States of America

10   9   8   7   6   5

To Nishi and Premal

operae pretium est

Contents

Preface

$\mathbf{H}$aving taught probability for the last 12 years or so, I finally decided it was time to put down my "experiences" with the subject in the form of a book. However, there are already so many excellent texts in probability[†] out there that writing yet another one was not something that really interested me. Therefore, I decided to write something that was a bit different.

Probability is a most fascinating and unique subject. However, one of its uniquenesses lies in the way common sense and intuition often fail when applied to apparently simple problems. The primary objective of this book is to examine some of the "classic" problems of probability that stand out, either because they have contributed to the field, or because they have been of historical significance. I also include in this definition problems that are of a counterintuitive nature. Not all the "classic" problems are old: *Problem 33: Parrondo's Perplexing Paradox*, for example, was discovered as recently as 1996. The book has considerable coverage of the history of the probability, although it is not a book on the history of the subject. The approach I have adopted here is to try to offer insights into the subject through its rich history. This book is targeted primarily to readers who have had at least a basic course in probability. Readers in the history of probability might also find it useful.

I have worked hard to make the presentation as clear as possible so that the book can be accessible to a wide audience. However, I have also endeavored to treat each problem in considerable depth and have provided mathematical proofs where needed. Thus, in the discussion of *Problem 16: The Buffon Needle Problem*, the reader will find much more than the conventional discussion found in most textbooks. I discuss alternative proofs by Joseph Barbier, which lead to more profound and general results. I also discuss the choice of random variables for which a uniform distribution is possible, which then naturally leads to a discussion on invariance. Likewise, the discussion of *Problem 19: Bertrand's Chords* involves much more than stating there are three well-known possible solutions to the problem. I discuss the implications of the indeterminacy of the problem, as well as the contributions made by Henri Poincaré and Edwin Jaynes. The same can be said of most of the problems discussed in the book. The reader will also find treatments of the limit and central theorems of probability. My hope is that the historical approach I have adopted will make these often misunderstood aspects of probability clearer.

Most of the problems are more or less of an elementary nature, although some are less so. *The reader is urged to refrain from focusing solely on the problems and their*

[†] For example, Paolella (2006, 2007), Feller (1968), Ross (1997), Ross and Pekoz (2006), Kelly (1994), Ash (2008), Schwarzlander (2011), Gut (2005), Brémaud (2009), and Foata and Fuchs (1998).

solutions, as it is in the discussion that the "meaty" parts will be found. Moreover, the selection of problems here are not necessarily the most important problems in the field of probability. Although a few such as *Problem 8: Jacob Bernoulli and his Golden Theorem, Problem 10: de Moivre, Gauss, and the Normal Curve, and Problem 14: Bayes, Laplace, and Philosophies of Probability* are important, many others are more interesting than they have been decisive to the field. Thus, few readers who are not very conversant with the history of probability are probably aware that scientists like Galileo and Newton ever used probability in their writings. Nor would anybody contend that these two men made any fundamental contributions to the theory of probability. Yet, I am hopeful that the manner in which these scientists tackled probability problems will awaken some interest in the reader.

I have refrained from giving extensive biographies of the mathematicians discussed in this book because my focus is mainly on the problems they solved and how they solved them. Readers eager to know more about these scientists should consult the excellent *Dictionary of Scientific Biography,* edited by C.-C. Gillispie.

If I had a criticism of the book, it would be the unevenness in the treatment of the problems. Thus, the reader will find more than 20 pages devoted to *Problem 4: The Chevalier de Méré Problem II: The Problem of Points,* but less than 5 pages spent on *Problem 15: Leibniz's Error.* This is inevitable given the historical contexts and implications of the problems. Nonetheless, I would be remiss to claim that there was no element of subjectivity in my allocation of space to the problems. Indeed, the amount of material partly reflects my personal preferences, and the reader will surely find some of the problems I treated briefly deserved more, or vice versa. There is also some unevenness in the level of difficulty of the problems since they are arranged in chronological order. For example, *Problem 23: Borel's Paradox and Kolmogorov's Axioms* is harder than many of the earlier and later problems because it contains some measure-theoretic concepts. However, I also wanted the reader to grasp the significance of Kolmogorov's work, and that would have been almost impossible to do without some measure theory.

Most of the citations were obtained from the original works. All the French translations were done by me, unless otherwise indicated. Although this book is about classic problems, I have also tried to include as many classic citations and references as possible.

I hope the book will appeal to both students, those interested in the history of probability, and all those who may be captivated by the charms of probability.

Prakash Gorroochurn
pg2113@columbia.edu
March 2012

Acknowledgments

My foremost thanks go to my friend and colleague Bruce Levin for carefully reading an almost complete version of the manuscript and for making several excellent suggestions. Bruce has been a constant source of encouragement, motivation, and inspiration, and I hope these few words can do justice to him. Arindam Roy Choudhury also read several chapters from an earlier version, and Bill Stewart discussed parts of **Problem 25** with me. In addition, I am grateful to the following for reading different chapters from the book: Nicholas H. Bingham, Steven J. Brams, Ronald Christensen, Joseph C. Gardiner, Donald A. Gillies, Michel Henry, Collin Howson, Davar Khoshnevisan, D. Marc Kilgour, Peter Killeen, Peter M. Lee, Paul J. Nahin, Raymond Nickerson, Roger Pinkham, and Sandy L. Zabell. I would also like to thank the following for help with the proofreading: Amy Armento, Jonathan Diah, Guqian Du, Rebecca Gross, Tianxiao Huang, Wei-Ti Huang, Tsun-Fang Hsueh, Annie Lee, Yi-Chien Lee, Keletso Makofane, Stephen Mooney, Jessica Overbey, Bing Pei, Lynn Petukhova, Nicolas Rouse, John Spisack, Lisa Strug, and Gary Yu. Finally, I thank the anonymous reviewers for their helpful suggestions.

My thanks also go to Susanne Steitz-Filler, editor at Wiley. Susanne has always been so attentive to the slightest concerns I have had, and she has made my experience with Wiley extremely positive.

Finally, I am of course forever indebted to my mother and my late father.

Problem 1

Cardano and Games of Chance (1564)

Problem. How many throws of a fair die do we need in order to have an even chance of at least one six?

Solution. Let A be the event "a six shows in one throw of a die" and p_A its probability. Then $p_A = 1/6$. The probability that a six does not show in one throw is $q_A = 1 - 1/6 = 5/6$. Let the number of throws be n. Therefore, assuming independence between the throws,

$$\Pr\{\text{a six shows at least once in } n \text{ throws}\} = 1 - \Pr\{\text{a six does not show at all in } n \text{ throws}\}$$

$$= 1 - \overbrace{q_A \times \cdots \times q_A}^{n \text{ terms}}$$
$$= 1 - (5/6)^n.$$

We now solve $1 - (5/6)^n \geq 1/2$ obtaining $n \geq \ln(1/2)/\ln(5/6) = 3.8$, so the number of throws is 4.

1.1 Discussion

In the history of probability, the physician and mathematician Gerolamo Cardano (1501–1575) (Fig. 1.1) was among the first to attempt a systematic study of the calculus of probabilities. Like those of his contemporaries, Cardano's studies were primarily driven by games of chance. Concerning his gambling for 25 years, he famously said in his autobiography (Cardano, 1935, p. 146)

> . . .and I do not mean to say only from time to time during those years, but I am ashamed to say it, everyday.

Classic Problems of Probability, Prakash Gorroochurn.
© 2012 John Wiley & Sons, Inc. Published 2012 by John Wiley & Sons, Inc.

Figure 1.1   Gerolamo Cardano (1501–1575).

Cardano's works on probability were published posthumously in 1663, in the famous 15-page *Liber de ludo aleae*[†] (Fig. 1.2) consisting of 32 small chapters (Cardano, 1564).

Cardano was undoubtedly a great mathematician of his time but stumbled on the question in **Problem 1**, and several others too. In this case, he thought the number of throws should be three. In Chapter 9 of his book, Cardano states regarding a die:

> One-half of the total number of faces always represents equality[‡]; thus the chances are equal that a given point will turn up in three throws. . .

Cardano's mistake stems from a prevalent general confusion between the concepts of probability and expectation. Let's now dig deeper into Cardano's

---

[†] The Book on Games of Chance. An English translation of the book and a thorough analysis of Cardano's connections with games of chance can be found in Ore's *Cardano: The Gambling Scholar* (Ore, 1953). More bibliographic details can be found in Gliozzi (1980, pp. 64–67) and Scardovi (2004, pp. 754–758).

[‡] Cardano frequently uses the term "equality" in the *Liber* to denote half of the total number of sample points in the sample space. See Ore (1953, p. 149).

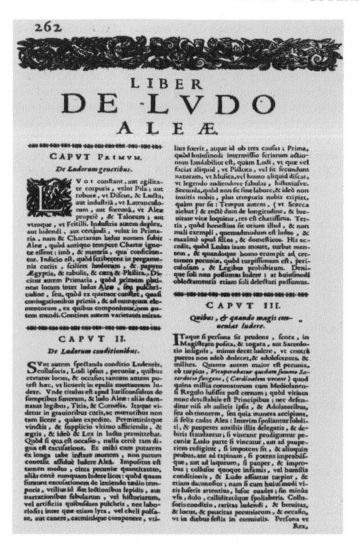

Figure 1.2   First page of the *Liber de ludo aleae*, taken from the *Opera Omnia* (Vol. I) (Cardano, 1564).

reasoning. In the *Liber*, Cardano frequently makes use of an erroneous principle, which Ore calls a "reasoning on the mean" (ROTM) (Ore, 1953, p. 150),[†] to deal with various probability problems. According to the ROTM, if an event has a probability p in one trial of an experiment, then in n independent trials the event will occur np times on average, which is then wrongly taken to represent the *probability* that the event will occur in n trials. For the question in **Problem 1**, we have $p = 1/6$ so that,

---

[†] See also Williams (2005).

with $n = 3$ throws, the event "at least a six" is wrongly taken to occur an average $np = 3$ $(1/6) = 1/2$ of the time (i.e., with probability $1/2$).

Using modern notation, let us see why the ROTM is wrong. Suppose an event has a probability p of occurring in a single repetition of an experiment. Then in n independent and identical repetitions of that experiment, the expected number of the times the event occurs is np. Thus, for the die example, the expectation for the number of times a six appears in three throws is $3 \times 1/6 = 1/2$. However, an expectation of $1/2$ in three throws is not the same as a probability of $1/2$ in three throws. These facts can formally be seen by using a binomial model.[†] Let X be the number of sixes in three throws. Then X has a binomial distribution with parameters $n = 3$ and $p = 1/6$, that is, $X \sim B(3, 1/6)$, and its probability mass function is

$$\Pr\{X = x\} = \binom{n}{x} p^x (1 - p)^{n-x} = \binom{3}{x}\left(\frac{1}{6}\right)^x \left(\frac{5}{6}\right)^{3-x} \quad \text{for } x = 0, 1, 2, 3.$$

From this formula, the probability of one six in three throws is

$$\Pr\{X = 1\} = \binom{3}{1}\left(\frac{1}{6}\right)^1 \left(\frac{5}{6}\right)^2 = .347,$$

and the probability of at least one six is

$$\begin{aligned}
\Pr\{X \geq 1\} &= 1 - \Pr\{X = 0\} \\
&= 1 - (5/6)^3 \\
&= .421.
\end{aligned}$$

Finally, the expected value of X is

$$\mathcal{E}X = \sum_{x=0}^{n} x \Pr\{X = x\} = \sum_{x=0}^{n} x \binom{n}{x} p^x (1 - p)^{n-x},$$

which can be simplified to give

$$\mathcal{E}X = np = (3)(1/6) = .5.$$

Thus, we see that although the expected number of sixes in three throws is $1/2$, neither the probability of one six or at least one six is $1/2$.

Cardano has not got the recognition that he perhaps deserves for his contributions to the field of probability, for in the *Liber de ludo aleae* he touched on many rules and problems that were later to become classics. Let us now outline some of these.

In Chapter 14 of the *Liber*, Cardano gives what some would consider the first definition of classical (or mathematical) probability:

> So there is one general rule, namely, that we should consider the whole circuit, and the number of those casts which represents in how many ways the favorable result can occur,

---

[†] For the origin of the binomial model, see **Problem 8**.

and compare that number to the rest of the circuit, and according to that proportion should the mutual wagers be laid so that one may contend on equal terms.

Cardano thus calls the "circuit" what is known as the sample space today, that is, the set of all possible outcomes when an experiment is performed. If the sample space is made up of r outcomes that are favorable to an event, and s outcomes that are unfavorable, and if all outcomes are equally likely, then Cardano correctly defines the odds in favor of the event by $r : s$. This corresponds to a probability of $r/(r+s)$. Compare Cardano's definition to

- The definition given by Leibniz (1646–1716) in 1710 (Leibniz, 1969, p. 161):

  If a situation can lead to different advantageous results ruling out each other, the estimation of the expectation will be the sum of the possible advantages for the set of all these results, divided into the total number of results.

- Jacob Bernoulli's (1654–1705) statement from the *Ars Conjectandi* (Bernoulli, 1713, p. 211)[†]:

  . . .if the integral and absolute certainty, which we designate by letter α or by unity 1, will be thought to consist, for example, of five probabilities, as though of five parts, three of which favor the existence or realization of some events, with the other ones, however, being against it, we will say that this event has $3/5\alpha$, or $3/5$, of certainty.

- De Moivre's (1667–1754) definition from the *De Mensura Sortis* (de Moivre, 1711; Hald, 1984):

  If p is the number of chances by which a certain event may happen, and q is the number of chances by which it may fail, the happenings as much as the failings have their degree of probability; but if all the chances by which the event may happen or fail were equally easy, the probability of happening will be to the probability of failing as p to q.

- The definition given in 1774 by Laplace (1749–1827), with whom the formal definition of classical probability is usually associated. In his first probability paper, Laplace (1774b) states:

  The probability of an event is the ratio of the number of cases favorable to it, to the number of possible cases, when there is nothing to make us believe that one case should occur rather than any other, so that these cases are, for us, equally possible.

However, although the first four definitions (starting from Cardano's) all anteceded Laplace's, it is with the latter that the classical definition was fully appreciated and began to be formally used. In modern notation, if a sample space consists of N equally likely outcomes, of which n_A are favorable to an event A, then Laplace's classical definition of the probability of the event A is

$$\Pr\{A\} \equiv \frac{n_A}{N}.$$

[†] The translation that follows is taken from Oscar Sheynin's translations of Chapter 4 of the *Ars Conjectandi* (Sheynin, 2005).

One of Cardano's other important contributions to the theory of probability is "Cardano's formula."[†] Suppose an experiment consists of t equally likely outcomes of which r are favorable to an event. Then the odds in favor of the event in one trial of the experiment are $r : (t - r)$.[‡] Cardano's formula then states that, in n independent and identical trials of the experiment, the odds in favor of the event occurring n times are $r^n : (t^n - r^n)$.[§] While this is an elementary result nowadays, Cardano had some difficulty establishing it.[**] At first he thought it was the odds that ought to be multiplied. Cardano calculated the odds against obtaining at least one 1 appearing in a toss of three dice as 125 to 91. Cardano then proceeded to obtain the odds against obtaining at least one 1 in two tosses of three dice as $(125/91)^2 \approx 2{:}1$. Thus, on the last paragraph of Chapter 14 of the *Liber*, Cardano writes (Ore, 1953, p. 202)

> Thus, if it is necessary for someone that he should throw an ace twice, then you know that the throws favorable for it are 91 in number, and the remainder is 125; so we multiply each of these numbers by itself and get 8281 and 15,625, and the odds are about 2 to 1.[++] Thus, if he should wager double, he will contend under an unfair condition, although in the opinion of some the condition of the one offering double stakes would be better.

However, in the very next chapter entitled *On an Error Which Is Made About This*, Cardano realizes that it is not the odds that must be multiplied. He comes to understand this by considering an event with odds 1:1 in one trial of an experiment. His multiplication rule for the odds would still give an odds of $(1/1)^3 = 1{:}1$ for three trials of the experiment, which is clearly wrong. Cardano thus writes (Ore, 1953, pp. 202–203)

> But this reasoning seems to be false, even in the case of equality, as, for example, the chance of getting one of any three chosen faces in one cast of one die is equal to the chance of getting one of the other three, but according to this reasoning there would be an even chance of getting a chosen face each time in two casts, and thus in three, and four, which is most absurd. For if a player with two dice can with equal chances throw an even and an odd number, it does not follow that he can with equal fortune throw an even number in each of three successive casts.

Cardano thus correctly calls his initial reasoning "most absurd," and then gives the following correct reasoning (Ore, 1953):

[†] Not to be confused with the other Cardano's formula having to do with the general solution of a "reduced" cubic equation (i.e., a cubic equation with no second-degree term). Cardano also provided methods to convert the general cubic equation to the reduced one. These results appeared in Cardano's opus *Ars Magna* (The Great Art) and had been communicated to him previously by the mathematician Niccolò Tartaglia of Brescia (1499–1557) after swearing that he would never disclose the results. A bitter dispute thereby ensued between Cardano and Tartaglia, and is nicely documented in Hellman's *Great Feuds in Mathematics* (Hellman, 2006, pp. 7–25).

[‡] Thus, the odds *against* the event in one trial are $(t - r){:}\ r$.

[§] This is the same as saying that, if an event has probability p $(=r/t)$ of occurring in one trial of an experiment, then the probability that it will occur in all of n independent and identical trials of the experiment is p^n.

[**] See also Katz (1998, p. 450).

[++] The odds calculated by Cardano are the odds *against* the event in question.

Therefore, in comparisons where the probability is one-half, as of even faces with odd, we shall multiply the number of casts by itself and subtract one from the product, and the proportion which the remainder bears to unity will be the proportion of the wagers to be staked. Thus, in 2 successive casts we shall multiply 2 by itself, which will be 4; we shall subtract 1; the remainder is 3; therefore a player will rightly wager 3 against 1; for if he is striving for odd and throws even, that is, if after an even he throws either even or odd, he is beaten, or if after an odd, an even. Thus he loses three times and wins once.

Cardano thus realizes that it is the probability, not the odds, that ought to be multiplied.[†] However, in the very next sentence following his previous correct reasoning, he makes a mistake again when considering three consecutive casts for an event with odds 1:1. Cardano wrongly states that the odds against the event happening in three casts are $1{:}(3^2 - 1) = 1{:}8$, instead of $1{:}(2^3 - 1) = 1{:}7$. Nevertheless, further in the book, Cardano does give the correct general rule (Ore, 1953, p. 205):

> Thus, in the case of one die, let the ace and the deuce be favorable to us; we shall multiply 6, the number of faces, into itself: the result is 36; and two multiplied into itself will be 4; therefore the odds are 4 to 32, or, when inverted, 8 to 1.

> If three throws are necessary, we shall multiply 3 times; thus, 6 multiplied into itself and then again into itself gives 216; and 2 multiplied into itself and again into 2, gives 8; take away 8 from 216: the result will be 208; and so the odds are 208 to 8, or 26 to 1. And if four throws are necessary, the numbers will be found by the same reasoning, as you see in the table; and if one of them be subtracted from the other, the odds are found to be 80 to 1.

In the above, Cardano has considered an event with probability 1/3, and correctly gives the odds against the event happening twice as $(3^2 - 1){:}1 = 8{:}1$, happening thrice as $(3^3 - 1){:}1 = 26{:}1$, and so on. Cardano thus finally reaches the following correct rule: if the odds in favor of an event happening in one trial of an experiment are $r : (t - r)$, then in n independent and identical trials of the experiment, the odds against the event happening n times are $(t^n - r^n) : r^n$.

Cardano also anticipated the law of large numbers (see **Problem 8**), although he never explicitly stated it. Ore writes (1953, p. 170)

> It is clear…that he [Cardano] is aware of the so-called law of large numbers in its most rudimentary form. Cardano's mathematics belongs to the period antedating the expression by means of formulas, so that he is not able to express the law explicitly in this way, but he uses it as follows: when the probability for an event is p then by a large number n of repetitions the number of times it will occur does not lie far from the value $m = np$.

Moreover, in Chapter 11 of the *De ludo aleae*, Cardano investigated the *Problem of Dice* (see **Problem 3**); in the *Practica arithmetice* (Cardano, 1539), he discussed the *Problem of Points* (see **Problem 4**), the *Gambler's Ruin* (see **Problem 5**) (Coumet, 1965a), and the *St Petersburg Problem* (see **Problem 11**) (Dutka, 1988); finally, in the *Opus novum de proportionibus* (Cardano, 1570), Cardano also made use of what later came to be known as Pascal's Arithmetic Triangle (see **Problem 4**) (Boyer, 1950).

[†] Thus for the 125:91 example, the correct odds against in two trials are $(216^2 - 91^2)/91^2 \approx 4.63{:}1$ and for the 1:1 example, the correct odds in favor in three trials are $1^3{:}(2^3 - 1^3) = 1{:}7$.

However, in none of these problems did Cardano reach the level of mathematical sophistication and maturity that was later to be evidenced in the hands of his successors.

We make a final comment on Cardano's investigations in probability. From the *Liber de ludo aleae* it is clear that Cardano is unable to disassociate the unscientific concept of luck from the mathematical concept of chance. He identifies luck with some supernatural force that he calls the "authority of the Prince" (Ore, 1953, p. 227). In Chapter 20 entitled *On Luck in Play*, Cardano states (Ore, 1953, pp. 215–216)

> In these matters, luck seems to play a very great role, so that some meet with unexpected success while others fail in what they might expect. . .
>
> If anyone should throw with an outcome tending more in one direction than it should and less in another, or else it is always just equal to what it should be, then, in the case of a fair game there will be a reason and a basis for it, and it is not the play of chance; but if there are diverse results at every placing of the wagers, then some other factor is present to a greater or less extent; there is no rational knowledge of luck to be found in this, though it is necessarily luck.

Cardano thus believes that there is some external force that is responsible for the fluctuations of outcomes from their expectations. He fails to recognize such fluctuations are germane to chance and not because of the workings of supernatural forces. Gigerenzer et al. (1989, p. 12) thus write

> . . .He [Cardano] thus relinquished his claim to founding the mathematical theory of probability. Classical probability arrived when luck was banished; it required a climate of determinism so thorough as to embrace even variable events as expressions of stable underlying probabilities, at least in the long run.

Problem 2

Galileo and a Discovery Concerning Dice (1620)

Problem. *Suppose three dice are thrown and the three numbers obtained added. The total scores of 9, 10, 11, and 12 can all be obtained in six different combinations. Why then is a total score of 10 or 11 more likely than a total score of 9 or 12?*

Solution. Table 2.1 shows each of the six possible combinations (unordered arrangements) for the scores of 9–12. Also shown is the number of ways (permutations or ordered arrangements) in which each combination can occur.

For example, reading the first entry under the column 12, we have a 6-5-1. This means that, to get a total score of 12, one could get a 6, 5, 1 in any order. Next to the 6-5-1 is the number 6. This is the number of different orders in which one can obtain a 6, 5, 1. Hence, we see that the scores of 9–12 can all be obtained using six combinations for each. However, because different combinations can be realized in a different number of ways, the total number of ways for the scores 9, 10, 11, and 12 are 25, 27, 27, and 25, respectively. Hence, scores of 10 or 11 are more likely than scores of 9 or 12.

2.1 Discussion

Almost a century after Cardano's times, this problem was asked by the Grand Duke of Tuscany to the renowned physicist and mathematician Galileo Galilei (1564–1642) (Fig. 2.1). The throwing of three dice was part of the game of *passadieci*, which involved adding up the three numbers and getting at least 11 points to win. Galileo gave the solution in his probability paper *Sopra le scoperte dei dadi*[†] (Galilei, 1620) (see Fig. 2.2). In his paper, Galileo states (David, 1962, p. 193)

[†] On a Discovery Concerning Dice.

Classic Problems of Probability, Prakash Gorroochurn.
© 2012 John Wiley & Sons, Inc. Published 2012 by John Wiley & Sons, Inc.

Table 2.1 Combinations and Number of Ways Scores of 9–12 that Can Be Obtained When Three Dice are Thrown

	Score							
	12		11		10		9	
	6-5-1	6	6-4-1	6	6-3-1	6	6-2-1	6
	6-4-2	6	6-3-2	6	6-2-2	3	5-3-1	6
	6-3-3	3	5-5-1	3	5-4-1	6	5-2-2	3
	5-5-2	3	5-4-2	6	5-3-2	6	4-4-1	3
	5-4-3	6	5-3-3	3	4-4-2	3	4-3-2	6
	4-4-4	1	4-4-3	3	4-3-3	3	3-3-3	1
Total no. of ways		25		27		27		25

Figure 2.1 Galileo Galilei (1564–1642).

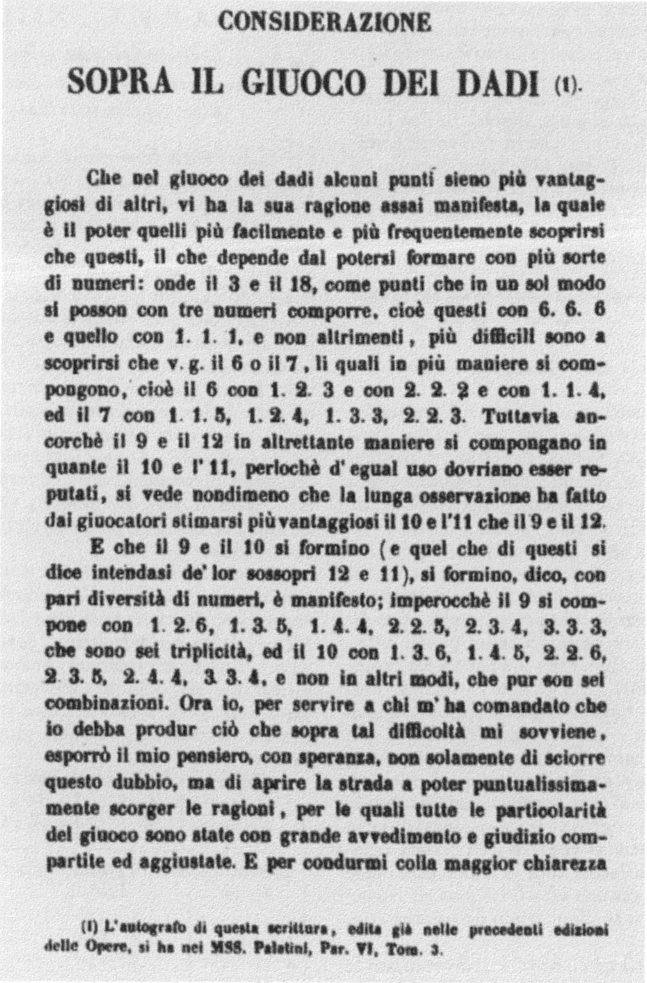

CONSIDERAZIONE

SOPRA IL GIUOCO DEI DADI (I).

Che nel giuoco dei dadi alcuni punti sieno più vantaggiosi di altri, vi ha la sua ragione assai manifesta, la quale è il poter quelli più facilmente e più frequentemente scoprirsi che questi, il che depende dal potersi formare con più sorte di numeri: onde il 3 e il 18, come punti che in un sol modo si posson con tre numeri comporre, cioè questi con 6. 6. 6 e quello con 1. 1. 1, e non altrimenti, più difficili sono a scoprirsi che v. g. il 6 o il 7, li quali in più maniere si compongono, cioè il 6 con 1. 2. 3 e con 2. 2. 2 e con 1. 1. 4, ed il 7 con 1. 1. 5, 1. 2. 4, 1. 3. 3, 2. 2. 3. Tuttavia ancorchè il 9 e il 12 in altrettante maniere si compongano in quante il 10 e l' 11, perlochè d' egual uso dovriano esser reputati, si vede nondimeno che la lunga osservazione ha fatto dai giuocatori stimarsi più vantaggiosi il 10 e l'11 che il 9 e il 12.

E che il 9 e il 10 si formino (e quel che di questi si dice intendasi de' lor sossopri 12 e 11), si formino, dico, con pari diversità di numeri, è manifesto; imperocchè il 9 si compone con 1. 2. 6, 1. 3. 5, 1. 4. 4, 2. 2. 5, 2. 3. 4, 3. 3. 3, che sono sei triplicità, ed il 10 con 1. 3. 6, 1. 4. 5, 2. 2. 6, 2. 3. 5, 2. 4. 4, 3. 3. 4, e non in altri modi, che pur son sei combinazioni. Ora io, per servire a chi m' ha comandato che io debba produr ciò che sopra tal difficoltà mi sovviene, esporrò il mio pensiero, con speranza, non solamente di sciorre questo dubbio, ma di aprire la strada a poter puntualissimamente scorger le ragioni, per le quali tutte le particolarità del giuoco sono state con grande avvedimento e giudizio compartite ed aggiustate. E per condurmi colla maggior chiarezza

(I) L'autografo di questa scrittura, edita già nelle precedenti edizioni delle Opere, si ha nei MSS. Palatini, Par. VI, Tom. 3.

Figure 2.2 Extract from Galileo's article "Sopra le scoperte dei dadi," taken from *Le Opere di Galileo XIV* (Galilei, 1620). The *Sopra* here appears under the name "Considerazione sopra il giuoco dei dadi."

But because the numbers in the combinations in three-dice throws are only 16, that is, 3, 4, 5, etc. up to 18, among which one must divide the said 216 throws, it is necessary that to some of these numbers many throws must belong; and if we can find how many belong to each, we shall have prepared the way to find out what we want to know, and it will be enough to make such an investigation from 3 to 10, because what pertains to one of these numbers, will also pertain to that which is the one immediately greater.

Galileo then proceeds to use a method similar to the one in the solution provided previously. The almost casual way in which he counts the number of favorable cases from the total number of equally possible outcomes indicates that the use of the

Succellio autem geminata , vt bonorum bis punctorum accedit ex circuitibus , inuicem ductis,videlicet tribus millibus fexcentis ictibus, cuius æqualitas eft dimidium,ictus fcilicet mille octingenti. In totidem enim poteft contingere, & non contingere. Et non fallit totus circuitus, nifi quia in vno poteft geminari, & bis , & ter. Hæc igitur cognitio eft fecundum coniecturam, & proximiorem ; & non eft ratio recta in his : Attamen contingit , quòd in multis circuitibus res fuccedit proxima coniecturæ.

CAPVT XII.

De trium Alearum iactu.

TErna puncta fimilia fiunt, nifi vno modo, vt in precedenti; ideòque funt fex. Puncta verò bina fimilia , & tertium difpar funt triginta ; & vnumquodque contingit tribus modis, erunt nonaginta. Puncta verò ex,tribus diffimilibus funt viginti , & variantur fex modis,erunt igitur cum iactu viginti , & circuitus ex omnibus ducenti fexdecim , & æqualitas in cétum octo, & ponam fimplices, & varios terminatos pro exemplo. Simplices , ergo fex geminati cuius puncti quinque modis. Cum ergo fint puncta fex , erunt modi triginta , feu ictuum varietates. Proponitur,& variatio,triplex;vt fint nonaginta. Sed viginti , qui funt omnes diffimiles , cum varientur modis fex, erunt centum,& viginti. Puncta ergo fimilia funt pars centefima octaua æqualitatis, geminata autem cum tria fint, erunt trigefima fexta eiufdem ; & vt in duabus Aleis ad vnguem , eft decima octaua. Ita hic iactus cum in decem octo mure ris fit, ad centum octo fexta pars eft , quare comparatus ad illum, triplo frequentius continget. Eaque eft geminorum lex, dicemus,aut de ratione pignoris iuxta hoc. Duo verò puncta inæqualia , vt vnum ; ac duo , fic diftinguemus , quoniam fi copulabitur vnum fiet tribus modis, fi duo, totidem erunt, ergo iam fex. Quatuor autem modis aliis contingit:At hi fex variantur finguli differentiis , erunt igitur viginti quatuor , vt cum reliquis fex

CAPVT XIII.

De Numeris compofitis,tam vfque ad fex, quam vltra , & tam in duabus Aleis , quàm in tribus.

IN duabus Aleis duodecim,& vndecim conftant eadem ratione , qua bis , fex , atque fex, & quinque. Decem autem ex bis quinque, & fex , & quatuor , hoc autem variatur dupliciter , erit igitur totum duodecima pars circuitus , & fexta æqualitatis. Rurfus ex nouem,& quinque ,& quattior,& fex, ac tribus,vt fit nona pars circuitus æqualitatis duplum nonæ partis.Octo autem puncta funt ex bis quatuor,tribus,& quinque,ac fex, & duobus. Totum quinque feptimâ fermè circuitus pars, & duæ feptimæ æqualitatis. Septem autem , ex fex, & vno quinque , ac duobus quatuor, ac tribus. Omnia igitur puncta funt fex, tertia pars æqualitatis , & fexta circuitus. At fex vt octo, & quinque , vt nouem, quatuor, vt decem, tria vt vndecim , & duo; vt duodecim.

Sed in Ludo fritilli vndecim puncta , adiicere decet , quia vna Alea poteft oftendi erunt igitur duorum punctorum iactus duodecim , & ita bes æqualitatis , & triens circuitus. Tria autem tredecim , quatuor autem quatuordecim , quinque quindecim,dextaris æqualitatis ; & à toto circuitu quincunx. Sex autem fexdecim , & valde propè æqualitatem.

Confenfus fortis in duabus Aleis.

| 2 | 12. | 1 | 3 | 11 | 2 | 4 | 10 | 3. Æqual. |
| 5 | 9 | 4 | 6 | 8 | 5 | 7 | 8 | 18. Ad Frit. |

Confenfus fortis in tribus Aleis tum Frit.

Sortis		Fritilli.
		3 115
3	18 1	4 125
4	17 3	5 126
5	16 6	6 133
6	15 10	7 33
7	14 15	8 36
8	13 21	9 37
9	12 25	10 36.
10	11 27	11 38
		12 26

Circuitus 216.
Æqualitas 108.

Figure 2.3 Cardano's solution of the problem considered by Galileo, as it appears in Chapter 13 of the *Liber de ludo aleae* (Cardano, 1564). The bottom left column on the right page has the two last rows reading 9, 12, 25 and 10, 11, 27. These correspond, respectively, to a total of 25 ways of obtaining a total of 9 or 12 with three dice, and a total of 27 ways of obtaining a total of 10 or 11 with three dice.

classical definition of probability was common at that time. Unbeknownst to Galileo, the same problem had actually already been successfully solved by Cardano almost a century earlier. The problem appeared in Chapter 13 of Cardano's *Liber de ludo aleae* (Ore, 1953, p. 198), which was published 21 years after Galileo's death (see Fig. 2.3).

Problem 3

The Chevalier de Méré Problem I: The Problem of Dice (1654)

Problem. *When a die is thrown four times, the probability of obtaining at least one six is a little more than 1/2. However, when two dice are thrown 24 times, the probability of getting at least one double-six is a little less than 1/2. Why are the two probabilities not the same, given the fact that Pr{double-six for a pair of dice} = 1/36 = 1/6·Pr{a six for a single die}, and you compensate for the factor of 1/6 by throwing 6 · 4 = 24 times when using two dice?*

Solution. Both probabilities can be calculated by using the multiplication rule of probability. In the first case, the probability of no six in one throw is $1 - 1/6 = 5/6$. Therefore, assuming independence between the throws,

$$\Pr\{\text{at least one six in 4 throws}\} = 1 - \Pr\{\text{no six in all 4 throws}\}$$
$$= 1 - (5/6)^4$$
$$= .518.$$

In the second case, the probability of no double-six in one throw of two dice is $1 - (1/6)^2 = 35/36$. Therefore, again assuming independence,

$$\Pr\{\text{at least one double-six in 24 throws}\} = 1 - \Pr\{\text{no double-six in all 24 throws}\}$$
$$= 1 - (35/36)^{24}$$
$$= .491.$$

Classic Problems of Probability, Prakash Gorroochurn.
© 2012 John Wiley & Sons, Inc. Published 2012 by John Wiley & Sons, Inc.

3.1 Discussion

It is fairly common knowledge that the gambler Antoine Gombaud (1607–1684), better known as the Chevalier de Méré,[†] had been winning consistently by betting even money that a six would come up at least once in four rolls with a single die. However, he had now been losing other bets, when in 1654 he met his friend, the amateur mathematician Pierre de Carcavi (1600–1684). This was almost a quarter century after Galileo's death. De Méré had thought the odds were favorable on betting that he could throw at least one *sonnez* (i.e., double-six) with 24 throws of a pair of dice. However, his own experiences indicated that 25 throws were required.[‡] Unable to resolve the issue, the two men consulted their mutual friend, the great mathematician, physicist, and philosopher Blaise Pascal (1623–1662).[§] Pascal himself had previously been interested in the games of chance (Groothuis, 2003, p. 10). Pascal must have been intrigued by this problem and, through the intermediary of Carcavi,[**] contacted the eminent mathematician, Pierre de Fermat (1601–1665),[++] who was a lawyer in Toulouse. Pascal knew Fermat through the latter's friendship with Pascal's father, who had died 3 years earlier. The ensuing correspondence, albeit short, between Pascal and Fermat is widely believed to be the starting point of the systematic

---

[†] Leibniz describes the Chevalier de Méré as "a man of penetrating mind who was both a player and a philosopher" (Leibniz, 1896, p. 539). Pascal biographer Tulloch also notes (1878, p. 66): "Among the men whom Pascal evidently met at the hotel of the Duc de Roannez [Pascal's younger friend], and with whom he formed something of a friendship, was the well-known Chevalier de Méré, whom we know best as a tutor of Madame de Maintenon, and whose graceful but flippant letters still survive as a picture of the time. He was a gambler and libertine, yet with some tincture of science and professed interest in its progress." Pascal himself was less flattering. In a letter to Fermat, Pascal says (Smith, 1929, p. 552): "...he [de Méré] has ability but he is not a geometer (which is, as you know, a great defect) and he does not even comprehend that a mathematical line is infinitely divisible and he is firmly convinced that it is composed of a finite number of points. I have never been able to get him out of it. If you could do so, it would make him perfect." The book by Chamaillard (1921) is completely devoted to the Chevalier de Méré.

[‡] Ore (1960) believes that the difference in the probabilities for 24 and 25 throws is so small that it is unlikely that de Méré could have detected this difference through observations. On the other hand, Olofsson (2007, p. 177) disagrees. With 24 throws, the casino would consistently make a profit of 2% (51–49%) and other gamblers would also pay for it. Furthermore, he points out that, if de Méré started with 100 pistoles (gold coins used in Europe in the seventeenth and eighteenth centuries), he would go broke with probability .97 before doubling, if the true probability is .49. If the true probability is .51, de Méré would double before going broke with very high probability. Thus, Olofsson contends it is possible to detect a difference between .49 and .51 through actual observations, and de Méré had enough time at his disposal for this.

[§] Of the several books that have been written on Pascal, the biographies by Groothuis (2003) and Hammond (2003) are good introductions to his life and works.

[**] Carcavi had been an old friend of Pascal's father and was very close to Pascal.

[++] Fermat is today mostly remembered for the so-called "Fermat Last Theorem", which he conjectured in 1637 and which was not proved until 1995 by Andrew Wiles (1995). The theorem states that no three positive integers a, b, c can satisfy the equation $a^n + b^n = c^n$ for any integer n greater than 2. A good introduction to Fermat's Last Theorem can be found in Aczel (1996). The book by Mahoney (1994) is an excellent biography of Fermat, whose probability work appears on pp. 402–410 of the book.

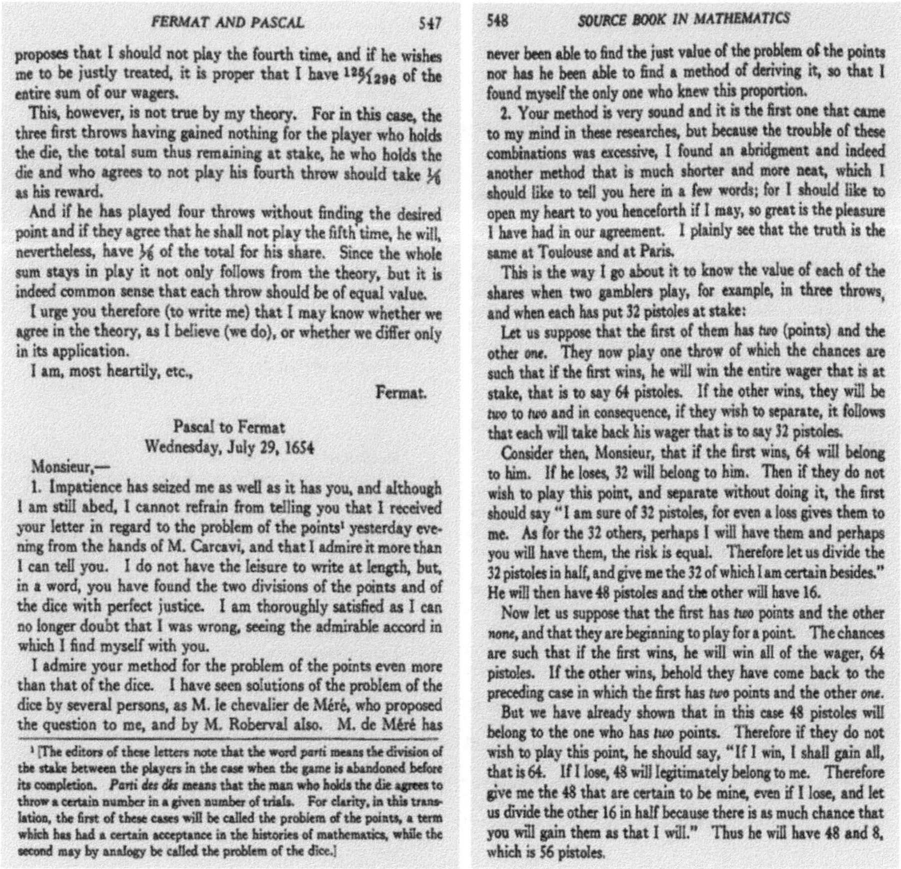

proposes that I should not play the fourth time, and if he wishes me to be justly treated, it is proper that I have $^{125}\!/_{296}$ of the entire sum of our wagers.

This, however, is not true by my theory. For in this case, the three first throws having gained nothing for the player who holds the die, the total sum thus remaining at stake, he who holds the die and who agrees to not play his fourth throw should take ⅙ as his reward.

And if he has played four throws without finding the desired point and if they agree that he shall not play the fifth time, he will, nevertheless, have ⅙ of the total for his share. Since the whole sum stays in play it not only follows from the theory, but it is indeed common sense that each throw should be of equal value.

I urge you therefore (to write me) that I may know whether we agree in the theory, as I believe (we do), or whether we differ only in its application.

I am, most heartily, etc.,

Fermat.

Pascal to Fermat
Wednesday, July 29, 1654

Monsieur,—

1. Impatience has seized me as well as it has you, and although I am still abed, I cannot refrain from telling you that I received your letter in regard to the problem of the points[1] yesterday evening from the hands of M. Carcavi, and that I admire it more than I can tell you. I do not have the leisure to write at length, but, in a word, you have found the two divisions of the points and of the dice with perfect justice. I am thoroughly satisfied as I can no longer doubt that I was wrong, seeing the admirable accord in which I find myself with you.

I admire your method for the problem of the points even more than that of the dice. I have seen solutions of the problem of the dice by several persons, as M. le chevalier de Méré, who proposed the question to me, and by M. Roberval also. M. de Méré has

[1] [The editors of these letters note that the word *parti* means the division of the stake between the players in the case when the game is abandoned before its completion. *Parti des dés* means that the man who holds the die agrees to throw a certain number in a given number of trials. For clarity, in this translation, the first of these cases will be called the problem of the points, a term which has had a certain acceptance in the histories of mathematics, while the second may by analogy be called the problem of the dice.]

never been able to find the just value of the problem of the points nor has he been able to find a method of deriving it, so that I found myself the only one who knew this proportion.

2. Your method is very sound and it is the first one that came to my mind in these researches, but because the trouble of these combinations was excessive, I found an abridgment and indeed another method that is much shorter and more neat, which I should like to tell you here in a few words; for I should like to open my heart to you henceforth if I may, so great is the pleasure I have had in our agreement. I plainly see that the truth is the same at Toulouse and at Paris.

This is the way I go about it to know the value of each of the shares when two gamblers play, for example, in three throws, and when each has put 32 pistoles at stake:

Let us suppose that the first of them has *two* (points) and the other *one*. They now play one throw of which the chances are such that if the first wins, he will win the entire wager that is at stake, that is to say 64 pistoles. If the other wins, they will be *two* to *two* and in consequence, if they wish to separate, it follows that each will take back his wager that is to say 32 pistoles.

Consider then, Monsieur, that if the first wins, 64 will belong to him. If he loses, 32 will belong to him. Then if they do not wish to play this point, and separate without doing it, the first should say "I am sure of 32 pistoles, for even a loss gives them to me. As for the 32 others, perhaps I will have them and perhaps you will have them, the risk is equal. Therefore let us divide the 32 pistoles in half, and give me the 32 of which I am certain besides." He will then have 48 pistoles and the other will have 16.

Now let us suppose that the first has *two* points and the other *none*, and that they are beginning to play for a point. The chances are such that if the first wins, he will win all of the wager, 64 pistoles. If the other wins, behold they have come back to the preceding case in which the first has *two* points and the other *one*.

But we have already shown that in this case 48 pistoles will belong to the one who has *two* points. Therefore if they do not wish to play this point, he should say, "If I win, I shall gain all, that is 64. If I lose, 48 will legitimately belong to me. Therefore give me the 48 that are certain to be mine, even if I lose, and let us divide the other 16 in half because there is as much chance that you will gain them as that I will." Thus he will have 48 and 8, which is 56 pistoles.

Figure 3.1 Extract from first extant letter from Pascal to Fermat, taken from Smith's *A Source Book in Mathematics* (Smith, 1929).

development of the theory of probability. In the first extant letter[†] Pascal addressed to Fermat, dated July 29, 1654, Pascal says (Figs. 3.1 and 3.2) (Smith, 1929, p. 552)

He [De Méré] tells me that he has found an error in the numbers for this reason:

If one undertakes to throw a six with a die, the advantage of undertaking to do it in 4 is as 671 is to 625.

If one undertakes to throw double sixes with two dice the disadvantage of the undertaking is 24.

[†] Unfortunately, the very first letter Pascal wrote to Fermat, as well as a few other letters between the two men, no longer exists. However, see the most unusual book by Rényi (1972), in which he fictitiously reconstructs four letters between Pascal and Fermat.

proposes that I should not play the fourth time, and if he wishes me to be justly treated, it is proper that I have $^{125}\!/_{1296}$ of the entire sum of our wagers.

This, however, is not true by my theory. For in this case, the three first throws having gained nothing for the player who holds the die, the total sum thus remaining at stake, he who holds the die and who agrees to not play his fourth throw should take $\frac{1}{6}$ as his reward.

And if he has played four throws without finding the desired point and if they agree that he shall not play the fifth time, he will, nevertheless, have $\frac{1}{6}$ of the total for his share. Since the whole sum stays in play it not only follows from the theory, but it is indeed common sense that each throw should be of equal value.

I urge you therefore (to write me) that I may know whether we agree in the theory, as I believe (we do), or whether we differ only in its application.

I am, most heartily, etc.,

Fermat.

Pascal to Fermat
Wednesday, July 29, 1654

Monsieur,—

1. Impatience has seized me as well as it has you, and although I am still abed, I cannot refrain from telling you that I received your letter in regard to the problem of the points[1] yesterday evening from the hands of M. Carcavi, and that I admire it more than I can tell you. I do not have the leisure to write at length, but, in a word, you have found the two divisions of the points and of the dice with perfect justice. I am thoroughly satisfied as I can no longer doubt that I was wrong, seeing the admirable accord in which I find myself with you.

I admire your method for the problem of the points even more than that of the dice. I have seen solutions of the problem of the dice by several persons, as M. le chevalier de Méré, who proposed the question to me, and by M. Roberval also. M. de Méré has

---

[1] [The editors of these letters note that the word *parti* means the division of the stake between the players in the case when the game is abandoned before its completion. *Parti des dés* means that the man who holds the die agrees to throw a certain number in a given number of trials. For clarity, in this translation, the first of these cases will be called the problem of the points, a term which has had a certain acceptance in the histories of mathematics, while the second may by analogy be called the problem of the dice.]

IV. FIELD OF PROBABILITY

FERMAT AND PASCAL ON PROBABILITY

(Translated from the French by Professor Vera Sanford, Western Reserve University, Cleveland, Ohio.)

Italian writers of the fifteenth and sixteenth centuries, notably Pacioli (1494), Tartaglia (1556), and Cardan (1545), had discussed the problem of the division of a stake between two players whose game was interrupted before its close. The problem was proposed to Pascal and Fermat, probably in 1654, by the Chevalier de Méré, a gambler who is said to have had unusual ability "even for the mathematics." The correspondence which ensued between Fermat and Pascal, was fundamental in the development of modern concepts of probability, and it is unfortunate that the introductory letter from Pascal to Fermat is no longer extant. The one here translated, written in 1654, appears in the *Œuvres de Fermat* (ed. Tannery and Henry, Vol. II, pp. 288–314, Paris, 1894) and serves to show the nature of the problem. For a biographical sketch of Fermat, see page 213; of Pascal, page 67. See also pages 165, 213, 214, and 326.

Monsieur,

If I undertake to make a point with a single die in eight throws, and if we agree after the money is put at stake, that I shall not cast the first throw, it is necessary by my theory that I take $\frac{1}{6}$ of the total sum to be impartial because of the aforesaid first throw.

And if we agree after that that I shall not play the second throw, I should, for my share, take the sixth of the remainder that is $\frac{5}{36}$ of the total.

If, after that, we agree that I shall not play the third throw, I should to recoup myself, take $\frac{1}{6}$ of the remainder which is $^{25}\!/_{216}$ of the total.

And if subsequently, we agree again that I shall not cast the fourth throw, I should take $\frac{1}{6}$ of the remainder or $^{125}\!/_{1296}$ of the total, and I agree with you that that is the value of the fourth throw supposing that one has already made the preceding plays.

But you proposed in the last example in your letter (I quote your very terms) that if I undertake to find the six in eight throws and if I have thrown three times without getting it, and if my opponent

---

Figure 3.2   Extract from first extant letter from Fermat to Pascal, taken from Smith's *A Source Book in Mathematics* (Smith, 1929).

But nonetheless, 24 is to 36 (which is the number of faces of two dice) as 4 is to 6 (which is the number of faces of one die).

This is what was his great scandal which made him say haughtily that the theorems were not consistent and that arithmetic was demented. But you can easily see the reason by the principles which you have.

De Méré was thus distressed that his observations were in contradiction with his mathematical calculations. To Fermat, however, the *Problem of Dice* was an elementary exercise that he solved without trouble, for Pascal says in his July 29 letter (Smith, 1929, p. 547)

I do not have the leisure to write at length, but, in a word, you have found the two divisions...of the dice with perfect justice. I am thoroughly satisfied as I can no longer doubt that I was wrong, seeing the admirable accord in which I find myself with you.

On the other hand, de Méré's erroneous mathematical reasoning was based on the incorrect *Old Gambler's Rule* (Weaver, 1982, p. 47), which uses the concept of the *critical value* of a game. The critical value C of a game is the smallest number of plays such that the probability the gambler will win at least one play is 1/2 or more. Let us now explain how the Old Gambler's Rule is derived. Recall Cardano's "reasoning on the mean" (ROTM, see **Problem 1**): if a gambler has a probability p of winning one play of a game, then in n independent plays the gambler will win an average of np times, which is then wrongly equated to the *probability* of winning in n plays. Then, by setting the latter probability to be half, we have

$$C \times p = \frac{1}{2}.$$

By substituting $p = 1/36$ in the above formula, Cardano had obtained a wrong answer of $C = 18$ for the number of throws. Furthermore, given a first game with (p_1, C_1), then a second game which has probability of winning p_2 in each play must have critical value C_2, where

$$C_1 p_1 = C_2 p_2 \quad \text{or} \quad C_2 = \frac{C_1 p_1}{p_2} \quad \text{(Old Gambler's Rule)}. \tag{3.1}$$

That is, the Old Gambler's Rule states that *the critical values of two games are in inverse proportion to their respective probabilities of winning.* Using $C_1 = 4, p_1 = 1/6$, and $p_2 = 1/36$, we get $C_2 = 24$. But we have seen that, with 24 throws, the probability of at least one double-six is .491, which is less than 1/2. So $C_2 = 24$ cannot be a critical value (the correct critical value is shown later to be 25), and the Old Gambler's Rule cannot be correct. It was thus the belief in the validity of the Old Gambler's Rule that made de Méré wrongly think that, with 24 throws, he should have had a probability of 1/2 for at least one double-six.

Digging further, let us see how the erroneous Old Gambler's Rule should be corrected. By definition, $C_1 = \lceil x_1 \rceil$, the smallest integer greater than or equal to x_1, such that $(1 - p_1)^{x_1} = .5$, that is, $x_1 = \ln(.5)/\ln(1 - p_1)$. With obvious notation, for the second game: $C_2 = \lceil x_2 \rceil$, where $x_2 = \ln(.5)/\ln(1 - p_2)$. Thus, the true relationship should be

$$x_2 = \frac{x_1 \ln(1 - p_1)}{\ln(1 - p_2)}. \tag{3.2}$$

We see that Eqs. (3.1) and (3.2) are quite different from each other. Even if p_1 and p_2 were very small, so that $\ln(1 - p_1) \approx -p_1$ and $\ln(1 - p_2) \approx -p_2$, we would get $x_2 = x_1 p_1/p_2$ approximately. This is still different from Eq. (3.1) because the latter uses the integers C_1 and C_2, instead of the real numbers x_1 and x_2.

The Old Gambler's Rule was later investigated by the renowned French mathematician Abraham de Moivre (1667–1754), who was a close friend to Isaac Newton. In his *Doctrine of Chances* (de Moivre, 1718, p. 14), de Moivre solves $(1 - p)^x = 1/2$ and obtains $x = -\ln(2)/\ln(1 - p)$. For small p,

PROBLEM V.

TO find in how many Trials an Event will Probably Happen, or how many Trials will be requisite to make it indifferent to lay on its Happening or Failing; supposing that a is the number of Chances for its Happening in any one Trial, and b the number of Chances for its Failing.

SOLUTION.

LET x be the number of Trials; therefore by what has been already demonstrated in the Introduction $\overline{a+b}^x - b^x = b^x$, or $\overline{a+b}^x = 2b^x$; therefore $x = \frac{Log:2}{Log:a+b - Log:b}$.

Moreover, let us reassume the Equation $\overline{a+b}^x = 2b^x$ and making $a, b :: 1, q$, the Equation will be changed into this $\overline{1+\frac{1}{q}}^x = 2$: let therefore $1+\frac{1}{q}$ be raised actually to the Power x by Sir *Isaac Newton*'s Theorem, and the Equation will be $1 + \frac{x}{q} + \frac{x \times x - 1}{1 \times 2 q q} + \frac{x \times x - 1 \times x - 2}{1 \times 2 \times 3 q^3}$ &c. = 2. In this Equation, if $q = 1$, then will x be likewise = 1; if q be infinite, then will x also be infinite. Suppose q infinite, then the Equation will be reduced to $1 + \frac{x}{q} + \frac{xx}{2qq} + \frac{x^3}{6q^3}$ &c. = 2: But the first part of this Equation is the number whose Hyperbolic Logarithm is $\frac{x}{q}$, therefore $\frac{x}{q}$ = Log: 2: But the Hyperbolic Logarithm of 2 is 0.693 or nearly 0.7; Wherefore $\frac{x}{q}$ = 0.7, and $x = 0.7q$ very near.

Figure 3.3   Extract from de Moivre's derivation of the gambling rule, taken from the first edition of the *Doctrine of Chances* (de Moivre, 1718, p. 14). Note that de Moivre defines q to be b/a or $(1/p) - 1$, whereas in the text we use $q = 1 - p$.

Table 3.1   Critical Values Obtained Using the Old Gambling Rule, de Moivre's Gambling Rule and the Exact Formula for Different Values of p, the Probability of the Event of Interest

Value of p	Critical value C using the Old Gambling Rule $C = C_1 p_1/p$ (assuming $C_1 = 4$ for $p_1 = 1/6$)	Critical value C using the de Moivre's Gambling Rule $C = \lceil .693/p \rceil$	Critical value C using the exact formula $C = \lceil -\ln 2/\ln(1 - p) \rceil$
1/216	144	150	150
1/36	24	25	25
1/6	4	5	4
1/4	3	3	3
1/2	2	2	1

$$x \approx \frac{.693}{p} \quad \text{(De Moivre's Gambling Rule)} \tag{3.3}$$

(see Fig. 3.3). Let us see if we obtain the correct answer when we apply de Moivre's Gambling Rule for the two-dice problem. Using $x \approx .693/p$ with $p = 1/36$ gives $x \approx 24.95$ and we obtain the correct critical value $C = 25$. The formula works only because p is small enough and is valid only for such cases.[†] The other formula that could be used, and that is valid for *all* values of p, is $x = -\ln(2)/\ln(1 - p)$. For the two-dice problem, this exact formula gives $x = -\ln(2)/\ln(35/36) = 24.60$ so that $C = 25$. Table 3.1 compares critical values obtained using the Old Gambler's Rule, de Moivre's Gambling Rule, and the exact formula.

We next use de Moivre's Gambling Rule to solve the following classic problem:

A gambler has a probability of 1/N of winning a game, where N is large. Show that she must play the game about (2/3)N times in order to have a probability of at least 1/2 of winning at least once.

To solve this problem, note that $p = 1/N$ is small so that we can apply de Moivre's Gambling Rule in Eq. (3.3):

$$x \approx \frac{.693}{1/N} \approx \frac{2}{3}N,$$

as required.

As a final note on the dice problem, Pascal himself never provided a solution to it in his known communication with Fermat. He had undoubtedly thought the problem was very simple. However, he devoted much of his time and energy on the next classic problem.

[†] For example, if we apply de Moivre's Gambling Rule to the one-die problem, we obtain $x = .693/(1/6) = 4.158$ so that $C = 5$. This cannot be correct because we showed in the solution that we need only 4 tosses.

Problem 4

The Chevalier de Méré Problem II: The Problem of Points (1654)

Problem. *Two players A and B play a fair game such that the player who wins a total of 6 rounds first wins a prize. Suppose the game unexpectedly stops when A has won a total of 5 rounds and B has won a total of 3 rounds. How should the prize be divided between A and B?*

Solution. The division of the prize is determined by the relative probabilities of A and B winning the prize, *had they continued the game.* Player A is one round short, and player B three rounds short, of winning the prize. The maximum number of hypothetical remaining rounds is $(1 + 3) - 1 = 3$, each of which could be equally won by A or B. The sample space for the game is $\Omega = \{A_1, B_1A_2, B_1B_2A_3, B_1B_2B_3\}$. Here B_1A_2, for example, denotes the event that B would win the first remaining round and A would win the second (and then the game would have to stop since A is only one round short). However, the four sample points in Ω are not equally likely. Event A_1 occurs if any one of the following four equally likely events occurs: $A_1A_2A_3$, $A_1A_2B_3$, $A_1B_2A_3$, and $A_1B_2B_3$. Event B_1A_2 occurs if any one of the following two equally likely events occurs: $B_1A_2A_3$ and $B_1A_2B_3$. In terms of equally likely sample points, the sample space is thus

$$\Omega = \{A_1A_2A_3, A_1A_2B_3, A_1B_2A_3, A_1B_2B_3, B_1A_2A_3, B_1A_2B_3, B_1B_2A_3, B_1B_2B_3\}$$

There are in all eight equally likely outcomes, only one of which ($B_1B_2B_3$) results in B hypothetically winning the game. Player A thus has a probability 7/8 of winning. The prize should therefore be divided between A and B in the ratio 7:1.

Classic Problems of Probability, Prakash Gorroochurn.
© 2012 John Wiley & Sons, Inc. Published 2012 by John Wiley & Sons, Inc.

4.1 Discussion

We note that the sample space for this game is not based on how many rounds each player has already won. Rather it depends on the maximum number of *remaining* rounds that could be played.

Problem 4, also known as the *Problem of Points*[†] or the *Division of Stakes Problem*, was another problem de Méré asked Pascal (Fig. 4.1) in 1654. The problem had already been known hundreds of years before the times of these mathematicians.[‡] It had appeared in Italian manuscripts as early as 1380 (Burton, 2006, p. 445). However, it first came in print in Fra Luca Pacioli's (1494) *Summa de arithmetica, geometrica, proportioni, et proportionalita*.[§] Pacioli's incorrect answer was that the prize should be divided in the same ratio as the total number of rounds the players had won. Thus, for **Problem 4**, the ratio is 5:3. A simple counterexample shows why Pacioli's reasoning cannot be correct. Suppose players A and B need to win 100 rounds to win a game, and when they stop A has won one round and B has won none. Then Pacioli's rule would give the whole prize to A even though he is a single round ahead of B and would have needed to win 99 more rounds had the game continued![**]

Cardano had also considered the *Problem of Points* in the *Practica arithmetice* (Cardano, 1539). His major insight was that the division of stakes should depend on how many rounds each player *had yet to win*, not on how many rounds they had *already won*. However, in spite of this, Cardano was unable to give the correct division ratio: he concluded that, if players A and B are a and b rounds short of winning, respectively, then the division ratio between A and B should be $b(b+1):a(a+1)$. In our case, $a=1$, $b=3$, giving a division ratio of 6:1.

At the opening of his book *Recherches sur la Probabilité des Jugements en Matières Criminelles et Matière Civile*, the distinguished mathematician Siméon Denis Poisson (1781–1840) pronounced these famous words (Poisson, 1837, p. 1):

> A problem relating to the games of chance, proposed to an austere Jansenist [Pascal] by a man of the world [de Méré], was at the origin of the calculus of probabilities. Its objective was to determine the proportion into which the stake should be divided between the players, when they decided to stop playing. . .

Poisson's words echo the still widely held view today that probability essentially sprung from considerations of games of chance during the Enlightenment. We should

---

[†] The *Problem of Points* is also discussed in Todhunter (1865, Chapter II), Hald (1990, pp. 56–63), Petkovic (2009, pp. 212–214), Paolella (2006, pp. 97–99), Montucla (1802, pp. 383–390), de Sá (2007, pp. 61–62), Kaplan and Kaplan (2006, pp. 25–30), Isaac (1995, p. 55), and Gorroochurn (2011).

[‡] For a full discussion of the *Problem of Points* before Pascal, see Coumet (1965b) and Meusnier (2004).

[§] Everything about arithmetic, geometry, and proportion.

[**] The correct division ratio between A and B here is approximately 53:47, as we shall soon see.

mention, however, that several other authors have a different viewpoint. For example, Maistrov (1974, p. 7) asserts:

> Up to the present time there has been a widespread false premise that probability theory owes its birth and early development to gambling.

Maistrov explains that gambling existed since ancient and medieval times, but probability did not develop then. He contends that its development occurred in the sixteenth and seventeenth century owing to economic developments, resulting in an increase in monetary transactions and trade. This seems to indeed be the case. In a recent work, Courtebras is of a similar opinion when he says (Courtebras, 2008, p. 51)

> ...behind the problems in games of chance which enable the elaboration of easily quantifiable solutions, new attitudes are being formed toward a world characterized by the development of towns and money, of productions and exchanges.

It is also important to note that the initial works in the field, from the times of Cardano until those before Jacob Bernoulli, were primarily concerned with problems of fairness or *equity*. Historian Ernest Coumet has thus noted that the *Problem of Points* has a judicial origin (Coumet, 1970). With the Renaissance came the legalization of both games of chance and aleatory contracts. By law, the latter had to be fair to either party. The *Problem of Points* is thus a model for the repartition of gains in arbitrary situations of uncertainty, mainly characterized by the notion of equity. As Gregersen (2011, p. 25) explains

> ...The new theory of chances was not, in fact, simply about gambling but also about the legal notion of a fair contract. A fair contract implied equality of expectations, which served as the fundamental notion in these calculations. Measures of chance or probability were derived secondarily from these expectations.
>
> Probability was tied up with questions of law and exchange in one other crucial respect. Chance and risk, in aleatory contracts, provided a justification for lending at interest, and hence a way of avoiding Christian prohibitions against usury. Lenders, the argument went, were like investors; having shared the risk, they deserved also to share in the gain. For this reason, ideas of chance had already been incorporated in a loose, largely nonmathematical way into theories of banking and marine insurance.

An additional interesting point is that, throughout their correspondence, although both Pascal and Fermat (Fig. 4.2) were deeply engaged in calculating probabilities, they never actually used the word "probability" in their investigations. Instead they talked about division ratios and used such terms as "value of the stake" or "value of a throw" to express a player's probability of winning. The term "probability" as a numerical measure was actually first used in Antoine Arnauld's (1612–1694) and Pierre Nicole's (1625–1695) widely influential *La Logique, ou l'Art de Penser*[++] (Arnauld and Nicole, 1662, pp. 469–470).

[†] Also known as *Logique de Port-Royal*, or the *Port-Royal Logic*. Arnauld and Nicole were both friends of Pascal and it is likely that Pascal contributed in their book. Arnauld was the main author of the book and is often called "The Great Arnauld."

As a final observation, before we dig deeper into Pascal's and Fermat's solutions to the *Problem of Points*, we comment on the significance of their communication. It is generally believed that the theory of probability really started through the correspondence of these mathematicians, although this notion has occasionally been disputed by some. For example, in his classic work *Cardano: the Gambling Scholar* (1953), Ore made a detailed study of Cardano's *Liber de ludo aleae* and pointed out (p. viii)

> ...I have gained the conviction that this pioneer work on probability is so extensive and in certain questions so successful that it would seem much more just to date the beginnings of probability theory from Cardano's treatise rather than the customary reckoning from Pascal's discussions with his gambling friend de Méré and the ensuing correspondence with Fermat, all of which took place at least a century after Cardano began composing his *De ludo aleae.*

Burton seems to share the same opinion, for he says (Burton, 2006, p. 445)

> For the first time, we find a transition from empiricism to the theoretical concept of a fair die. In making it, Cardan [Cardano] probably became the real father of modern probability theory.

On the other hand, Edwards (1982) is quite categorical:

> ...in spite of our increased awareness of the earlier work of Cardano (Ore, 1953) and Galileo (David, 1962) it is clear that before Pascal and Fermat no more had been achieved than the enumeration of the fundamental probability set in various games with dice or cards.

Although Cardano was the first to study probability, it was Pascal's and Fermat's work however that provided the first impetus for a systematic study and development of the mathematical theory of probability. Likewise, contemporaries of Cardano such as Pacioli, Tartaglia, and Peverone did consider probability calculations involving various games of chance. However, as Gouraud (1848, p. 3) explains

> ...but these crude essays, consisting of extremely erroneous analyses and having all remained equally sterile, do not merit the consideration of either critiques or history...

Let us now discuss Pascal's and Fermat's individual contributions to the *Problem of Points*. Pascal was at first unsure of his own solution to the problem, and turned to a friend, the mathematician Gilles Personne de Roberval (1602–1675). Roberval was not of much help, and Pascal then asked for the opinion of Fermat, who was immediately intrigued by the problem. A beautiful account of the ensuing correspondence between Pascal and Fermat can be found in a recent book by Keith Devlin, *The Unfinished Game: Pascal, Fermat and the Seventeenth Century Letter That Made the World Modern* (Devlin, 2008). An English translation of the extant letters can be found in Smith (1929, pp. 546–565). In a letter dated August 24, 1654, Pascal says (Smith, 1929, p. 554)

> I wish to lay my whole reasoning before you, and to have you do me the favor to set me straight if I am in error or to endorse me if I am correct. I ask you this in all faith and sincerity for I am not certain even that you will be on my side.

Figure 4.1   Blaise Pascal (1623–1662).

Fermat made use of the fact that the solution depended not on how many rounds each player had already won but *on how many each player must still win to win the prize*. This is the same observation Cardano had previously made, although he had been unable to solve the problem correctly.

The solution we provided earlier is based on Fermat's idea of extending the unfinished game. Fermat also enumerated the different sample points like in our solution and reached the correct division ratio of 7:1.

Pascal seems to have been aware of Fermat's method of enumeration (Edwards, 1982), at least for two players, and also believed there was a better method. For he says in his first extant letter, dated July 29, 1654 (Smith, 1929, p. 548)

> Your method is very sound and it is the first one that came to my mind in these researches, but because the trouble of these combinations was excessive, I found an abridgment and indeed another method that is much shorter and more neat, which I should like to tell you here in a few words; for I should like to open my heart to you henceforth if I may, so great is the pleasure I have had in our agreement. I plainly see that the truth is the same at Toulouse and at Paris[‡‡].

[†] Pascal was residing in Paris while Fermat was in Toulouse.

Figure 4.2   Pierre de Fermat (1601–1665).

Let us now examine the "shorter and more neat" method Pascal referred to previously. In the July 29 letter, he continues

> This is the way I go about it to know the value of each of the shares when two gamblers play, for example, in three throws and when each has put 32 pistoles at stake:

> Let us suppose that the first of them has two (points) and the other one. They now play one throw of which the chances are such that if the first wins, he will win the entire wager that is at stake, that is to say 64 pistoles. If the other wins, they will be two to two and in consequence, if they wish to separate, it follows that each will take back his wager that is to say 32 pistoles.

> Consider then, Monsieur, that if the first wins, 64 will belong to him. If he loses, 32 will belong to him. Then if they do not wish to play this point, and separate without doing it, the first should say "I am sure of 32 pistoles, for even a loss gives them to me. As for the 32 others, perhaps I will have them and perhaps you will have them, the risk is equal.

Therefore, let us divide the 32 pistoles in half, and give me the 32 of which I am certain besides." He will then have 48 pistoles and the other will have 16.

Now let us suppose that the first has two points and the other none, and that they are beginning to play for a point. The chances are such that if the first wins, he will win all of the wager, 64 pistoles. If the other wins, behold they have come back to the preceding case in which the first has two points and the other one.

But we have already shown that in this case 48 pistoles will belong to the one who has two points. Therefore if they do not wish to play this point, he should say, "If I win, I shall gain all, that is 64. If I lose, 48 will legitimately belong to me. Therefore give me the 48 that are certain to be mine, even if I lose, and let us divide the other 16 in half because there is as much chance that you will gain them as that I will." Thus he will have 48 and 8, which is 56 pistoles.

Let us now suppose that the first has but *one* point and the other none. You see, Monsieur, that if they begin a new throw, the chances are such that if the first wins, he will have two points to none, and dividing by the preceding case, 56 will belong to him. If he loses, they will be point for point, and 32 pistoles will belong to him. He should therefore say, "If you do not wish to play, give me the 32 pistoles of which I am certain, and let us divide the rest of the 56 in half." From 56 take 32, and 24 remains. Then divide 24 in half, you take 12 and I take 12 which with 32 will make 44.

By these means, you see, by simple subtractions that for the first throw, he will have 12 pistoles from the other; for the second, 12 more; and for the last 8.

Pascal thus considers a fair game worth a total of 64 pistoles and which stops once a player wins a total of three rounds. In the first case, he assumes player A has already won two rounds and player B has won one round. Then the game suddenly stops. How should the division be done? Pascal reasons that if A were to win one more round (with probability 1/2) then he would win a total of three rounds. Player A would then win the 64 pistoles. On the other hand, if A was to lose (with probability 1/2), they would have won two rounds each and A would win 32 pistoles. Player A's *expected* win is therefore $(1/2)(64) + (1/2)(32) = 48$ pistoles. This gives a division ratio of 48:16 or 3:1 between A and B.

Pascal then considers a second case. Suppose A has won two rounds and B none, when the game stops. Then, if A were to win the next round, he would win 64 pistoles. If A was to lose, then A and B would have won two and one rounds, respectively. They would then be in the first case, with A winning 48 pistoles. Player A's expected win for the second case is thus $(1/2)(64) + (1/2)(48) = 56$, which gives a division ratio of 56:8 or 7:1.

Finally, for the third case, Pascal assumes A has won one round and B none. If A wins the next round, then A would have won two rounds and B none. This leads us to the second case with A winning 56 pistoles. If A loses the next round, then A and B would have won one round each, and A would win 32 pistoles. Player A's expected win for the third case is therefore $(1/2)(56) + (1/2)(32) = 44$ pistoles giving a division ratio of 44:20 or 11:5.

Pascal's reasoning may be generalized as follows. Suppose the total prize is 1 dollar and the person who wins a total of R fair rounds first collects the prize. Let the game stop suddenly when A is short to win by a rounds and B is short to win by b rounds, and let A's expected win be $e_{a,b}$. Then, A can either win the next round with probability 1/2 (in which case his expected win would be $e_{a-1,b}$) or lose the next round with probability 1/2 (in which case his expected win would be $e_{a,b-1}$)

$$\left.\begin{aligned}
e_{a,b} &= \frac{1}{2}(e_{a-1,b} + e_{a,b-1}), & a, b &= 1, 2, \ldots, R. \\
e_{a,0} &= 0, & a &= 1, 2, \ldots, R-1. \\
e_{0,b} &= 1, & b &= 1, 2, \ldots, R-1.
\end{aligned}\right\} \tag{4.1}$$

Note that A's expected win $e_{a,b}$ is also his *probability* $p_{a,b}$ of winning, since the prize is 1 dollar.[†]

The elegance of Pascal's method becomes obvious when we apply it to the original question posed in **Problem 4**. Let the total prize be 1 dollar. Instead of looking at the case where A and B have won a total of five and three rounds first, let us assume A and B have won five and four rounds instead. If A wins the next round, he will win 1 dollar and if he loses he will win 1/2 dollar. His expected win is thus $(1/2)(1) + (1/2)$ $(1/2) = 3/4$ dollar. Let us now look at the case when A and B have won five and three rounds. If A wins, he will win 1 dollar. If A loses, A and B will be at five and four wins, which is the previous case (in which A wins 3/4 dollar). Thus A's expected win is $(1/2)(1) + (1/2)(3/4) = 7/8$. The division ratio is therefore 7:1, which is the same answer as before.

Pascal's solution is noteworthy for several reasons. It does not need the kinds of enumeration required in Fermat's method. The latter would have been very tedious in those days even if the number of remaining rounds was small ($\sim$10). Pascal's method is recursive in nature and it uses the concept of expectation. Although the latter concept is usually attributed to Huygens, Pascal was actually the first to use it. Hacking correctly points out (2006)

> Not until the correspondence between Fermat and Pascal do we find expectation well understood.

Pascal realized that his recursive method would quickly become unwieldy for large a and b. Moreover, he was unable to use it when player A is $b - 1$ rounds short and player B is b rounds short. Therefore, he resorted to the Arithmetic Triangle[‡] for a solution (Pascal, 1665). This triangle has ones along its two outermost diagonals, and

---

[†] See p. 44

[‡] The Arithmetic Triangle was known well before Pascal and had also been used by Cardano in his *Opus novum* (Cardano, 1570; Boyer, 1950). It is called *Yang Hui's Triangle* in China in honor of the Chinese mathematician Yang Hui (1238–1298) who used it in 1261. Others have called it *Halayudha's Triangle* since the Indian writer Halayudha used it in the tenth century. The triangle was first called *Pascal's Triangle* by Montmort (1708, p. 80) in his *Essay d'Analyse sur les Jeux de Hazard*, see Samueli and Boudenot (2009, pp. 38–39). For a modern treatment of the Arithmetic Triangle, see Edwards (2002) and Hald (1990, pp. 45–54).

Value of $a+b-1$				1				
1				1		1		
2			1		2		1	
3		1		3		3		1
4	1		4		6		4	1
$\vdots$		$\vdots$		$\vdots$		$\vdots$		$\vdots$

Figure 4.3   A modern representation of the Arithmetic Triangle.

each inner entry is the sum of the two nearest top entries (see Figs. 4.3 and 4.4). Moreover, for a given row n, the ith entry counting from left is denoted by $\binom{n}{i}$ and has a special property. It gives the number of ways of choosing i objects out of n identical objects. In the binomial expansion of $(x+y)^n$, $\binom{n}{i}$ is equal to the coefficient of x^i and is thus called a binomial coefficient, that is,

$$(x+y)^n = \sum_{i=0}^{n} \binom{n}{i} x^i y^{n-i},$$

where n is a natural number. This is the *binomial theorem*, due to Pascal (1665) and can be extended to the *generalized binomial theorem*, due to Newton (1665):

$$(1+x)^\alpha = 1 + \alpha x + \frac{\alpha(\alpha-1)}{2!} x^2 + \cdots + \frac{\alpha(\alpha-1)\cdots(\alpha-i+1)}{i!} x^i + \cdots,$$

where α is any rational number and $|x| < 1$.

Pascal was able to relate the *Problem of Points* to the Arithmetic Triangle. He correctly identified the value of $a+b-1$ with each row of the triangle, such that the corresponding entries give the number of ways A can win $0, 1, 2, \ldots$, rounds. Thus, for row $a+b-1$, the jth entry counting from left is $\binom{a+b-1}{j}$, the number of ways A can win j rounds out of $a+b-1$. Now, suppose player A is short by a rounds and player B by b rounds. Player A wins if he wins any of the remaining $a, a+1, \ldots, a+b-1$ rounds. Pascal showed that the number of ways this can happen is given by the sum[†]

$$\binom{a+b-1}{a} + \binom{a+b-1}{a+1} + \cdots + \binom{a+b-1}{a+b-1}$$

$$\equiv \binom{a+b-1}{0} + \binom{a+b-1}{1} + \cdots + \binom{a+b-1}{b-1}, \qquad (4.2)$$

[†] In the following, we will use the fact that $\binom{n}{r} \equiv \binom{n}{n-r}$.

which is the sum of the first b entries in the Arithmetic Triangle for row $a + b - 1$. Similarly, player B wins if he wins any of the remaining $b, b + 1, \ldots, a + b - 1$ rounds. The number of ways this can happen is given by the sum

$$\binom{a + b - 1}{b} + \binom{a + b - 1}{b + 1} + \cdots + \binom{a + b - 1}{a + b - 1}, \tag{4.3}$$

which is the sum of the last a entries in the Arithmetic Triangle for row $a + b - 1$. Pascal was thus able to give the general division rule for a fair game between A and B from the entries of the Arithmetic Triangle, counting from the left:

(sum of the first b entries of row $a + b - 1$) : (sum of the last a entries of row $a + b - 1$)

Although Pascal solved only the case when A and B were $b - 1$ and b rounds short in his correspondence with Fermat, he was able to prove the general division rule above in his *Traité du Triangle Arithmétique*[†] (Pascal, 1665).[‡] Applying this simple rule to our question in **Problem 4**, we have $a = 1, b = 3, a + b - 1 = 3$, and a division of stakes of $(1 + 3 + 3):1 = 7:1$ between A and B, as required. Using Eq. (4.2) we can also obtain the probability of A winning a fair game:

$$p^*_{a,b} = \frac{1}{2^{a+b-1}} \left\{ \binom{a + b - 1}{a} + \binom{a + b - 1}{a + 1} + \cdots + \binom{a + b - 1}{a + b - 1} \right\}.$$

Let us now generalize Pascal's idea when players A and B have probabilities p and q $(= 1 - p)$ of winning each round. Suppose A and B are a and b rounds, respectively, short of winning the prize, when the game suddenly stops. If the game had continued, the maximum number of possible more rounds would have been $a + b - 1$. Player A wins the prize by winning any of $a, a + 1, \ldots, a + b - 1$ rounds. Now A can win j rounds out of $a + b - 1$ rounds in $\binom{a + b - 1}{j}$ ways, so A's probability of winning the prize is

$$p_{a,b} = \sum_{j=a}^{a+b-1} \binom{a + b - 1}{j} p^j q^{a+b-1-j}. \tag{4.4}$$

Equation (4.4) first appeared in the second edition of Pierre Rémond de Montmort's[§] (1678–1719) *Essay d'Analyse sur les Jeux de Hazard*[**] (Montmort,

---

[†] Unfortunately, the English translation of Pascal's *Traité* in Smith (1929, pp. 67–79) is not complete, as only one application of the Arithmetic Triangle is included. In particular, the application of the triangle to the *Problem of Points* is missing. However, the full *Traité* in French can be found in the *Oeuvres Complètes de Pascal, Vol. II* (Pascal, 1858, pp. 415–439). The *Traité* also marks the beginning of modern combinatorial theory.

[‡] In Pascal's *Oeuvres Completes*, Vol. II (Pascal, 1858, pp. 434–436).

[§] See **Problem 7** for more on Montmort.

[**] An Analytical Essay on the Games of Chance.

1713, pp. 244–245) as the first formula for the *Problem of Points* and corresponds to a division ratio of $p_{a,b} : (1 - p_{a,b})$ between A and B. The solution had been communicated to Montmort by John Bernoulli (1667–1748) in a letter that is reproduced in the *Essay* (Montmort, 1713, p. 295).

Having received Fermat's method of enumeration for two players, Pascal makes two important observations in his August 24 letter. First, he states that his friend Roberval believes there is a fault in Fermat's reasoning and that he has tried to convince Roberval that Fermat's method is indeed correct. Roberval's argument was that, in **Problem 4**[†] for example, it made no sense to consider three hypothetical additional rounds, because in fact the game could end in one, two, or perhaps three rounds. The difficulty with Roberval's reasoning is that it leads us to write the sample space as $\Omega = \{A_1, B_1A_2, B_1B_2A_3, B_1B_2B_3\}$. Since there are three ways out of four for A to win, a naöve application of the classical definition of probability results in the wrong division ratio of 3:1 between A and B[‡] (instead of the correct 7:1). The problem here is that the sample points in Ω above are not all equally likely, so that the classical definition cannot be applied. It is thus important to consider the *maximum* number of hypothetical rounds, namely three, for us to be able to write the sample space in terms of equally likely sample points, that is,

$$\Omega = \{A_1A_2A_3, A_1A_2B_3, A_1B_2A_3, A_1B_2B_3, B_1A_2A_3, B_1A_2B_3, B_1B_2A_3, B_1B_2B_3\},$$

from which the correct division ratio of 7:1 can be deduced.

Pascal's second observation concerns his belief that Fermat's method was not applicable to a game with three players. In the August 24, 1654 letter, Pascal says (Smith, 1929, p. 554)

> When there are but two players, your theory which proceeds by combinations is very just. But when there are three, I believe I have a proof that it is unjust that you should proceed in any other manner than the one I have.

Let us now explain how Pascal made a slip when dealing with the *Problem of Points* with three players. Pascal considers the case of three players A, B, C who are respectively 1, 2, and 2 rounds short of winning. In this case, the maximum of further rounds before the game has to finish is $(1 + 2 + 2) - 2 = 3$.[§] With three maximum rounds, there are $3^3 = 27$ possible combinations in which the three players can win each round. Pascal correctly enumerates all the 27 ways but now makes a mistake: he counts the number of favorable combinations that lead to A

[†] Recall that, in this fair game, two players A and B are, respectively, one and three rounds short of winning a prize, when the game stops.

[‡] This is exactly the same wrong line of reasoning that d'Alembert later used when calculating the probability of at least one head when a coin is tossed twice. See **Problem 12**.

[§] The general formula is: maximum number of remaining rounds = (sum of the number of rounds each player is short of winning) − (number of players) + 1.

Table 4.1   Possible Combinations When A, B, and C are 1, 2, and 2 Rounds Short of Winning the Game, Respectively

$A_1A_2A_3$ ✓	$B_1A_2A_3$ ✓	$C_1A_2A_3$ ✓
$A_1A_2B_3$ ✓	$B_1A_2B_3$ ✓	$C_1A_2B_3$ ✓
$A_1A_2C_3$ ✓	$B_1A_2C_3$ ✓	$C_1A_2C_3$ ✓
$A_1B_2A_3$ ✓	$B_1B_2A_3$ X	$C_1B_2A_3$ ✓
$A_1B_2B_3$ ✓	$B_1B_2B_3$	$C_1B_2B_3$
$A_1B_2C_3$ ✓	$B_1B_2C_3$	$C_1B_2C_3$
$A_1C_2A_3$ ✓	$B_1C_2A_3$ ✓	$C_1C_2A_3$ X
$A_1C_2B_3$ ✓	$B_1C_2B_3$	$C_1C_2B_3$
$A_1C_2C_3$ ✓	$B_1C_2C_3$	$C_1C_2C_3$

The ticks and crosses indicate the combinations that Pascal incorrectly chose to correspond to A winning the game. However, the crosses cannot be winning combinations for A because $B_1B_2A_3$ results in B winning and $C_1C_2A_3$ results in C winning.

winning the game as 19. As can be seen in Table 4.1, there are 19 combinations (denoted by ticks and crosses) for which A wins at least one round. But out of these, only 17 lead to A winning the game (the ticks), because in the remaining two (the crosses) either B or C wins the game first. Similarly, Pascal incorrectly counts the number of favorable combinations leading to B and C winning as 7 and 7, respectively, instead of 5 and 5.

Pascal thus reaches an incorrect division ratio of 19:7:7. Now Pascal again reasons incorrectly and argues that, out of the 19 favorable cases for A winning the game, six of these (namely $A_1B_2B_3$, $A_1C_2C_3$, $B_1A_2B_3$, $B_1B_2A_3$, $C_1A_2C_3$, and $C_1C_2A_3$) result in either both A and B winning the game or both A and C winning the game. So he argues the net number of favorable combinations for A should be $13 + (6/2) = 16$. Likewise he changes the number of favorable combinations for B and C, finally reaching a division ratio of 16:5½:5½. But he correctly notes that the answer cannot be right, for his own recursive method gives the correct ratio of 17:5:5. Thus, Pascal at first wrongly believed Fermat's method of enumeration could not be generalized to more than two players. Fermat was quick to point out the error in Pascal's reasoning. In his September 25, 1654 letter, Fermat explains (Smith, 1929, p. 562)

> In taking the example of the three gamblers of whom the first lacks one point, and each of the others lack two, which is the case in which you oppose, I find here only 17 combinations for the first and 5 for each of the others; for when you say that the combination *acc* is good for the first, recollect that everything that is done after one of the players has won is worth nothing. But this combination having made the first win on the first die, what does it matter that the third gains two afterwards, since even when he gains thirty all this is superfluous? The consequence, as you have well called it "this fiction," of extending the game to a certain number of plays serves only to make the rule easy and (according to my opinion) to make all the chances equal; or better, more intelligibly to reduce all the fractions to the same denomination.

Fermat thus correctly argues that a combination such as $A_1C_2C_3$ cannot be counted as favorable to both A and C, for once A has won the first round, he has won the game and the remaining rounds do not matter. Similarly, both $B_1B_2A_3$ and $C_1C_2A_3$ cannot be counted as wins for A because B and C have already won the game, respectively, by winning the first two rounds, and the third round does not matter. Thus, there are only 17 winning combinations for A (the ticks). Similarly, there are five winning combinations for B (namely $B_1B_2A_3$, $B_1B_2B_3$, $B_1B_2C_3$, $B_1C_2B_3$, and $C_1B_2B_3$), and five winning combinations for C (namely $B_1C_2C_3$, $C_1B_2C_3$, $C_1C_2A_3$, $C_1C_2B_3$, $C_1C_2C_3$). Thus, the correct division ratio should be 17:5:5.

In the same letter, Fermat further continues and gives an alternative to the method of counting that can be used to get the correct probability of A winning:

The first may win in a single play, or in two or in three.

> If he wins in a single throw, it is necessary that he makes the favorable throw with a three-faced die at the first trial. A single die will yield three chances. The gambler then has 1/3 of the wager because he plays only one third. If he plays twice, he can gain in two ways,— either when the second gambler wins the first and he the second, or when the third wins the throw and when he wins the second. But two dice produce 9 chances. The player then has 2/9 of the wager when they play twice. But if he plays three times, he can win only in two ways, either the second wins on the first throw and the third wins the second, and he the third; or when the third wins the first throw, the second the second, and he the third; for if the second or the third player wins the two first, he will win the wager and the first player will not. But three dice give 27 chances of which the first player has 27 of the chances when they play three rounds. The sum of the chances which makes the first gambler win is consequently 1/3, 2/9, and 2/17, which makes 17/27.

Fermat's reasoning is based on the waiting time for a given number of "successes." Let us generalize Fermat's idea using modern notation. Note that both Pascal and Fermat considered only fair games, but here we shall assume A and B's probabilities of winning one round are p and $q = 1 - p$, respectively. Now A is a rounds short of winning and the maximum number of possible more rounds is $a + b - 1$. He can win either on the ath round, or the $(a+1)$th round, $\dots$, or the $(a+b-1)$th round. This means that, respectively, he can either win $(a - 1)$ rounds out of a total of $(a - 1)$ rounds and then win the ath round, or win $(a - 1)$ rounds out of a total of a rounds and then win the $(a + 1)$th round, $\dots$ or win $(a - 1)$ rounds out of a total of $(a + b - 2)$ rounds and then win the $(a + b - 1)$th round. Thus A's probability of winning can also be written as

$$
\begin{aligned}
p_{a,b} &= \binom{a-1}{a-1}p^{a-1}\cdot p + \binom{a}{a-1}p^{a-1}q\cdot p + \binom{a+1}{a-1}p^{a-1}q^2\cdot p + \cdots + \binom{a+b-2}{a-1}p^{a-1}q^{b-1}\cdot p \\
&= p^a + p^a\binom{a}{1}q + p^a\binom{a+1}{2}q^2 + \cdots + p^a\binom{a+b-2}{b-1}q^{b-1} \\
&= p^a\sum_{j=0}^{b-1}\binom{a-1+j}{j}q^j.
\end{aligned}
$$

$$(4.5)$$

The reader might recognize that this second reasoning is based on the *negative binomial distribution.*[†] The latter is also sometimes called the Pascal distribution although it is Fermat who first actually made use of it for the case $p = 1/2$. Equation (4.5) first appeared in the second edition of Montmort's *Essay d'Analyse sur les Jeux de Hazard*[‡] (Montmort, 1713, p. 245) as the second formula for the *Problem of Points.*[§]

Both Fermat and Pascal therefore reached the same answer, albeit using different approaches. But whose method was more groundbreaking? Which of the two was the more significant figure in the foundation of the calculus of probabilities? David is clearly in favor of Fermat. For example, she writes (David, 1962, p. 88).

> Had Pascal left it at this, one would have marvelled at the quick intelligence which was able to generalise from such small arithmetical examples. The continuation of this present letter, however, and of succeeding letters does to a certain extent throw doubt as to whether he really understood what he was about.

Similarly, Devlin (2008, p. 62) says

> A detailed description of Pascal's recursive solution to the problem of the points quickly becomes too technical for this book. But my main reason for not including it is that Fermat's solution is simply much better.

On the other hand, Rényi (1972, p. 82) states

> The witty solution of the problem of rightful distribution (the Problem of Points) comes certainly from Pascal, and this, a recursive procedure making superfluous the actual counting of the individual cases, is in itself a significant contribution to the foundation of probability theory.

[†] In general, consider a set of Bernoulli trials (see **Problem 8**) with probability of success p and probability of failure $q = 1 - p$. Let X be the number of failures before the rth success occurs. Then X is said to have a negative binomial distribution with parameters p and r, and $\Pr\{X = x\} = \binom{x + r - 1}{r - 1} p^{r-1} q^x \cdot p =$ $\binom{x + r - 1}{x} p^r q^x$ for $x = 0, 1, 2, \ldots$. To understand why X is said to have a *negative* binomial distribution, note that $\binom{r + x - 1}{x}(-1)^x \equiv \binom{-r}{x}$, so that $\Pr\{X = x\} = \binom{-r}{x} p^r (-q)^x$. $\Pr\{X = x\}$ is thus the $(x + 1)$th term in the expansion of $p^r(1 - q)^{-r}$, which is a binomial expression with a *negative* index. The negative binomial distribution is discussed in more details in Chapter 5 of Johnson et al. (2005)

[‡] In the first edition of 1708, Montmort discussed the *Problem of Points*, but only for a fair game (Montmort, 1708, pp. 165–178).

[§] For the *Generalized Problem of Points*, see pp. 86–88.

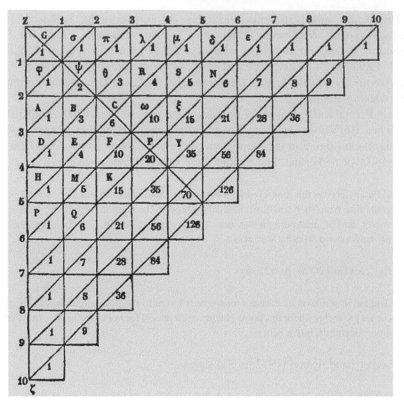

Figure 4.4   Pascal's Arithmetic Triangle, taken from the *Oeuvres Completes Vol. 2* (Pascal, 1858, p. 416).

Edwards (2003, pp. 40–52) further adds

> In section 3 [of the *Traité du Triangle Arithmétique*][†] Pascal breaks new ground, and this section, taken together with his correspondence with Fermat, is the basis of his reputation as the father of probability theory.

The correspondence between Pascal and Fermat clearly shows the latter as the consummate mathematician very much at ease with the calculus of probabilities. Fermat was fully conversant with the existing probability principles of the time and knew how to apply them to the old *Problem of Points*. On the other hand, Pascal's approach was more trailblazing and general, and his genius lies in that he was able to devise new ways to deal with old problems. Moreover, Pascal was the first to have solved the initial significant problem in the calculus of probabilities, namely the *Problem of Points*.

[†] The *Traité* must be taken into account in a discussion of the relative contributions of Pascal and Fermat toward the foundation of the probability calculus, for Fermat had already received a copy of it from Pascal prior to his correspondence with the latter.

We have explained how Pascal, one of the greatest mathematicians of his times, made a mistake when dealing with three players in the *Problem of Points*. This slip should perhaps not be judged too harshly. Indeed, such is the character of the calculus of probabilities that simple-looking problems can deceive even the sharpest minds. In his celebrated *Essai Philosophique sur les Probabilités* (Laplace, 1814a, English edition, p. 196), the eminent French mathematician Pierre-Simon Laplace (1749–1827) said

... the theory of probabilities is at bottom only common sense reduced to calculus.

There is no doubt that Laplace is right, but the fact remains that blunders and fallacies continue to persist in the field of probability, often when "common sense" is applied to problems. In another famous essay on probability, written more than a century ago, the English mathematician Morgan Crofton (1826–1915) made these very pertinent remarks (Crofton, 1885):

From its earliest beginnings, a notable feature in our subject has been the strange and insidious manner in which errors creep in – often misleading the most acute minds, as is the case of D'Alembert[†] – and the difficulty of detecting them, even when one is assured of their presence by the evident incorrectness of the result. This is probably in many cases occasioned by the poverty of language obliging us to use one term in the same context for different things – thus introducing the fallacy of ambiguous idle; e.g., the same word "probability" referring to the same event may sometimes mean its probability *before* a certain occurrence, sometimes *after;* thus the chance of a horse winning the Derby is different after the Two Thousand from what it was before. Again, it may mean the probability of the event according to one source of information, as distinguished from its probability taking everything into account; for instance, as astronomer thinks he can notice in a newly-discovered planet a rotation from east to west; the probability that this is the case is of course that of his observations in like cases turning out to be correct, if we had no other sources of information; but the actual probability is less, because we know at least that the vast majority of the planets and satellites revolve from west to east. It is easy to see that such employment of terms in the same context must prove a fruitful source of fallacies; and yet, without worrisome repetitions, it cannot always be avoided.

Our final discussion concerns Pascal's further use of mathematical expectation in a famous argument favoring belief in God.[‡] The argument appears in Pascal's *Pensées*

[†] D'Alembert will be further discussed in **Problems 11, 12**, and **13**.

[‡] Religion had an indelible impact on Pascal's short life. Pascal went through his first religious conversion to Jansenism in 1646 at the age of 23, while two male nurses were attending to his injured father for a period of three months. Jansenism was a distinct movement within the Catholic Church characterized by deep prayer, recognition of human sinfulness, asceticism, and predestination. The movement was in a bitter feud with the Jesuit establishment and the Pope. In 1647, Pascal was very ill and his physicians recommended avoiding stress. Pascal then turned to worldly pleasures. However, following a nearly fatal accident on the Pont de Neuilly, Pascal underwent an ecstatic second conversion on the night of November 23, 1654, the so-called "night of fire." This was soon after his probability correspondence with Fermat. Thereafter, Pascal gave up mathematics and retreated to the Jansenists at the Port-Royal, only to return to mathematics once in 1658. He died in 1662 at the age of 39 as one of the greatest "could-have-been" in the history of mathematics, but having done enough to still rank as one of the greatest. See Groothuis (2003) for more details.

Table 4.2   Decision Matrix in Pascal's Wager

Decision	True state of nature	
	God exists (prob. p)	God does not exist (prob. $1 - p$)
Wager for	$U = \infty$	$U = y$
Wager against	$U = x$	$U = z$

(Pascal, 1670, Fig. 4.5) and is commonly known as *Pascal's Wager*.[†] The Wager is not an argument that God *exists*. Rather it uses a decision-theoretic approach to argue that it is prudentially more rational to *believe* in God. Hacking has given Pascal's Wager an even higher importance than Pascal's work on probability with Fermat. He writes (Hacking, 1972):

> Pascal's wager was a decisive turning point in probability theory. Pascal is celebrated for the probability correspondence with Fermat, but that is not his most profound contribution to probability. Pascal's thought showed that the mathematics of games of chance had quite general applications. He made it possible for the doctrine of chances (a theory involving "objective" frequencies, whose primitive concept is expectation, or average winnings) to become the art of conjecture (a "subjective" theory about degrees of confidence).

According to Hacking, Pascal gave three versions of his wager, although Jordan identifies a fourth version within Pascal's writings (Jordan, 2006, p. 24). Here we shall focus on what Jordan dubs the "canonical version." Pascal's announces his wager as follows (Palmer, 2001, p. 327):

> ...even though there were an infinite number of chances, of which only one were in your favour, you would still be right to wager one in order to win two; and you would be acting wrongly, being obliged to play, in refusing to stake one life against three in a game, where out of an infinite number of chances there is one in your favour, if there were an infinity of infinitely happy life to be won. But here there is an infinity of infinitely happy life to be won, one chance of winning against a finite number of chances of losing, and what you are staking is finite. That leaves no choice; wherever there is infinity, and where there are not infinite chances of losing against that of winning, there is no room for hesitation, you must give everything. And thus, since you are obliged to play, you must be renouncing reason if you hoard your life rather than risk it for an infinite gain, just as likely to occur as a loss amounting to nothing.

Using modern notation, Pascal's argument can be represented in the decision matrix in Table 4.2. Here, p is the probability that God exists and is assumed to be extremely small (but nonzero).

The various U's represent *utilities* (or "values") for various decisions given the true states of nature. For example, $U = x$ means the utility of wagering against God

---

[†] Pascal's Wager is also discussed in Hacking (1972), Jordan (2006), Elster (2003, pp. 53–74), Palmer (2001, pp. 286–302), Oppy (2006, Chapter 5), Erickson and Fossa (1998, p. 150), Sorensen (2003, p. 228), and Joyce (1999, Chapter 1).

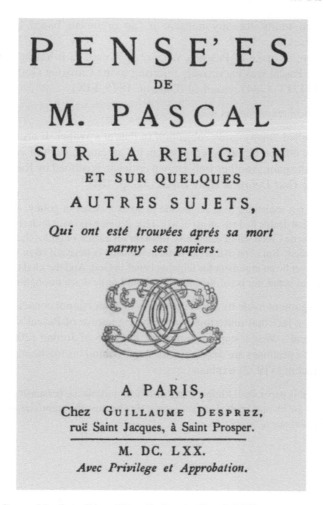

PENSE'ES

DE

M. PASCAL

SUR LA RELIGION

ET SUR QUELQUES

AUTRES SUJETS,

Qui ont esté trouvées aprés sa mort parmy ses papiers.

A PARIS,

Chez GUILLAUME DESPREZ,

ruë Saint Jacques, à Saint Prosper.

M. DC. LXX.

Avec Privilege et Approbation.

Figure 4.5   Cover of the first edition of Pascal's *Pensées* (Pascal, 1670).

(i.e., believing that God does not exist) given that He exists. Pascal contends that wagering for God when He exists brings the infinite happiness of an eternity in heaven and has infinite utility; on the other hand, x, y, and z are all finite. Then, even if the probability p that God exists is extremely small but nonzero, the expected utility of wagering for God is

$$\mathscr{E}U_{\text{for}} = (\infty)(p) + (y)(1-p) = \infty.$$

On the other hand, the expected utility of wagering against God is

$$\mathscr{E}U_{\text{against}} = (x)(p) + (y)(1-p) = p(x-y) + y \ll \mathscr{E}U_{\text{for}}.$$

Thus, the expected utility of wagering for God is infinitely bigger than the expected utility of wagering against God. Hacking calls the decision criterion based on the expected utility for any nonzero p "an argument based on dominating expectation" (Hacking, 1972).

What are we to make of Pascal's Wager? Sure enough it has attracted a lot of criticism. Since Pascal was exclusively referring to the Christian God in his Wager, Denis Diderot (1713–1784) remarked (Diderot, 1875, LIX)

> An Imam could reason just as well.

This is the so-called "many-gods" objection to Pascal's Wager. It argues that Pascal does not include other religious alternatives in his argument, each of which could lay claim to an infinite gain. Another well-known criticism is echoed by Richard Dawkins (b. 1941) in *The God Delusion* (Dawkins, 2006, p. 104):

> Believing is not something you can decide to do as a matter of policy. At least, it is not something I can decide to do as an act of will. I can decide to go to church and I can decide to recite the Nicene Creed, and I can decide to swear on a stack of bibles that I believe every word inside them. But none of that can make me actually believe it if I don't. Pascal's wager could only ever be an argument for feigning belief in God. And the God that you claim to believe in had better not be of the omniscient kind or he'd see through the deception.

The essential argument made by Dawkins here is that one cannot consciously decide to believe in God, a fact that undermines the whole premise of Pascal's Wager. Other critiques of Pascal's Wager can be found in Chapter 5 of Jordan (2006). However, although Pascal's premises are debatable, the *logic* behind his mathematical argument is valid. As Hacking (1972) explains

> Nowhere in this paper shall I imply that Pascal's arguments are persuasive. But this is not because they are invalid (in the logician's sense) but because the premisses of the arguments are, at best, debatable.

Problem 5

Huygens and the Gambler's Ruin (1657)

Problem. *Two players A and B, having initial amounts of money a and (t − a) dollars, respectively, play a game in which the probability that A wins a round is p and the probability that B wins is q = 1 − p. Each time A wins, he gets one dollar from B, otherwise he gives B one dollar. What is the probability that A will eventually get all of B's money?*

Solution. Let $w_{a,t}$ be the probability that A eventually wins all the t dollars (i.e., ruins B), starting with a dollars. Then, conditioning on the first round, we can obtain the following difference equation for $w_{a,t}$:

$$w_{a,t} = pw_{a+1,t} + qw_{a-1,t} \tag{5.1}$$

with $w_{t,t} \equiv 1, w_{0,t} \equiv 0$. To solve the above difference equation, we let $w_{a,t} = \lambda^a$, where λ is a constant. Therefore, Eq. (5.1) becomes

$$\lambda^a = p\lambda^{a+1} + q\lambda^{a-1}$$
$$p\lambda^2 - \lambda + q = 0$$
$$\lambda = \frac{1 \pm \sqrt{1 - 4pq}}{2p}$$
$$= \frac{1 \pm \sqrt{1 - 4p + 4p^2}}{2p}$$
$$= \frac{1 \pm (1 - 2p)}{2p}$$
$$= \frac{q}{p}, 1.$$

Classic Problems of Probability, Prakash Gorroochurn.
© 2012 John Wiley & Sons, Inc. Published 2012 by John Wiley & Sons, Inc.

In the case $p \neq q$, we have two distinct roots so that $w_{a,t} = C + D(q/p)^a$, where C and D are arbitrary constants. Substituting $w_{t,t} \equiv 1, w_{0,t} \equiv 0$, we have

$$C + D(q/p)^t = 1$$

$$C + D = 0,$$

giving $C = -D = -[(q/p)^t - 1]^{-1}$. Therefore, for $p \neq q$, we have

$$w_{a,t} = \frac{(q/p)^a - 1}{(q/p)^t - 1}.$$

For $p = q$, we consider the above expression in the limit as $z = q/p$ approaches 1:

$$w_{a,t} = \lim_{z \to 1} \frac{z^a - 1}{z^t - 1}$$

$$= \lim_{z \to 1} \frac{az^{a-1}}{tz^{t-1}}$$

$$= \frac{a}{t}.$$

Hence, we have

$$w_{a,t} = \begin{cases} \dfrac{(q/p)^a - 1}{(q/p)^t - 1}, & p \neq q, \\[2ex] \dfrac{a}{t}, & p = q. \end{cases} \tag{5.2}$$

5.1 Discussion

The groundbreaking work of Pascal and Fermat, decisive as it was, did not get published in any of the journals of the time. Pascal had communicated to the *Académie Parisienne*[†] in 1654, that he would soon publish the foundations of a brand new science: *Alea geometria* or the Mathematics of Chance (Pascal, 1858, pp. 391–392). However, this project never came to fruition. Pascal was busy with other projects and fighting inner religious battles. As for Fermat, he was not interested in publishing his mathematical works himself and has rightly been called the "Prince of Amateurs."[‡] Destiny had left

---

[†] Founded in 1635 by the mathematician and music theorist Marin Mersenne (1588–1648), the *Académie Parisienne* would in 1666 grow into the *Académie des Sciences*. See Droesbeke and Tassi (1990, pp. 24–25).

[‡] See, for example, Bell's *Men of Mathematics* (Bell, 1953, pp. 60–78). Samueli and Boudenot (2009, p. 36) report that Fermat published only one manuscript during his lifetime, although his mathematical works were quite extensive.

Figure 5.1   Christiaan Huygens (1629–1695).

it to the prominent Dutch astronomer and mathematician Christiaan Huygens (1629–1695) to popularize his two famous contemporaries' work (Fig. 5.1).

In 1655, a few months after Pascal and Fermat's correspondence, Huygens visited France. There he became aware of the *Problem of Points*, although he did not meet any of the two Frenchmen. However, Huygens met the mathematicians Claude Mylon (1618–1660), who was a friend of Carcavi, and Gilles de Roberval, with whom Pascal had previously discussed de Méré's questions. Back in the Netherlands, Huygens wrote a small manuscript on probability, which he sent to his mentor, the Dutch mathematician Franciscus van Schooten (1615–1660). The latter translated Huygens' manuscript into Latin and included it in his book *Exercitationum mathematicarum*. Huygens' treatise came out in 1657 under the name *De ratiociniis in aleae ludo*[†] (Fig. 5.2) and, at 18 pages, was the first *published* book on

---

[†] On the Calculations in Games of Chance. An English translation of this book was published in 1692 by the Scottish physician John Arbuthnott (1667–1735) under the title "Of the Laws of Chance, or, a Method of Calculation of the Hazards of Game, Plainly demonstrated, And applied to Games as present most in Use." However, Arbuthnott's work was not just a translation; he included a preface, solutions to Huygens' problems, and comments of his own.

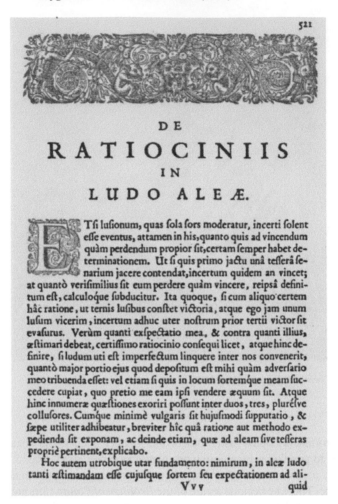

Figure 5.2 First page of Huygens' *De ratiociniis in aleae ludo*, taken from van Schooten's *Exercitationum mathematicarum* (Huygens, 1657).

probability (Huygens, 1657). Although neither Pascal nor Fermat is mentioned in the book, Huygens is quick to acknowledge in the preface (David, 1962, p. 114):

> It should be said, also, that for some time some of the best mathematicians of France[†] have occupied themselves with this kind of calculus so that no one should attribute to me the honour of the first invention. This does not belong to me. But these savants, although they put each other to the test by proposing to each other many questions difficult to solve, have hidden their methods. I have had therefore to examine and to go deeply for myself into this matter by beginning with the elements, and it is impossible for me for this reason to affirm

---

[†] Although he did so indirectly, Huygens thus became the first to acknowledge Pascal and Fermat as the founders of the probability calculus. See Pichard (2004, p. 18).

Figure 5.3   The last four problems in Huygens' *De ratiociniis in aleae ludo*, taken from van Schooten's *Exercitationum mathematicarum* (Huygens, 1657). The very last problem (Problem 5) is the famous *Gambler's Ruin Problem*.

that I have even started from the same principle. But finally I have found that my answers in many cases do not differ from theirs.

It appears from the above that Huygens did not get the solutions to the problems Pascal and Fermat had previously discussed, and had to work them out for himself. However, historian Ernest Coumet has cast some doubts on this premise (Coumet, 1981, p. 129). Referring to a 1656 letter from Huygens to van Schooten, Coumet advances the hypothesis that Huygens might have received hints on the solutions of these problems from Pascal through the intermediary of de Méré. Moreover, it is known that Carcavi did later communicate the essential results, but not the detailed solutions, of Pascal and Fermat to Huygens.[†]

The *De ratiociniis* contained 14 propositions and concluded with five exercises for the reader. It is the fifth exercise that is of interest here (see Fig. 5.3):

[†] See Courtebras (2008, p. 76).

A and B taking 12 Pieces of Money each, play with 3 Dice on this Condition, That if the Number 11 is thrown, A shall give B one Piece, but if 14 shall be thrown, then B shall give one to A; and he shall win the Game that first gets all the Pieces of Money. And the Proportion of A's Chance to B's is found to be, as 244,140,625 to 282,429,536,481.

We now describe, using modern notation, the solution that Huygens later provided in Volume 14 of the *Oeuvres Complètes* (Huygens, 1920). The reader should note that Huygens was the first to have *formalized* the use of mathematical expectations[†]. In *Proposition III* of his book, Huygens (1657) explicitly states

If I have p chances of obtaining a and q chances of obtaining b, and if every chance can happen as easily, it is worth to me as much as $(pa + qb)/(p + q)$.

In modern terms, if the probability of a player winning an amount a is $p/(p + q)$ and the probability of her winning an amount b is $q/(p + q)$, then the expected value of her win W is

$$\mathscr{E}W = \sum_i w_i \Pr\{W = w_i\}$$

$$= a \cdot \frac{p}{p + q} + b \cdot \frac{q}{p + q}$$

$$= \frac{pa + qb}{p + q}.$$

Similarly, let $I = 1$ if a player wins a game, the probability of the latter event being π, and let $I = 0$ otherwise. Then the expectation of I is

$$\mathscr{E}I = 1(\pi) + 0(1 - \pi) = \pi.$$

The player's expectation to win in a game, or more simply her expectation, is thus her probability to win.

Coming back to Huygens' fifth exercise, let $\mathscr{E}_i$ be A's expectation of ruining B when he has won a net amount i from B. Let α be the probability of obtaining a score of 14, and β the probability of a score of 11, when three dice are thrown, conditional on a throw of either a 14 or an 11. Huygens first considers the hypothetical cases of A ruining B whenever A leads by two points, and of B ruining A whenever A lags by two points. *This is equivalent to saying that A and B both start with two units of money and whoever wins all four units ruins his opponent.* The problem then is to find the value of $\mathscr{E}_0$, which is A's expectation of ruining B at the start of the game (and also at any time A and B have won equal amounts). We have $\mathscr{E}_2 \equiv 1$, $\mathscr{E}_{-2} \equiv 0$, and

$$\mathscr{E}_1 = \alpha(1) + \beta\mathscr{E}_0 = \alpha + \beta\mathscr{E}_0,$$

$$\mathscr{E}_{-1} = \alpha\mathscr{E}_0 + \beta(0) = \alpha\mathscr{E}_0,$$

$$\mathscr{E}_0 = \alpha\mathscr{E}_1 + \beta\mathscr{E}_{-1}.$$

[†] Although Pascal was the first to have actually used it, as we noted in **Problem 4**.

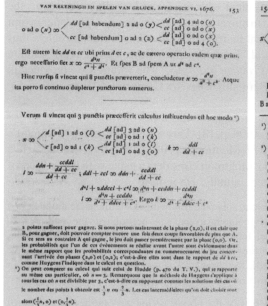

Figure 5.4   Extract from Huygens' proof of the formula for the Gambler's Ruin Problem, taken from the *Oeuvres Completes*, Vol. 14 (Huygens, 1920).

Huygens also makes use of a probability tree, the first indeed, to illustrate the last set of relationships.[†] Eliminating $\mathscr{E}_{-1}$ and $\mathscr{E}_1$, we have

$$\mathscr{E}_0 = \frac{\alpha^2}{\alpha^2 + \beta^2}.$$

This implies that B's probability of ruining A is $\beta^2/(\alpha^2 + \beta^2)$, giving a relative probability for A and B of α^2/β^2. Huygens then considers the case when both players start with four and eight units of money each, and obtains probability ratios of α^4/β^4 and α^8/β^8, respectively (see Fig. 5.4). Huygens states that for the case of n units of money the relative probability is α^n/β^n, without proving this claim by induction. By using $\alpha/\beta = 5/9$[‡] and $n = 12$, we obtain a probability ratio of

$$\left(\frac{\alpha}{\beta}\right)^{12} = \left(\frac{5}{9}\right)^{12} = \frac{244,140,625}{282,429,536,481},$$

which is the answer Huygens gave in the *De ratiociniis* (see Fig. 5.3).

[†] See also Edwards (1983) and Shoesmith (1986).

[‡] This is the ratio of the probability of obtaining a 14 to the probability of obtaining an 11 with three dice.

Although Huygens is usually credited with the *Gambler's Ruin Problem*,[†] the problem was actually communicated to him at a later stage by Fermat. The latter in turn had been asked the problem by Pascal in 1656. The *Gambler's Ruin Problem* thus seems to actually originate from Pascal.[‡] Edwards (1983) traces Pascal's own solution to the difference equation from the *Problem of Points*.[§] The solution we provided in **Problem 5** was first given by Jacob Bernoulli in the *Ars Conjectandi* (Bernoulli, 1713, p. 99).

An even more difficult related problem at the time was the *Problem of the Duration of the Play* (see, e.g., Roux, 1906, p. 18). This was first treated in Montmort's *Essay d'Analyse sur les Jeux de Hazard* (Montmort, 1708, p.184), then in de Moivre's *De Mensura Sortis* (de Moivre, 1711). Montmort considers a game with two players of equal skill, each having three counters initially. He writes

> We can, by the method in this Problem, solve the following: Determine the duration of a game where one would play always by counting excess wins according to the conditions we have explained, p.178. One will find, for example, that if we play according to these rules with three counters each the odds that the game will end in seven rounds are 37 to 27; & in five rounds an odds of 7 to 9. One will find in the formula $\frac{1}{4} + \frac{3}{4^2} + \frac{3^2}{4^3} + \frac{3^3}{4^4} + \frac{3^4}{4^5} +$ &c. the reason for the odds that the game will finish in a certain number of rounds ...

Thus by adding the first three terms of the above series we obtain probability of 37/64 giving odds of 37 to 27; by adding the first two terms we obtain a probability of 7/16, giving odds of 7 to 9.

The general *Problem of the Duration of the Play* can be answered by finding the probability mass function of N, the number of games before either A or B is ruined. Suppose A and B initially have a and $t - a$ dollars, and their probabilities of winning one game are p and $q = 1 - p$, respectively. Let $u_a^{(n)}$ be the probability that A is ruined on the nth game, starting with a dollars. Then, conditioning on the first game, we obtain

$$u_a^{(n)} = pu_{a+1}^{(n-1)} + qu_{a-1}^{(n-1)}, \tag{5.3}$$

where

$$u_0^{(n)} \equiv u_t^{(n)} \equiv 0, \qquad \text{for } n \geq 1,$$
$$u_0^{(0)} \equiv 1, \tag{5.4}$$
$$u_a^{(0)} \equiv 0, \qquad \text{for } a > 0.$$

[†] The gambler's ruin problem is also discussed in Ibe (2009, p. 220), Mosteller (1987, p. 54), Rosenthal (2006, p. 75), Andel (2001, p. 29), Karlin and Taylor (1975, pp. 92–94), Ethier (2010, Chapter 7), and Hassett and Stewart (2006, p. 373).

[‡] However, historian Ernest Coumet (1965a), cites an interesting paragraph from Cardano's *Practica arithmeticae generalis omnium copiosissima & utilissima*, Chapter LXI, S 17 (Cardano, 1539), which discusses a problem very similar to the *Gambler's Ruin*. See p. 109 (footnote) for more.

[§] See **Problem 4**.

We now follow Pierre-Simon Laplace (1749–1827) who solved the problem in its full generality through the method of probability generating functions (Laplace, 1812, p. 225).[†] The reader might wish to review this method in **Problem 9** before moving on. We first define the probability generating function of the number of games (N_1) before A is ruined:

$$G_a(s) \equiv \mathcal{E} s^{N_1} = \sum_{n=0}^{\infty} u_a^{(n)} s^n.$$

Equation (5.3) becomes

$$\sum_{n=0}^{\infty} u_a^{(n)} s^n = p \sum_{n=1}^{\infty} u_{a+1}^{(n-1)} s^n + q \sum_{n=1}^{\infty} u_{a-1}^{(n-1)} s^n$$

$$= ps \sum_{n=0}^{\infty} u_{a+1}^{(n)} s^n + qs \sum_{n=0}^{\infty} u_{a-1}^{(n)} s^n$$

Therefore, we have

$$G_a(s) = ps G_{a+1}(s) + qs G_{a-1}(s), \tag{5.5}$$

where, from Eq. (5.4), $G_0(s) \equiv 1$ and $G_t(s) \equiv 0$. To solve this difference equation, we let $G_a(s) = \xi^a$. Equation (5.5) then becomes

$$1 = ps\xi + qs\xi^{-1} \Rightarrow \xi_1, \xi_2 = \frac{1 \pm \sqrt{1 - 4pqs^2}}{2ps}.$$

Using these two roots, we can write the general solution of Eq. (5.5) as

$$G_a(s) = C_1 \xi_1^a + C_2 \xi_2^a. \tag{5.6}$$

Since $G_0(s) \equiv 1$ and $G_t(s) \equiv 0$, we have

$$\begin{cases} C_1 + C_2 = 1, \\ C_1 \xi_1^t + C_2 \xi_2^t = 0. \end{cases}$$

Solving for C_1 and C_2 and substituting in Eq. (5.6), we finally obtain

$$G_a(s) = \mathcal{E} s^{N_1} = \left(\frac{q}{p}\right)^a \frac{\xi_1^{t-a} - \xi_2^{t-a}}{\xi_1^t - \xi_2^t}.$$

Similarly, the generating function for the number of games (N_2) before B is ruined is

$$\mathscr{E}\, s^{N_2} = \frac{\xi_1^{a} - \xi_2^{a}}{\xi_1^{t} - \xi_2^{t}}.$$

Since $\Pr\{N = n\} = \Pr\{N_1 = n\} + \Pr\{N_2 = n\}$, the probability generating function of the number of games until either A or B is ruined is

$$\mathscr{E}\, s^{N} = \mathscr{E}\, s^{N_1} + \mathscr{E}\, s^{N_2} = \frac{(q/p)^{a}(\xi_1^{t-a} - \xi_2^{t-a}) + (\xi_1^{a} - \xi_2^{a})}{\xi_1^{t} - \xi_2^{t}}. \tag{5.7}$$

A closed-form expression for $\Pr\{N = n\}$ could be obtained by extracting the coefficient of s^n in the generating function above, but is complex.[†] However, Eq. (5.7) can be differentiated to obtain the expected duration of the game[‡]:

$$\mathscr{E}N = \frac{d}{ds}\, \mathscr{E}\, s^{N}\bigg|_{s=1} = \begin{cases} \dfrac{1}{q-p}\left[a - \dfrac{t\{1 - (q/p)^{a}\}}{1 - (q/p)^{t}}\right], & p \neq q, \\[2ex] a(t - a), & p = q = 1/2. \end{cases}$$

[†] See Feller (1968, pp. 352–354).

[‡] An alternative method would be to solve the following difference equation for $\mathscr{E}N = D_a$: $D_a = pD_{a+1} + qD_{a-1} + 1$, subject to $D_0 = 0$ and $D_a = 0$.

Problem 6

The Pepys–Newton Connection (1693)

Problem. *A asserts that he will throw at least one six with six dice. B asserts that he will throw at least two sixes by throwing 12 dice. C asserts that he will throw at least three sixes by throwing 18 dice. Which of the three stands the best chance of carrying out his promise?*

Solution. The solution involves the application of the binomial distribution. Let X be the number of sixes when n balanced dice are thrown independently, $n = 1, 2, \ldots$ Then $X \sim B(n, 1/6)$ and

$$\Pr\{X = x\} = \binom{n}{x}(1/6)^x(5/6)^{n-x}, \quad x = 0, 1, 2, \ldots, n.$$

For A, $X \sim B(6, 1/6)$ and

$$\begin{aligned} \Pr\{X \geq 1\} &= 1 - \Pr\{X = 0\} \\ &= 1 - (5/6)^6 \\ &= .665. \end{aligned}$$

For B, $X \sim B(12, 1/6)$ and

$$\begin{aligned} \Pr\{X \geq 2\} &= 1 - \Pr\{X = 0\} - \Pr\{X = 1\} \\ &= 1 - (5/6)^{12} - 12(1/6)(5/6)^{11} \\ &= .619. \end{aligned}$$

For C, $X \sim B(18, 1/6)$ and

$$\Pr\{X \geq 3\} = 1 - \Pr\{X = 0\} - \Pr\{X = 1\} - \Pr\{X = 2\}$$
$$= 1 - (5/6)^{18} - 18(1/6)(5/6)^{17} - 153(1/6)^2(5/6)^{16}$$
$$= .597.$$

Thus, A is more likely than B who, in turn, is more likely than C of carrying out his promise.

6.1 Discussion

The above problem is of historical significance because it was posed to none other than Sir Isaac Newton (1643–1727) (Fig. 6.2) by naval administrator, Member of

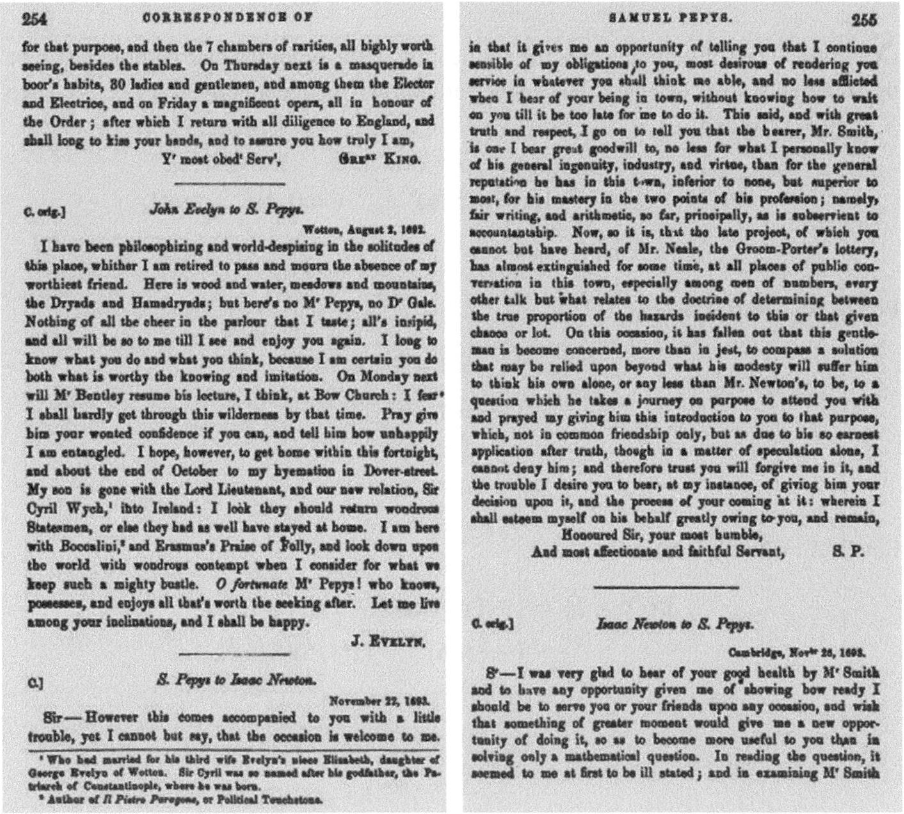

Figure 6.1   Pepys' November 22, 1693 accompanying letter to Newton, taken from Pepys (1866).

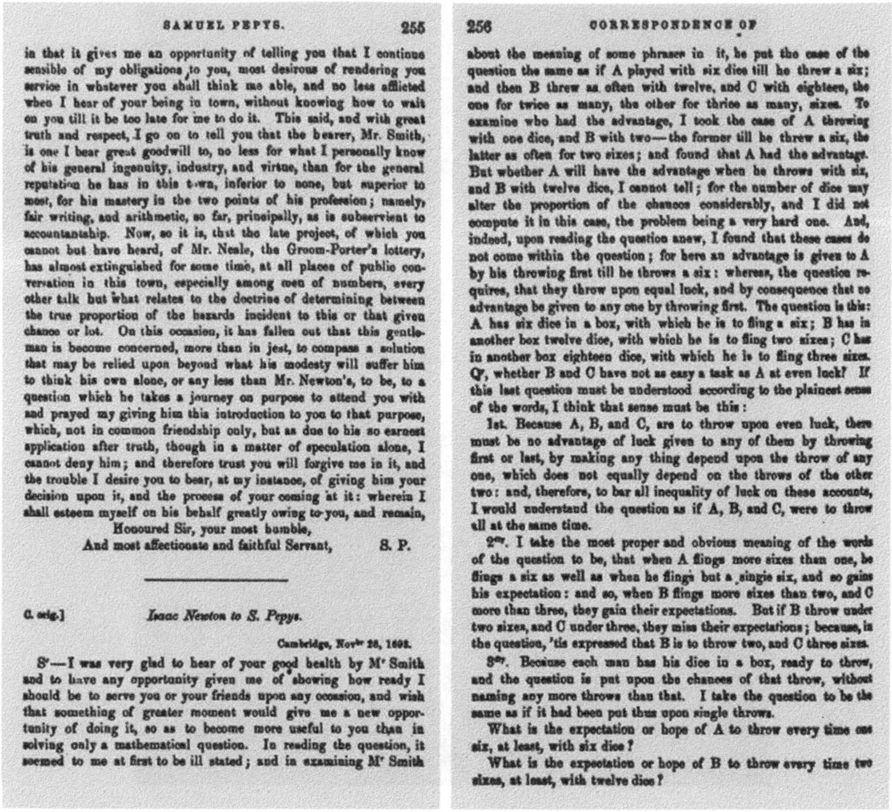

in that it gives me an opportunity of telling you that I continue sensible of my obligations to you, most desirous of rendering you service in whatever you shall think me able, and no less afflicted when I hear of your being in town, without knowing how to wait on you till it be too late for me to do it. This said, and with great truth and respect, I go on to tell you that the bearer, Mr. Smith, is one I bear great goodwill to, no less for what I personally know of his general ingenuity, industry, and virtue, than for the general reputation he has in this town, inferior to none, but superior to most, for his mastery in the two points of his profession; namely, fair writing, and arithmetic, so far, principally, as is subservient to accountantship. Now, so it is, that the late project, of which you cannot but have heard, of Mr. Neale, the Groom-Porter's lottery, has almost extinguished for some time, at all places of public conversation in this town, especially among men of numbers, every other talk but what relates to the doctrine of determining between the true proportion of the hazards incident to this or that given chance or lot. On this occasion, it has fallen out that this gentleman is become concerned, more than in jest, to compass a solution that may be relied upon beyond what his modesty will suffer him to think his own alone, or any less than Mr. Newton's, to be, to a question which he takes a journey on purpose to attend you with and prayed my giving him this introduction to you to that purpose, which, not in common friendship only, but as due to his so earnest application after truth, though in a matter of speculation alone, I cannot deny him; and therefore trust you will forgive me in it, and the trouble I desire you to bear, at my instance, of giving him your decision upon it, and the process of your coming at it: wherein I shall esteem myself on his behalf greatly owing to you, and remain,

Honoured Sir, your most humble,

And most affectionate and faithful Servant,   S. P.

G. orig.]   *Isaac Newton to S. Pepys.*

Cambridge, Nov^br 26, 1693.

S^r—I was very glad to hear of your good health by Mr Smith and to have any opportunity given me of showing how ready I should be to serve you or your friends upon any occasion, and wish that something of greater moment would give me a new opportunity of doing it, so as to become more useful to you than in solving only a mathematical question. In reading the question, it seemed to me at first to be ill stated; and in examining Mr Smith

about the meaning of some phrases in it, he put the case of the question the same as if A played with six dice till he threw a six; and then B threw as often with twelve, and C with eighteen, the one for twice as many, the other for thrice as many, sixes. To examine who had the advantage, I took the case of A throwing with one dice, and B with two—the former till he threw a six, the latter as often for two sixes; and found that A had the advantage. But whether A will have the advantage when he throws with six, and B with twelve dice, I cannot tell; for the number of dice may alter the proportion of the chances considerably, and by consequence that no compute it in this case, the problem being a very hard one. And, indeed, upon reading the question anew, I found that these cases do not come within the question; for here an advantage is given to A by his throwing first till he throws a six: whereas, the question requires, that they throw upon equal luck, and by consequence that no advantage be given to any one by throwing first. The question is this: A has six dice in a box, with which he is to fling a six; B has in another box twelve dice, with which he is to fling two sixes; C has in another box eighteen dice, with which he is to fling three sixes. Q^r, whether B and C have not as easy a task as A at even luck? If this last question must be understood according to the plainest sense of the words, I think that sense must be this:

1st. Because A, B, and C, are to throw upon even luck, there must be no advantage of luck given to any of them by throwing first or last, by making any thing depend upon the throw of any one, which does not equally depend on the throws of the other two: and, therefore, to bar all inequality of luck on these accounts, I would understand the question as if A, B, and C, were to throw all at the same time.

2^ly. I take the most proper and obvious meaning of the words of the question to be, that when A flings more sixes than one, he flings a six as well as when he flings but a single six, and so gains his expectation: and so, when B flings more sixes than two, and C more than three, they gain their expectations. But if B throw under two sixes, and C under three, they miss their expectations; because, in the question, 'tis expressed that B is to throw two, and C three sixes.

3^ly. Because each man has his dice in a box, ready to throw, and the question is put upon the chances of that throw, without naming any more throws than that. I take the question to be the same as if it had been put thus upon single throws.

What is the expectation or hope of A to throw every time one six, at least, with six dice?

What is the expectation or hope of B to throw every time two sixes, at least, with twelve dice?

Figure 6.2   Extract from Newton's letter of November 26, 1693 to Pepys, taken from Pepys (1866).

Parliament, and diarist Samuel Pepys (1633–1703) in 1693.[†] The question was sent to Newton by Mr. Smith, the writing-master at Christ's Hospital. The accompanying letter from Pepys is shown in Fig. 6.1.

Pepys' original question required the probability for "exact" numbers of sixes, rather than "at least" a certain number of sixes. In his reply to Pepys (Fig 6.1), Newton made the latter change stating the original question was ambiguous. Newton's correct answer was that the probabilities are in decreasing order for A, B, and C, stating the following reason (Pepys, 1866, p. 257)

> If the question be thus stated, it appears by an easy computation, that the expectation of A is greater than that of B or C; that is, the task of A is the easiest; and the reason is because A has all the chances of sixes on his dyes for his expectation, but B and C have not all the chances on theirs; for when B throws a single six, or C but one or two sixes, they miss of their expectations.

[†] The Pepys–Newton connection is also discussed in David (1962, Chapter 12), Schell (1960), Pedoe (1958, pp. 43–48), and Higgins (1998, p. 161).

Figure 6.3   Sir Isaac Newton (1643–1727).

However, as Stigler (2006) has noted, although Newton's answer was correct, his reasoning was not completely beyond reproach. The ordering in the three probabilities is not true in general, and depends crucially on their being of the form $\Pr\{X \geq np\}$.[†] As Pepys was not convinced of the answer and thought that C would have the highest probability, Newton wrote back to him, giving details of the calculations he used and the exact probabilities for the first two cases, namely 31,031/46,656 and 1,346,704,211/2,176,782,336. David (1962, p. 128) notes that Newton stopped at 12 dice and that the "arithmetic would have been formidable for 18 dice and Huygens would have used logarithms."

The Pepys–Newton problem seems to be a rare venture Newton made into probability. David (1962, p. 126) states that

> ...Newton did not research in the probability calculus; he may have disapproved of gambling...

[†] Stigler gives an example, taken from Evans (1961), in which the ordering in not true: $\Pr\{X \geq 1 | n = 6, p = 1/4\} = .822 < \Pr\{X \geq 2 | n = 12, p = 1/4\} = .842$.

Mosteller (1987, p. 34) adds

As far as I know this is Newton's only venture into probability.

However, Newton did use probabilistic ideas and methods on some occasions in his writings (Sheynin, 1971). Gani (2004) states

The correspondence with Pepys, outlined earlier, provides convincing evidence that Newton was conversant with the calculus of probabilities of his time.

In any case, when asked mathematical questions, presumably of the probabilistic type, Newton used to recommend his good friend Abraham de Moivre (1667–1754): "Go to Mr. de Moivre; he knows these things better than I do." (Bernstein, 1996, p. 126). The feeling was mutual, for de Moivre dedicated the first edition of his epic *Doctrine of Chances* (de Moivre, 1718) to Newton:

The greatest help I have received in writing on this subject having been from your incomparable works, especially your method of series; I think it my duty publicly to acknowledge that the improvements I have made in the matter here treated of are principally derived from yourself. The great benefit which has accrued to me in this respect, requires my share in the general tribute of thanks due to you from the learned world. But one advantage, which is more particularly my own, is the honour I have frequently had of being admitted to your private conversation, wherein the doubts I have had upon any subject relating to Mathematics have been resolved by you with the greatest humanity and condescension. Those marks of your favour are the more valuable to me because I had no other pretence to them, but the earnest desire of understanding your sublime and universally useful speculations. . ..

Problem 7

Rencontres with Montmort (1708)

Problem. *Suppose you have a jar containing n balls numbered 1, 2, 3, ..., n, respectively. They are well mixed and then drawn one at a time. What is the probability that at least one ball is drawn in the order indicated by its label (e.g., the ball labeled 2 could be the second ball drawn from the jar)?*

Solution. There are $n!$ permutations (or ordered sequences) of the n balls $\{1, 2, \ldots, n\}$. Assume that the jth element in a permutation represents the jth ball drawn. Let P_j ($j = 1, 2, \ldots, n$) be the property that, in a given permutation, the jth ball drawn has label j, and let A_j denote the set of permutations with property P_j. We now compute the total number of permutations, L_n, for which at least one ball is drawn in the order indicated by its label. We have

$$L_n \equiv |A_1 \cup A_2 \cup \cdots \cup A_n|$$

$$= \sum_i |A_i| - \sum_{i<j} |A_i \cap A_j| + \sum_{i<j<k} |A_i \cap A_j \cap A_k|$$

$$- \cdots + (-1)^{n+1} |A_1 \cap A_2 \cap \cdots \cap A_n|.$$

In the above, $|S|$ denotes the number of elements in the set S, and we have used the principle of inclusion and exclusion.[†] Now $|A_i \cap A_j| = (n-2)!$ and there are $\binom{n}{2}$ such

---

[†] See, for example, Johnson et al. (2005, Chapter 10) and Charalambides (2002, p. 26).

Classic Problems of Probability, Prakash Gorroochurn.
© 2012 John Wiley & Sons, Inc. Published 2012 by John Wiley & Sons, Inc.

combinations, and similarly for the others. Therefore,

$$L_n = \binom{n}{1}(n-1)! - \binom{n}{2}(n-2)! + \binom{n}{3}(n-3)! - \cdots + (-1)^{n+1}(1)$$

$$= \frac{n!}{1!} - \frac{n!}{2!} + \frac{n!}{3!} - \cdots + (-1)^{n+1}\frac{n!}{n!}$$

$$= n!\left\{1 - \frac{1}{2!} + \frac{1}{3!} - \cdots + \frac{(-1)^{n+1}}{n!}\right\}.$$

Hence, the probability that at least one ball is drawn in the order indicated by its label is

$$p_n \equiv \frac{L_n}{n!} = 1 - \frac{1}{2!} + \frac{1}{3!} - \cdots + \frac{(-1)^{n+1}}{n!}. \tag{7.1}$$

7.1 Discussion

This problem is the best-known contribution to the theory of probability of the French mathematician Pierre Rémond de Montmort (1678–1719).[†] The latter included it in his treatise *Essay d'Analyse sur les Jeux de Hazard*[‡] (Montmort, 1708) (Fig. 7.1), a work largely inspired by the then deceased Jacob Bernoulli. The original problem went under the heading of the game of *Treize*[§] (see Fig. 7.2) and is also known as the *Problème des Rencontres*[**] (Montmort, 1708, pp. 54–64). It involves the throwing of 13 cards numbered $1, 2, \ldots, 13$. The cards are then drawn out singly and the problem asks for the probability that, once at least, the number on a card coincides with the number of the order in which it was drawn. In his 1865 classic *A History of the Mathematical Theory of Probability From the Time of Pascal to That of Laplace*, which still ranks as one of the best history books on probability ever written, the great mathematical historian Isaac Todhunter (1820–1884) (Fig. 7.5) notes, with his usual attention to detail (1865, p. 91):

[†] For a biography of Montmort, see Hacking (1980c).

[‡] An Analytical Essay on the Games of Chance.

[§] The comprehensive book by van Tenac (1847) gives a full description of various games of the time. The game of *Treize* appears on p. 98.

[**] Problem of Coincidences. The *Problème des Rencontres* is further discussed in Hald (1990, Chapter 19), Chuang-Chong and Khee-Meng (1992, pp. 160–163), MacMahon (1915, Chapter III), Riordan (1958, Chapter 3) and Dorrie (1965, pp. 19–21).

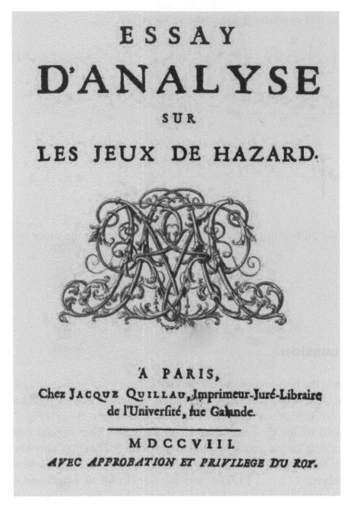

ESSAY
D'ANALYSE
SUR
LES JEUX DE HAZARD.

'A PARIS,
Chez JACQUE QUILLAU, Imprimeur-Juré-Libraire
de l'Univerfité, rue Galande.

MDCCVIIL
AVEC APPROBATION ET PRIVILEGE DU ROY.

Figure 7.1   Cover of the first edition of the *Essay d'Analyse sur les Jeux de Hazard.*

In his first edition Montmort did not give any demonstrations of his results; but in his second edition he gives two demonstrations which he had received from Nicolas Bernoulli. . .

(see Fig. 7.3). Indeed, the 1708 publication of the first edition of *Essay d'Analyse* prompted the reaction from three renowned mathematicians of the time, namely John Bernoulli (1667–1748) (Fig. 7.4) (archrival of the older brother Jacob), Nicholas Bernoulli[†] (1687–1759) (nephew of Jacob and John, and student of the former), and Abraham de Moivre (1667–1754). First was a somewhat harsh letter from John pointing out errors and omissions in Montmort's book, as well as general solutions to

---

[†] Also known as Nicholas I Bernoulli.

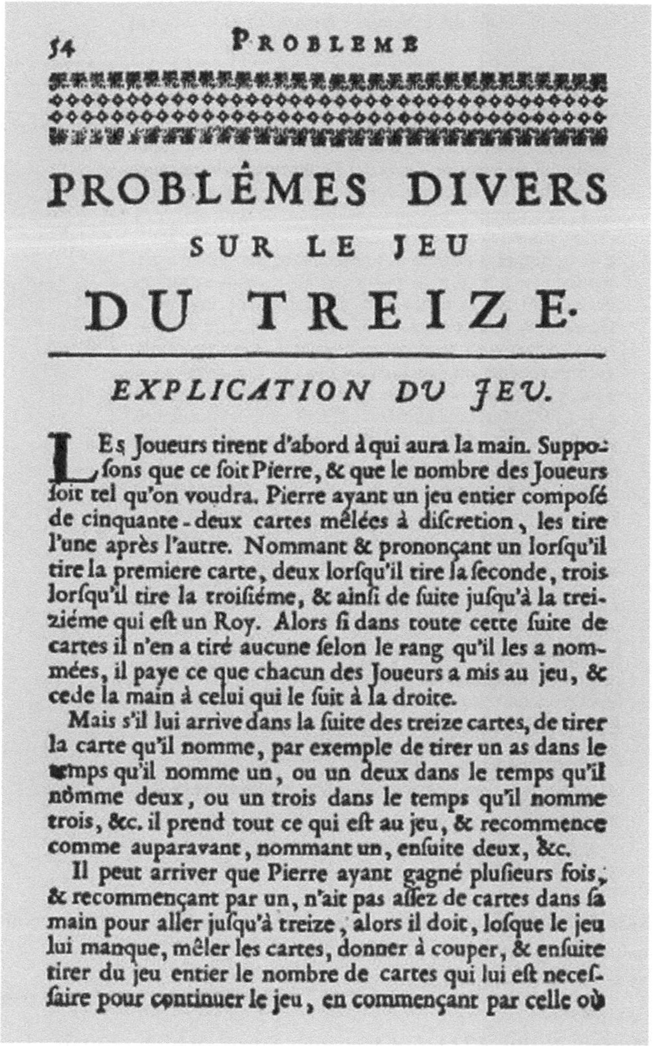

54 PROBLEME

PROBLÊMES DIVERS

SUR LE JEU

DU TREIZE.

EXPLICATION DU JEU.

LES Joueurs tirent d'abord à qui aura la main. Suppoſons que ce ſoit Pierre, & que le nombre des Joueurs ſoit tel qu'on voudra. Pierre ayant un jeu entier compoſé de cinquante-deux cartes mêlées à diſcretion, les tire l'une après l'autre. Nommant & prononçant un lorſqu'il tire la premiere carte, deux lorſqu'il tire la ſeconde, trois lorſqu'il tire la troiſiéme, & ainſi de ſuite juſqu'à la treiziéme qui eſt un Roy. Alors ſi dans toute cette ſuite de cartes il n'en a tiré aucune ſelon le rang qu'il les a nommées, il paye ce que chacun des Joueurs a mis au jeu, & cede la main à celui qui le ſuit à la droite.

Mais s'il lui arrive dans la ſuite des treize cartes, de tirer la carte qu'il nomme, par exemple de tirer un as dans le temps qu'il nomme un, ou un deux dans le temps qu'il nomme deux, ou un trois dans le temps qu'il nomme trois, &c. il prend tout ce qui eſt au jeu, & recommence comme auparavant, nommant un, enſuite deux, &c.

Il peut arriver que Pierre ayant gagné pluſieurs fois, & recommençant par un, n'ait pas aſſez de cartes dans ſa main pour aller juſqu'à treize, alors il doit, loſque le jeu lui manque, mêler les cartes, donner à couper, & enſuite tirer du jeu entier le nombre de cartes qui lui eſt neceſſaire pour continuer le jeu, en commençant par celle où

Figure 7.2 Montmort's "Jeux de Treize" from the first edition of the *Essay* (Montmort, 1708, p. 54).

some of the problems Montmort had previously considered. On the other hand, Nicholas' remarks were more constructive and started a fruitful correspondence between him and Montmort.† John's and Nicholas' input led to the publication by Montmort in 1713 of a greatly expanded and improved second edition of the *Essay d'Analyse*. In the new edition, Montmort reproduced the letters from both John and Nicholas, giving credit to both of them. However, Montmort was not equally

† It was through this correspondence that the so-called St Petersburg problem was first enunciated by Nicholas Bernoulli. This problem went on to gain immortal status through its later publication by Nicholas' younger cousin, Daniel Bernoulli. See **Problem 11**.

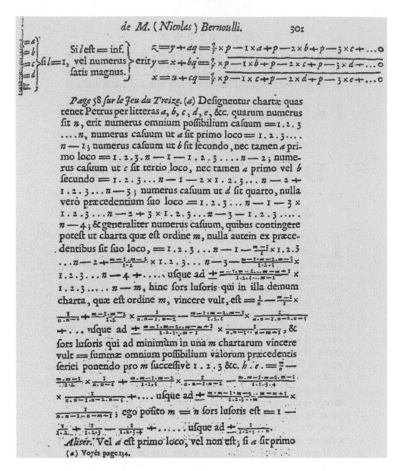

Figure 7.3   Nicholas Bernoulli's proof of the solution to the *Game of Treize*, as it appears in the second edition of Montmort's *Essay* (Montmort, 1713, p. 301).

kind to de Moivre. In his 1711 *De Mensura Sortis*, which contained many of the problems in Montmort's 1708 book, de Moivre had been quite critical of the latter (de Moivre, 1711; Hald, 1984):

> By thy exhortation, most distinguished sir, I have solved certain problems pertaining to the die, and have revealed the principles by which their solution may be undertaken; and now I have published them by command of the Royal Society. Huygens was the first that I know who presented rules for the solution of this sort of problems, which a French author has very recently well illustrated with various examples; but these distinguished gentlemen do not seem to have employed that simplicity and generality which the nature of the matter demands: moreover, while they take up many unknown quantities, to represent the various conditions of gamesters, they make their calculation too complex; and while they suppose that the skill of gamesters is always equal, they confine this doctrine of games within limits too narrow. . .

Figure 7.4   John Bernoulli (1667–1748).

In the second edition of his book, Montmort replied measure for measure (Montmort, 1713, p. xxvii):

> M. Moivre was right that I would need his book [De Mensura Sortis] to answer to the critique he made of mine in his Advertisement. His commendable intention to promote and elevate his work has made him demean mine and dispute the novelty of my methods. As he has thought it possible to attack me without me complaining, I believe I can answer him without giving him reason to complain against me. . .

However, by 1718, when the first edition of de Moivre's *Doctrine of Chances* came out, the two men seemed to have made up. In the preface of his book, de Moivre (1718, p. ii) writes

> Since the printing of my specimen [de Mensura Sortis], Mr. de Montmort, author of the Analyse des Jeux du Hazard, published a second edition of that book, in which he has particularly given proofs of his singular genius, and extraordinary capacity, which testimony I give both to truth, and with the friendship with which he is pleased to honour me.

Figure 7.5   Isaac Todhunter (1820–1884).

Coming back to the game of *Treize*, Montmort attacks the problem by first giving the solution for a few specific cases. Using the same reasoning as for these cases, he then gives the following recursive equation for p_n (Montmort, 1708, p. 58):

$$p_n = \frac{(n-1)p_{n-1} + p_{n-2}}{n}, \quad n \geq 2, \quad p_0 = 0, \ p_1 = 1. \tag{7.2}$$

To understand how Eq. (7.2) is derived, let us write $q_n = 1 - p_n$, where q_n is the probability of a *derangement* of n balls, that is, a permutation for which no ball in the ith position has label i ($i = 1, 2, \ldots, n$). Consider a particular derangement. Suppose the ith ball drawn has label j ($\neq i$). Then the jth ball drawn can either have label i with probability $1/n$ (and the remaining $(n-2)$ balls are also "deranged") or the jth ball drawn can have a label different from i with probability $(1 - 1/n)$ (and the remaining $(n-1)$ balls, i.e., with ball i excluded, are also "deranged"). Therefore, we have

$$q_n = \frac{1}{n}q_{n-2} + \left(1 - \frac{1}{n}\right)q_{n-1}.$$

Replacing $q_n = 1 - p_n$, we obtain Eq. (7.2). Using the latter formula, Montmort calculates $p_{13} = 109{,}339{,}663/172{,}972{,}800 \approx .632120558$. Montmort also gives the explicit formula we derived in our solution for p_n (Montmort, 1708, p. 59). Using a result from Leibniz,[†] he shows that

$$\lim_{n \to \infty} p_n = 1 - e^{-1} (\approx .632120558).$$

David (1962, p. 146) mentions that this was perhaps the first exponential limit in the calculus of probability. Coming back to p_n, we make two observations. First, its value settles down to .6321 (to 4 d.p.) as soon as $n = 7$. It remains the same (to 4 d.p.) even if $n = 10^6$. The reason for this behavior can be understood from the series of rapidly decreasing terms in Eq. (7.1). The second observation relates to the relatively large value of p_n itself, for $n \geq 7$. If the ball in jar experiment with $n = 7$ balls is repeated many times, in slightly less than 2 out of 3 repetitions, at least one ball will on average will be drawn in the order indicated by its label. These are pretty good odds.

In the 1751 memoir, *Calcul de la probabilité dans le jeu de rencontre*, the great mathematician Leonhard Euler (1707–1783) also solved the *Problème des Rencontres*, independently of Nicolas Bernoulli (Euler, 1751). Compared to his works in mathematical analysis, Euler's contributions in probability were modest, although he did write four papers on the Genoese Lottery. His other probability contributions are described in Chapter 12 of Debnath (2010).

[†] Namely the well-known formula $e^x = 1 + x + x^2/2! + x^3/3! + \ldots$ for all x.

Problem 8

Jacob Bernoulli and his Golden Theorem (1713)

Problem. *Show that the probability of an equal number of heads and tails when a fair coin is tossed 2m times is approximately $1/\sqrt{\pi m}$ for large m.*

Solution. Let X be the number of heads for $2m$ independent tosses of a fair coin. Then $X \sim B(2m, 1/2)$ and

$$\Pr\{X = x\} = \binom{2m}{x}\left(\frac{1}{2}\right)^x\left(\frac{1}{2}\right)^{2m-x}$$

$$= \frac{1}{2^{2m}}\binom{2m}{x}, \qquad x = 0, 1, \ldots, 2m.$$

The probability of an equal number of heads and tails is

$$\Pr\{X = m\} = \frac{1}{2^{2m}}\binom{2m}{m}$$

$$= \frac{1}{2^{2m}} \cdot \frac{(2m)!}{(m!)^2}.$$

Applying Stirling's formula $N! \sim \sqrt{2\pi}N^{N+1/2}\,e^{-N}$ for large N, we have

$$\Pr\{X = m\} \approx \frac{1}{2^{2m}} \cdot \frac{\sqrt{2\pi}(2m)^{2m+1/2}\,e^{-2m}}{\left[\sqrt{2\pi}\,m^{m+1/2}\,e^{-m}\right]^2}$$

$$= \frac{1}{2^{2m}} \cdot \frac{\sqrt{2\pi}(2m)^{2m+1/2}\,e^{-2m}}{2\pi m^{2m+1}e^{-2m}} \qquad\qquad (8.1)$$

$$= \frac{1}{\sqrt{\pi m}}.$$

Classic Problems of Probability, Prakash Gorroochurn.
© 2012 John Wiley & Sons, Inc. Published 2012 by John Wiley & Sons, Inc.

Table 8.1   The Probability of an Equal Number of Heads and Tails When a Fair Coin is Tossed $2m$ Times

Number of tosses of a fair coin ($2m$)	Prob. of an equal number of heads and tails
10	.1262
20	.0892
50	.0564
100	.0399
500	.0178
1,000	.0126
10,000	.0040
100,000	.0013
1,000,000	.0004

8.1 Discussion

The last formula implies that, for a fair coin, the probability of an equal number of heads and tails is inversely proportional to $\sqrt{m}$. Table 8.1 shows that, as more coins are tossed, it becomes less likely that there will be an equal number of heads and tails.

This last statement might appear as a direct contradiction of Bernoulli's Law of Large Numbers. Let us briefly explain this law before dealing with the apparent contradiction. Bernoulli's law is the simplest form of the *Weak Law of Large Numbers* (WLLN) and considers n independent tosses of a coin with probability of heads p such that the total number of heads is S_n.[†] Then Bernoulli's law states that the proportion of heads converges in probability to p as n becomes large, that is,

$$\lim_{n \to \infty} \Pr\left\{ \left| \frac{S_n}{n} - p \right| < \varepsilon \right\} = 1, \tag{8.2}$$

where $\varepsilon > 0$ is an arbitrarily small number.[‡] In other words, consider an experiment that consists of tossing a coin n times, yielding a sequence of heads and tails. Suppose the experiment is repeated a large number of times, resulting in a large number of sequences of heads and tails. Let us consider the proportion of heads for each sequence. Bernoulli's law implies that, for a *given* large n, *the fraction of sequences (or experiments) for which the proportions of heads is arbitrarily close to p is high, and increases with n.* That is, the more tosses we perform for a given experiment, the

---

[†] In general, independent and identically random variables each of which can take only two values are eponymously called Bernoulli random variables.

[‡] This limit statement itself means that for any $\varepsilon > 0$ and $\delta > 0$, no matter how small, an N can be found sufficient large so that, for all $n \geq N$, we have $\Pr\{|S_n/n - p| < \varepsilon\} \geq 1 - \delta$.

Table 8.2   As the Number of Tosses Increases, the Proportion of Heads Gets Closer to .5. Yet, at the same time, the absolute difference between the number of heads and tails keeps on increasing

Number of tosses of a fair coin $(2m)$	Proportion of heads	Number of heads	Difference in the Number of heads and tails
1,000	.400	400	200
5,000	.450	2,250	500
10,000	.530	5,300	600
50,000	.520	26,000	2,000
100,000	.490	49,000	2,000
500,000	.495	247,500	5,000
1,000,000	.497	497,000	6,000

greater the probability that the proportion of heads in the corresponding sequence will be very close to the true p. This fact is very useful in practice. For example, it gives us confidence in using the frequency interpretation of probability.[†]

Let us now return to the apparent contradiction. Applying Bernoulli's law to our case of a fair coin, we see that the proportion of heads in $2m$ tosses should be very close to 1/2 with increasing probability as m increases. Doesn't this then mean that the probability of an equal number of heads and tails should increase with m? The answer is no. In fact, it is still possible for the *proportion* of heads to be ever *closer* to 1/2 while the *difference* in the number of heads and tails is ever larger. Consider Table 8.2. As the number of tosses increases, we see that the proportion of heads gets closer to 1/2 in spite of the fact the absolute difference between the number of heads and tails increases. Thus, there is no contradiction with Table 8.1.

Digging further, we now prove that, as the number of tosses of a fair coin increases, it is increasingly likely that the absolute difference in the number of heads and tails becomes arbitrarily large, even as the proportion of heads converges to 1/2.

Let X be the number of heads in n independent tosses of a fair coin. Then $X \sim B$ $(n, 1/2)$ and the absolute difference between the number of heads and tails is $|X - (n - X)| = |2X - n|$. For any $\Delta > 0$, we have

$$\Pr\{|2X - n| > \Delta\} = 1 - \Pr\{|2X - n| \leq \Delta\}$$

$$= 1 - \Pr\{-\Delta \leq 2X - n \leq \Delta\}$$

$$= 1 - \Pr\left\{\frac{n}{2} - \frac{\Delta}{2} \leq X \leq \frac{n}{2} + \frac{\Delta}{2}\right\}.$$

[†] The various interpretations of probability are discussed in **Problem 14**.

We now use the normal approximation[†] $X \sim N(n/2, n/4)$ so that

$$\Pr\{|2X - n| > \Delta\} \approx 1 - \Pr\left\{\frac{-\Delta}{\sqrt{n}} \leq Z \leq \frac{\Delta}{\sqrt{n}}\right\},$$

where $Z \sim N(0, 1)$. As $n \to \infty$, we have

$$\Pr\{|2X - n| > \Delta\} \to 1 - \Pr\{Z = 0\} = 1 - 0 = 1.$$

Thus, it is increasingly likely that the absolute difference in the number of heads and tails exceeds any finite number, however large, as the number of tosses of a fair coin increases. This is in spite of the fact that the ratio of the number of heads to the total number of tosses is increasingly likely to be very close to 1/2. The absolute difference is likely to increase because this difference becomes ever more variable as the number of tosses increases:

$$\begin{aligned} \text{var}|2X - n| &= \text{var}(2X - n) \\ &= 4 \text{ var } X \\ &= 4(n/4) \\ &= n, \end{aligned}$$

which becomes infinitely large as $n \to \infty$.

The great Swiss mathematician Jacob Bernoulli[‡] (1654–1705) had every right to be proud of his WLLN, which he first enunciated in the fourth part of the *Ars Conjectandi*[§] (Bernoulli, 1713) (Figs. 8.1–8.3). Before we dig deeper into Bernoulli's celebrated law, let us recall Bernoulli's use of the binomial distribution. He is considered to have been the first to explicitly use it.[**] In the first part of the *Ars Conjectandi* (p. 45), Bernoulli considers a game consisting of n independent trials in each of which a player has the same probability p of bringing about a particular event. Such trials are now eponymously called *Bernoulli trials*. Further, Bernoulli considers

---

[†] In general, if $X \sim B(n, p)$, then for large n, $X \sim N(np, npq)$ approximately, where $q = 1 - p$. The history of this approximation is given in **Problem 10**.

[‡] Also known as James Bernoulli.

[§] The Art of Conjecturing. Bernoulli probably got the name of his book from Arnauld and Nicole's celebrated *La Logique, ou l'Art de Penser* (Arnauld and Nicole 1662), which in Latin is *Logica, sive Ars Cogitandi*.

[**] Both Pascal and Montmort had previously indirectly worked with the binomial distribution, but only for the symmetric case of $p = 1/2$.

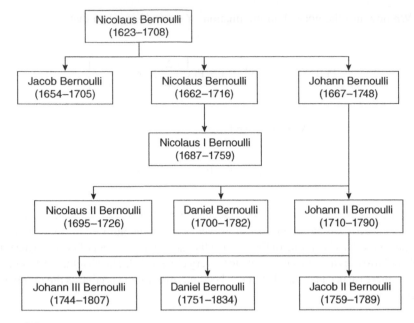

Figure 8.1 The Bernoulli family tree.

the probability that the player brings about the event a total of m times out of n, and gives the probability as

$$\frac{n(n-1)(n-2)\cdots(n-m+1)}{1\cdot 2\cdot 3\cdot \cdots \cdot m}p^m(1-p)^{n-m} \equiv \frac{n(n-1)(n-2)\cdots(m+1)}{1\cdot 2\cdot 3\cdot \cdots \cdot (n-m)}p^m(1-p)^{n-m}.$$

In modern notation, suppose an experiment consists of n Bernoulli trials each with "success" probability p, and let X be the number of successes. Then X is said to a have a $B(n, p)$ distribution, and the probability that $X=x$ is given by

$$p_X(x) = \Pr\{X = x\}$$
$$= \frac{n!}{x!(n-x)!}p^x(1-p)^{n-x}$$
$$= \binom{n}{x}p^x(1-p)^{n-x}, \quad x = 0, 1, \ldots, n.$$

Let us now come back to Bernoulli's WLLN, which he called his "golden theorem,"[†] On p. 227 of the *Ars Conjectandi*, Bernoulli says

[†] The name "law of large numbers" was first given later by Poisson (1781–1840).

JAKOB BERNOULLI um 1687

Figure 8.2   Jacob Bernoulli (1654–1705).

This is therefore the problem that I now wish to publish here, having considered it closely for a period of twenty years, and it is a problem of which the novelty as well as the high utility together with its grave difficulty exceed in value all the remaining chapters of my doctrine. Before I treat of this "Golden Theorem" I will show that a few objections, which certain learned men have raised against my propositions, are not valid.

Gouraud (1848, p. 19) eloquently puts the *Ars Conjectandi* in context as follows:

> . . .The calculus of probabilities had been walking like the chance whose name it claimed to erase from the dictionary of human knowledge. We had indeed, here and there, in France for the first time, on the occasion of a difficulty in a game, in Holland, twenty years later, in a matter of public economy, in England finally, after another twenty years, regarding a similar problem, evaluated fairly accurately some probabilities and some chances, but nobody had thought of defining what chance was and what probability was; we had not further explained under what law and on the basis of what philosophy, we were applying mathematics to such matters; we were lacking in principles as in definitions, and finally, if it was true that thanks to the genius of Fermat we possessed a method, that method itself still needed to be studied in radically greater depth to serve the new analysis. It is not

Figure 8.3   Cover of the *Ars Conjectandi*.

surprising that we had all waited to give the calculus of probabilities the share of attention it deserved, that it finally emerged from this state of experimentation, to show itself in the definite character a doctrine.

The first days of the XVIIIth century were destined to finally see the accomplishment of this great progress. Another few years and thanks to the fruitful vigils of one of the finest geniuses of modern analysis, the organization that it lacked was to be acquired.

Historian Ivor Grattan-Guinness has edited a book that includes Bernoulli's *Ars Conjectandi* as one of the landmark writings in western mathematics.[†]

Let us now discuss the relevant part of Bernoulli's *Ars Conjectandi* in greater detail and try to understand how Bernoulli obtained his celebrated theorem. First Bernoulli enunciates his concept of probability (Bernoulli, 1713)[‡]

> Certainty of some thing is considered either *objectively* and in itself and means none other than its real existence at present or in the future; or *subjectively*, depending on us, and consists in the measure of our knowledge of this existence... (p. 210)

[†] The book contains a full analysis of the *Ars Conjectandi* by Schneider (2005b).

[‡] The three sets of translations which follow are taken from Oscar Sheynin's translations of Chapter 4 of the *Ars Conjectandi* (Sheynin, 2005).

As to *probability*, this is the degree of certainty, and it differs from the latter as a part from the whole. Namely, if the integral and absolute certainty, which we designate by letter α or by unity 1, will be thought to consist, for example, of five probabilities, as though of five parts, three of which favor the existence or realization of some event, with the other ones, however, being against it, we will say that this event has $\frac{3}{5}\alpha$, or $\frac{3}{5}$, of certainty (p. 211).

In the first paragraph above, Bernoulli points to certainty being either logically or subjectively determined. The subjective interpretation is an *epistemic* concept,[†] that is, it relates to our degree of knowledge. In the second paragraph, Bernoulli defines probability as "the degree of certainty" that "differs from the latter (i.e., certainty) as a part from the whole." Bernoulli thus conceives probability as either objective or epistemic. Further in his book, Bernoulli introduces *a priori* and *a posteriori* probabilities (p. 224):

> ...what we are not given to derive a priori, we at least can obtain *a posteriori*, that is, can extract it from a repeated observation of the results of similar examples. Because it should be assumed that each phenomenon can occur and not occur in the same number of cases in which, under similar circumstances, it was previously observed to happen and not to happen. Actually, if, for example, it was formerly noted that, from among the observed three hundred men of the same age and complexion as Titius now is and has, two hundred died after ten years with the others still remaining alive, we may conclude with sufficient confidence that Titius also has twice as many cases for paying his debt to nature during the next ten years than for crossing this border.

The *a priori* probability Bernoulli refers to in the above is the classical definition. On the other hand the *a posteriori* probability is the frequency concept, by which Bernoulli plans to estimate the *a priori* probability. Bernoulli's "program" is thus enunciated as follows:

> To make clear my desire by illustration, I suppose that without your knowledge three thousand white pebbles and two thousand black ones are hidden in an urn, and that, to determine [the ratio of] their numbers by experiment, you draw one pebble after another (but each time returning the drawn pebble before extracting the next one so that their number in the urn will not decrease), and note how many times is a white pebble drawn, and how many times a black one. It is required to know whether you are able to do it so many times that it will become ten, a hundred, a thousand, etc., times more probable (i.e., become at last morally certain) that the number of the white and the black pebbles which you extracted will be in the same ratio, of 3 to 2, as the number of pebbles themselves, or cases, than in any other different ratio. (p. 226)

> Let the number of fertile cases be to the number of sterile cases precisely or approximately as r to s; or to the number of all the cases as r to $r + s$, or as r to t so that this ratio is contained between the limits $(r + 1)/t$ and $(r - 1)/t$. It is required to show that it is possible to take such a number of experiments that it will be in any number of times (for example, in c times) more likely that the number of fertile observations will occur between these limits rather than beyond them, that is, that the ratio of the number of fertile observations to the number of all of them will be not greater than $(r + 1)/t$ and not less than $(r - 1)/t$. (p. 236)

[†] See **Problem 14**.

Thus, Bernoulli's aim was to determine *how many times an experiment ought to be repeated before one can be "morally" certain that the relatively frequency of an event is close enough to its true probability, given that the true probability of the event is a known quantity.* In modern notation,[†] let h_n be the frequency of an event with probability $p = r/t$, where $t = r + s$ and n is the number of independent trials. Bernoulli's aim is to show that for any given small positive number $\varepsilon = 1/t$ and given large natural number c, we can find an n such that

$$\frac{\Pr\{|h_n - p| \le 1/t\}}{\Pr\{|h_n - p| > 1/t\}} > c, \tag{8.3}$$

which is equivalent to

$$\Pr\left\{|h_n - p| \le \frac{1}{t}\right\} > \frac{c}{c+1}.$$

That is, we can find an n such that h_n is within a tolerance $1/t$ of p with probability greater than $c/(c+1)$. Bernoulli first considers the expression $(p + q)^n = (r + s)^n t^{-n}$, where $q = 1 - p = s/t$. We have

$$(r + s)^n \equiv \sum_{x=0}^{n} \binom{n}{x} r^x s^{n-x} = \sum_{i=-kr}^{ks} f_i, \tag{8.4}$$

where

$$f_i \equiv \binom{kr + ks}{kr + i} r^{kr+i} s^{ks-i}, \quad i = -kr, -kr + 1, \ldots, ks.$$

In the above, $k = n/t$ is chosen to be a positive integer, for convenience. To prove Eq. (8.3), Bernoulli shows that the central term of the series in Eq. (8.4) plus k terms on each side is larger than c times the sum of the $k(r - 1)$ terms of the left tail and the $k(s - 1)$ terms of the right tail. Since $f_{-i}(k, r, s) \equiv f_i(k, s, r)$ for $i = 0, 1, \ldots, kr$, it suffices for Bernoulli to rigorously prove that

$$\frac{\sum_{i=1}^{k} f_i}{\sum_{i=k+1}^{ks} f_i} \ge c,$$

which he was able to do as long as

$$n \ge \max\left\{m_1 t + \frac{st(m_1 - 1)}{r + 1}, m_2 t + \frac{rt(m_2 - 1)}{s + 1}\right\}$$

[†] Bernoulli's original method is fully described in Stigler (1986, pp. 66–70), Uspensky (1937, Chapter VI), Henry (2004), Rényi (2007, p. 195), Schneider (2005b), Liagre (1879, pp. 85–90), Montessus (1908, pp. 50–57).

with

$$m_1 \geq \frac{\ln\{c(s-1)\}}{\ln(1+r^{-1})}, \quad m_2 \geq \frac{\ln\{c(r-1)\}}{\ln(1+s^{-1})}.$$

Next, Bernoulli gives an example with $r = 30$, $s = 20$, so that $p = 3/5$. He wants to have a high certainty that $|h_n - 3/5| \leq 1/(30 + 20)$, that is, $29/50 \leq h_n \leq 31/50$. Bernoulli chooses $c = 1000$ that corresponds to a moral certainty of $1000/1001$. Then, how large a sample size is needed so that

$$\Pr\left\{\left|h_n - \frac{3}{5}\right| \leq \frac{1}{50}\right\} > \frac{1000}{1001}$$

is satisfied? With these numbers, Bernoulli obtains $m_1 = 301$, $m_1 t + st(m_1 - 1)/(r + 1) = 24{,}750$, and $m_2 = 211$, $m_2 t + rt(m_2 - 1)/(s + 1) = 25{,}500$. Therefore, with a sample size of $n = 25{,}500$, there is more than a $1000/1001$ probability (i.e., it is morally certain) that $29/50 \leq h_n \leq 31/50$.

The sample size of $n = 25{,}500$ that Bernoulli calculates is disappointingly large and may have been a major reason why he did not finish the *Ars Conjectandi* before he died.[†] As Stigler (1986, p. 77) explains

> . . .Bernoulli's upper bound was a start but it must have been a disappointing one, both to Bernoulli and to his contemporaries. To find that 25,550 experiments are needed to learn the proportion of fertile to sterile cases within one part in fifty is to find that nothing reliable can be learned in a reasonable number of experiments. The entire population of Basel was then smaller than 25,550; Flamsteed's 1725 catalogue listed only 3,000 stars. The number 25,550 was more than astronomical; for all practical purposes it was infinite. I suspect that Jacob Bernoulli's reluctance to publish his deliberations of twenty years on the subject was due more to the magnitude of the number yielded by his first and only calculation than to any philosophical reservations he may have had about making inferences based upon experimental data.

> The abrupt ending of the Ars Conjectandi (Figure 2.1) would seem to support this- Bernoulli literally quit when he saw the number 25,550. . .

However, this should in no way belittle Bernoulli's major achievement. His law was the first limit theorem in probability. It was also the first attempt to apply the calculus of probability outside the realm of games of chance. In the latter, the calculation of odds through classical (or mathematical) reasoning had been quite successful because the outcomes were equally likely. However, this was also a limitation of the classical method. Bernoulli's law provided an empirical framework

---

[†] It was thus left to Nicholas Bernoulli (1687–1759), Jacob Bernoulli's nephew, to publish the *Ars Conjectandi* in 1713, 8 years after his uncle's death. Nicholas Bernoulli later refined Jacob Bernoulli's rather loose bound for the law of large numbers (Montmort, 1713, pp. 388–394), and also interested de Moivre in the problem, which culminated in de Moivre's discovery of the normal approximation to the binomial distribution (see **Problem 10** for more details). Meusnier (2006) writes "Concerning his uncle, his cousins, the historiography, the bernoullian probabilistic project in its different components, for all of it, Nicolas is an outstanding nephew, in all the meanings of the term."

that enabled the estimation of probabilities even in the case of outcomes that were not equally likely. Probability was no longer only a mathematically abstract concept. Rather, now it was a quantity that could be estimated with increasing confidence as the sample size became larger. It is not that the latter fact was completely unknown: most mathematicians before had tacitly assumed it to be "obvious." But Bernoulli was the first to rigorously prove that this should be so. Tabak (2004, p. 36) further adds

> In his book Jacob Bernoulli showed that there was a robust and well-defined structure for the class of independent random processes... Bernoulli succeeded in demonstrating the existence of a deep structure associated with events that until then had simply been described as unpredictable.

Nowadays, few textbooks give Bernoulli's original proof because there is a much simpler one using Chebyshev's inequality that was discovered in 1853,[†] much later than Bernoulli's times. Chebyshev's inequality states for any random variable X with mean μ and variance $\sigma^2 < \infty$, we have

$$\Pr\{|X - \mu| \geq \varepsilon\} \leq \frac{\sigma^2}{\varepsilon^2}$$

for any $\varepsilon > 0$. Setting $X = S_n/n$ we have $\mu = p$ and $\sigma^2 = pq/n$, so that

$$\Pr\left\{\left|\frac{S_n}{n} - p\right| \geq \varepsilon\right\} \leq \frac{pq}{n\varepsilon^2} \leq \frac{1}{4n\varepsilon^2} \to 0, \quad \text{as } n \to \infty,$$

and Bernoulli's WLLN is proved.[‡] Although Chebyshev's inequality is very useful, it is cruder than Bernoulli's result. To see this, let us write the above inequality as

$$\Pr\left\{\left|\frac{S_n}{n} - p\right| < \varepsilon\right\} \geq 1 - \frac{pq}{n\varepsilon^2}.$$

We now solve $1 - pq/(n\varepsilon^2) = 1000/1001$, where $p = 1 - q = 3/5$ and $\varepsilon = 1/50$, obtaining $n = 600,600$. This is much larger than the sample size of $n = 25,500$ from Bernoulli's result. In **Problem 10**, we shall see that the sample size can be greatly reduced by using the Central Limit Theorem (CLT).

We mentioned before that Bernoulli' law belongs to the class of weak laws. It is worth noting that there is a stronger form of Bernoulli's law, known as Borel's Strong Law of Large Numbers (SLLN).[§] The latter also considers n independent tosses of a

---

[†] The inequality was first proved in 1853 by Irenée-Jules Bienaymé (1796–1878) and later in 1867 by Pafnuty Lvovich Chebyshev (1821–1894). Chebyshev's inequality is thus often more appropriately called Bienaymé-Chebyshev's inequality.

[‡] Here we have used the fact that the maximum value of $pq = p(1 - p)$ is $1/4$.

[§] Due to the eminent French mathematician Emile Borel (1871–1956).

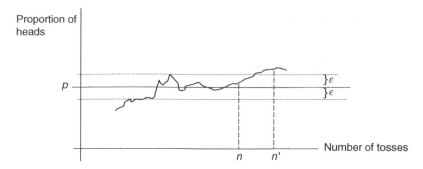

Figure 8.4   For a given large n, one particular sequence for which the proportion of heads is very close to p. However, for an even larger n (say n') the proportion of heads for that sequence drifts away from p. The question is: what do the weak and strong laws say about such a scenario? The weak law cannot guarantee it will not happen; on the other hand, the strong law says it has probability zero.

coin with probability of heads p and total number of heads S_n. Borel's SLLN states that the proportion of heads converges *almost surely* to p, that is,

$$\Pr\left\{\lim_{n\to\infty}\frac{S_n}{n}=p\right\}=1. \tag{8.5}$$

The reader might find it difficult to understand the difference between Eqs. (8.2) and (8.5). So let us spend some time to explain this. Again, consider an experiment that consists of tossing a coin a large number n times, yielding a sequence of heads and tails. The experiment is repeated a large number of times, resulting in a large number of sequences of heads and tails. We look at the proportion of heads for each sequence. Bernoulli's law states that, for a *given* large n, the fraction of sequences for which the proportions of heads is arbitrarily close to p is large, and increases with n. Thus, for a given large n and a given sequence, there is a high probability that the proportion of heads will be close to p. Now consider the following scenario: for a given large n, one particular sequence is among those for which the proportion of heads is very close to p, but for an even larger n (say n') the proportion of heads for that sequence drifts away from p (see Fig. 8.4). Can Bernoulli's law guarantee this will not happen? The answer is no. Bernoulli's law only tells us that most sequences will be close to p for a given large n, but cannot assure us that they will each remain close for even larger n. This is where the strong law comes in. It explicitly states that scenarios like the one we have just described above are so unlikely that they have a probability of zero. *That is, the strong law not only states that, for a given sequence, the proportion of heads will most likely be close to p for a reasonably large n, but also that the proportion of heads will most likely remain close to p for all larger values of n in that sequence.* In contrast to Bernoulli's WLLN, Borel's SLLN thus deals with the whole *infinite* sequence of heads and tails, and asserts that the set of sequences for which

$|S_j/j - p| > \varepsilon$ for $j > n$, where n is reasonably large, has probability zero. Eq. (8.5) can therefore also be written as[†]

$$\lim_{n \to \infty} \Pr\left\{ \left| \frac{S_j}{j} - p \right| > \varepsilon, \quad \text{for } j = n+1, n+2, \ldots \right\} = 0.$$

The reader should realize that just because the event $\{|S_j/j - p| > \varepsilon, j > n\}$, where n is large, has a probability of zero does not mean it is *impossible*.[‡] Rather, it means that the event is so rare that assigning it any nonzero probability, however small, would overestimate its chance of occurring. In fact, as a consequence of the first Borel-Cantelli lemma,[§] the probability of zero also means that, as the number of repetitions of the experiment of tossing the coin is increased without bound, the event $\{|S_j/j - p| > \varepsilon, j > n\}$ occurs over only *finitely many* such repetitions. Thus, we speak of "almost sure convergence" rather than "sure convergence" when we refer to the strong law of large numbers. Let us give an explicit example, again in the context of coin tossing, to make this point clear. Consider a coin-tossing experiment for which $X_i = 1$ if the ith toss is a head (with probability p) and $X_i = 0$ otherwise (with probability $1 - p$). In the long run, we would expect the proportion of heads to be p, that is, we would expect

$$\lim_{n \to \infty} \frac{X_1 + X_2 + \cdots + X_n}{n} = p, \quad \text{for all } \omega \in \Omega.$$

But this is not true for *all* $\omega \in \Omega$. For example, let $\omega^* = \{0, 0, \ldots\}$, the sequence of all tails. For this ω^*,

$$\lim_{n \to \infty} \frac{X_1 + X_2 + \cdots + X_n}{n} = 0.$$

The strong law, though, tells us that ω^* is so unlikely that $\Pr\{\omega^*\} = 0$. Therefore, we speak of "almost sure convergence" because we do not have convergence for all ω, but have convergence for *almost all* ω, such that the set of ω's for which convergence does not occur has probability zero.

One question that still needs to be answered in Borel's SLLN is, how do we model an *infinite* sequence of heads and tails? In Bernoulli's WLLN, we toss a coin a large (and finite) number of times n, repeat this experiment several times, and show convergence in probability as n increases, but still remains finite. On the other hand, in Borel's SLLN, the proportion of heads in the whole infinite sequence is p almost surely. One way of conceptualizing an infinite sequence of heads and tails is to identify

---

[†] This can be written more compactly as $\lim_{n \to \infty} \Pr\{\sup_{j \geq n} |S_j/j - p| > \varepsilon\} = 0$. The latter equation implies that almost sure convergence of S_n/n is the same as convergence in probability to zero of $\sup_{j \geq n} |S_j/j - p|$.

[‡] To give a simple example of why this is true, consider choosing a real number X uniformly in $[0, 1]$. All events of the form $\{X = x\}$ are so rare that they are assigned zero probability. Yet some $x = x_0$ gets chosen after all, showing that $\{X = x_0\}$ is not impossible. For more details, see Székely (1986, p. 198).

[§] The first Borel-Cantelli lemma states that if $A_1, A_2, \ldots$ is a sequence of *arbitrary* events such that $\sum_{j=1}^{\infty} \Pr\{A_j\} < \infty$, then $\Pr\{A_j$ occurs infinitely often$\} = 0$.

the sequence of heads (1's) and tails (0's) with the binary expansion of a real number in [0,1]. That is, the sequence $X_1, X_2, X_3, \ldots$ of heads and tails corresponds to the real number U in [0,1] such that

$$U = \frac{X_1}{2} + \frac{X_2}{2^2} + \frac{X_3}{2^3} + \cdots$$

For example, the sequence *HHTH* (i.e., 1,1,0,1) corresponds to $1/2 + 1/4 + 0/8 + 1/16 = 13/16$. Conversely, .23 is .001111101... in binary and corresponds to the sequence *TTHHHHHHTH*...[†]

To summarize, Bernoulli's WLLN is about weak convergence (or convergence *in probability*): it states that, for a given large number of tosses n, there is a large fraction of sequences for which the proportion of heads will be close to the true p, and that this fraction gets ever closer to one as n increases. However, this leaves out the possibility for the proportion of heads in a particular sequence to drift away from p for a larger n. On the other hand, Borel's SLLN is about strong (or almost sure) convergence: it states that, with probability one, the proportion of heads in a sequence will be close to p for a reasonably large n and remain close to p for *every* larger n. Borel's SLLN thus says more about the behavior of the sequences, and also implies Bernoulli's law.

Of course, several key developments and extensions were made by others since the pioneering works of Bernoulli and Borel. It is important to note that these investigations did not delve into the further detection of stability of proportions (and of means in general), but rather into *the conditions* under which such stability could be achieved.

Let us outline extensions to Bernoulli's law first. A major development came through Siméon Denis Poisson's[‡] (1781–1840) (Fig. 8.5) proof that the Bernoulli's law of large numbers still holds if the probability of heads is not held constant from trial to trial (Poisson, 1837). Then, in 1867, Pafnuty Lvovich Chebyshev (1821–1894) observed that it was not necessary for S_n to be the sum of *Bernoulli* random variables for the law of large numbers to hold (Chebyshev, 1867) (Fig 8.6). He showed that if $X_1, X_2, \ldots, X_n$ is a sequence of *any* independent random variables such that $\mathscr{E}X_i = \mu_i$ and $\mathrm{var}X_i = \sigma_i^2 \leq K < \infty$, where K is a uniform upper bound for each σ_i^2, then

$$\lim_{n\to\infty} \mathrm{Pr}\left\{ \left| \frac{1}{n}\sum_{i=1}^{n} X_i - \frac{1}{n}\sum_{i=1}^{n}\mu_i \right| < \varepsilon \right\} = 1.$$

[†] The reader might also wonder what type of numbers in [0, 1] correspond to sequences that do not converge to p almost surely. This issue is addressed on p. 198 when we discuss Borel's normal numbers.

[‡] Poisson's other significant contributions include the introduction of the concepts of a random variable, of a cumulative distribution function, and of the probability density function as the derivative of the distribution function (Sheynin, 1978). Moreover, he derived the so-called Poisson distribution (Poisson 1837, pp. 205–206). Poisson's proof is based on approximating the B(n,p) distribution as p tends to zero and n tends to infinity such that $np = \lambda$ remains constant. He showed that if $X \sim$ B(n,p) then, under the conditions just stated, $\mathrm{Pr}\{X \leq m\} = \sum_{x=0}^{m}(e^{-\lambda}\lambda^x)/x!$ The Poisson probability mass function is thus $\mathrm{Pr}\{X = x\} = (e^{-\lambda}\lambda^x)/x!$ for $\lambda > 0$ and $x = 0, 1, 2, \ldots$ However, a form of the Poisson distribution had already previously appeared in de Moivre's 1711 *De Mensura Sortis* (Hald, 1984, p. 231).

Figure 8.5   Siméon Denis Poisson (1781–1840).

To prove the last result, Chebyshev used his famous inequality, as follows. Applying Chebyshev's inequality

$$\Pr\left\{\left|\frac{1}{n}\sum_{i=1}^{n}X_i - \frac{1}{n}\sum_{i=1}^{n}\mu_i\right| \geq \varepsilon\right\} \leq \frac{\mathrm{var}[1/n\sum_{i=1}^{n}X_i]}{\varepsilon^2}$$

$$= \frac{\mathrm{var}\sum_{i=1}^{n}X_i}{n^2\varepsilon^2}$$

$$= \frac{\sum_{i=1}^{n}\sigma_i^2}{n^2\varepsilon^2}.$$

Since $\sigma_i^2 \leq K < \infty$, we have

$$\Pr\left\{\left|\frac{1}{n}\sum_{i=1}^{n}X_i - \frac{1}{n}\sum_{i=1}^{n}\mu_i\right| \geq \varepsilon\right\} \leq \frac{\sum_{i=1}^{n}\sigma_i^2}{n^2\varepsilon^2}$$

$$\leq \frac{nK}{n^2\varepsilon^2}$$

$$= \frac{K}{n\varepsilon^2}$$

$$\to 0,$$

as $n \to 0$.

Figure 8.6 Pafnuty Lvovich Chebyshev (1821–1894).

In 1906, Andrei Andreevich Markov (1856–1922) (Fig. 8.7), who was Cheby-shev's student, showed that independence of the X_i's was not necessary for the last equation to hold as long as $(\text{var} \sum_i X_i)/n^2 \to 0$ when $n \to \infty$ (Markov, 1906). In 1929, Aleksandr Yakovlevich Khintchine (1894–1959) (Fig. 8.9) proved that when the X_i's are independent and identically distributed (IID), the sufficient condition for the WLLN to hold is simply $\mathscr{E}|X_i| < \infty$ (Khintchine, 1929). However, the necessary and sufficient condition for the WLLN for any sequence of random variables was given in 1927 by Andrey Nikolaevich Kolmogorov (1903–1987) (Kolmogorov, 1927) (Figs. 8.8)

$$\lim_{n \to \infty} \mathscr{E}\left(\frac{\Lambda_n^2}{1 + \Lambda_n^2}\right) = 0,$$

where $\Lambda_n = (S_n - \mathscr{E} S_n)/n$ and $S_n = \sum_{i=1}^n X_i$.

Regarding Borel's SLLN, the first key generalization was provided by Francesco Paolo Cantelli (1875–1966) in 1917 (Cantelli, 1917) (Fig. 8.10). He showed that if

Figure 8.7 Andrei Andreevich Markov (1856–1922).

$X_1, X_2, \ldots, X_n$ is a sequence of independent random variables with finite fourth moment and $\mathscr{E}|X_k - \mathscr{E} X_k|^4 \leq C < \infty$, then

$$\Pr\left\{ \lim_{n \to \infty} \frac{S_n - \mathscr{E} S_n}{n} = 0 \right\} = 1.$$

Figure 8.8 Andrey Nikolaevich Kolmogorov (1903–1987).

Figure 8.9   Aleksandr Yakovlevich Khintchine (1894–1959).

In 1930, Kolmogorov provided another sufficient condition for a sequence $X_1, X_2, \ldots, X_n$ of independent random variables with finite variance to converge almost surely[†] (Kolmogorov, 1930):

$$\sum_k \frac{\operatorname{var} X_k}{k^2} < \infty.$$

Finally, Kolmogorov showed in his landmark book that $\mathscr{E}|X_i| < \infty$ was both necessary and sufficient for a sequence of IID random variables to converge almost surely to μ (Kolmogorov, 1933).

Having come thus far, it would be unfortunate if we did not comment on yet another important limit theorem, namely the *Law of the Iterated Logarithm* (LIL). Consider the sum $S_n = X_1 + X_2 + \cdots + X_n$ of IID random variables, where without

---

[†] In analogy with Chebyshev, Kolmogorov used his own inequality to prove his version of the strong law. Kolmogorov's inequality states that for any sequence $X_1, X_2, \ldots, X_n$ of independent random variables such that $\operatorname{var} X_k = \sigma_k{}^2 < \infty$, then $\Pr\{\max_{1 \le k \le n} |S_k - \mathscr{E}S_k| \ge \varepsilon\} \le (\sum_k \sigma_k^2)/\varepsilon^2$.

Figure 8.10   Francesco Paolo Cantelli (1875–1966).

loss of generality we assume each X_i has mean zero and variance one. The SLLN states that for large n, S_n/n gets close to zero and subsequently *remains close* to it. The Central Limit Theorem[†] states that for large n, $S_n/\sqrt{n}$ has an approximate N(0, 1) distribution, so that the probability that $S_n/\sqrt{n}$ exceeds any given value can be estimated. However, none of these give the magnitudes of fluctuations of $S_n/\sqrt{n}$ about zero. This where the LIL comes in, by giving bounds for the fluctuations. It was first proved by Khintchine (1924) for independent and identically distributed Bernoulli random variables and later generalized by Kolmogorov (1929). Using the above notation, Khintchine's LIL states that

$$\lim_{n \to \infty} \sup \frac{S_n}{\sqrt{2n \ln \ln n}} = 1.$$

The implication is that the magnitude of fluctuations of $S_n/\sqrt{n}$ is of the order $\sqrt{2 \ln \ln n}$. Further, suppose the constants c_1 and c_2 are such that $c_1 > \sqrt{2 \ln \ln n} > c_2$. Then the LIL implies that, with probability one, $S_n/\sqrt{n}$ will exceed $\pm c_2$ infinitely often but will exceed $\pm c_1$ only finitely often.

[†] To be further discussed in **Problem 10**.

Problem 9

De Moivre's Problem (1730)

Problem. *A fair die is thrown n independent times. Find the probability of obtaining a sum equal to t, where t is a natural number.*

Solution. Let X_j be the score on the die on the jth ($j = 1, 2, \ldots, n$) throw and let T_n be the sum of the scores obtained. Then

$$T_n = X_1 + X_2 + \cdots + X_n,$$

where, for all $j = 1, 2, \ldots, n$,

$$\Pr\{X_j = x\} = \frac{1}{6}, \quad \text{for } x = 1, 2, \ldots, 6.$$

We now take the probability generating function of T_n, remembering that the X_j's are independent and have the same distribution:

$$\mathscr{E}s^{T_n} = \mathscr{E}s^{X_1 + X_2 + \cdots + X_n}$$
$$= \mathscr{E}s^{X_1} \, \mathscr{E}s^{X_2} \cdots \mathscr{E}s^{X_n}$$
$$= \left(\mathscr{E}s^{X_j}\right)^n.$$

Now

$$\mathscr{E}s^{X_j} = \sum_{x=1}^{6} s^x \Pr\{X_j = x\}$$
$$= \frac{1}{6}(s + s^2 + \cdots + s^6)$$
$$= \frac{1}{6} \cdot \frac{s(1 - s^6)}{1 - s}, \quad |s| < 1.$$

Classic Problems of Probability, Prakash Gorroochurn.
© 2012 John Wiley & Sons, Inc. Published 2012 by John Wiley & Sons, Inc.

Therefore,

$$
\mathscr{E} s^{T_n} = \frac{1}{6^n} \left\{ \frac{s(1 - s^6)}{(1 - s)} \right\}^n
$$

$$
= \frac{1}{6^n} s^n \sum_{i=0}^{n} \binom{n}{i} (-s^6)^i \sum_{k=0}^{\infty} \binom{-n}{k} (-s)^k
$$

$$
= \frac{1}{6^n} \sum_{i=0}^{n} \binom{n}{i} (-1)^i s^{6i+n} \sum_{k=0}^{\infty} \binom{n+k-1}{k} s^k,
$$

where we have made use of the fact that $\binom{-n}{k} \equiv \binom{n+k-1}{k} (-1)^k$. Now, $\Pr\{T_n = t\}$ is the coefficient of s^t in $\mathscr{E} s^{T_n}$ and can be obtained by multiplying the coefficient of s^{6i+n} in the first summation above with the coefficient of $s^{t-(6i+n)}$ in the second summation, then summing over all i, and finally dividing by 6^n, that is,

$$
\Pr\{T_n = t\} = \frac{1}{6^n} \sum_{i=0}^{\lfloor (t-n)/6 \rfloor} \binom{n}{i} (-1)^i \binom{n + (t - 6i - n) - 1}{t - 6i - n}
$$

$$
= \frac{1}{6^n} \sum_{i=0}^{\lfloor (t-n)/6 \rfloor} (-1)^i \binom{n}{i} \binom{t - 6i - 1}{n - 1}. \tag{9.1}
$$

In the above, $\lfloor (t - n)/6 \rfloor$ is the greatest integer smaller than or equal to $(t - n)/6$ and is obtained by setting $n - 1 \leq t - 6i - 1$.

9.1 Discussion

This classic problem was first investigated in much detail but with no general formulas by Jacob Bernoulli in the *Ars Conjectandi* (Bernoulli, 1713, p. 30). In his analysis, Bernoulli first gives all the ways of producing a certain total score with three dice. He then proceeds to outline principles through which the total scores with more than three dice can be enumerated.

The first explicit formula was given by the eminent French mathematician Abraham de Moivre[†] (1667–1754) in the *De Mensura Sortis* (de Moivre, 1711 Hald, 1984). As part of Problem 5, de Moivre considers the problem of finding "the

---

[†] Although de Moivre was born in France, he emigrated to England when he was very young due to religious persecution. He spent the rest of his life and career in England. See Hacking (1980b) for more biographical details.

number of chances by which a given number of points may be thrown with a given number of dice." De Moivre considers n dice each with f faces such that the total required is $p + 1$. He then gives without proof the following formula for the number of ways of obtaining a total of $p + 1$:

$$\left(\frac{p}{1}\frac{p-1}{2}\frac{p-2}{3}\cdots\right) - \left(\frac{q}{1}\frac{q-1}{2}\frac{q-2}{3}\cdots\frac{n}{1}\right) + \left(\frac{r}{1}\frac{r-1}{2}\frac{r-2}{3}\cdots\frac{n}{1}\frac{n-1}{2}\right)$$

$$-\left(\frac{s}{1}\frac{s-1}{2}\frac{s-2}{3}\cdots\frac{n}{1}\frac{n-1}{2}\frac{n-2}{3}\right) + \cdots$$

(9.2)

where $p - f = q$, $q - f = r$, $r - f = s$, and so on. The above formula needs to be continued until a zero or negative factor is obtained. If we generalize the formula we derived in Eq. (9.1) to the case of a die with f faces, we obtain

$$\Pr\{T_n = t + 1 | f \text{ faces}\} = \frac{1}{f^n}\sum_{i=0}^{\infty}(-1)^i\binom{n}{i}\binom{t-fi}{n-1}.$$

When the latter is expanded, we obtain the same expression as in Eq. (9.2) above. While de Moivre did not prove his result in the *De Mensura Sortis*, he later did so in the *Miscellanea analytica* (de Moivre, 1730). He is thus the first to make use of probability generating functions, a technique that was to be later given its name and fully developed by Laplace (1812, 1814b, 1820). It is interesting to note that, in between de Moivre's two aforementioned publications, Montmort claimed that he had already derived the formula on p. 141 of the 1708 edition of his book (Montmort, 1713, p. 364). Although this is not actually true, Montmort seems to have been aware of the formula before it came out in de Moivre's *De Mensura Sortis* (Todhunter, 1865, pp. 84–85).

De Mensura Sortis is also noteworthy because it is there that de Moivre first gave his explicit definition of classical probability (de Moivre, 1711; Hald, 1984):

> If p is the number of chances by which a certain event may happen, and q is the number of chances by which it may fail, the happenings as much as the failings have their degree of probability; but if all the chances by which the event may happen or fail were equally easy, the probability of happening will be to the probability of failing as p to q.

De Moivre was also the first to explicitly define independence (de Moivre, 1718) and conditional probability (de Moivre, 1738). On p. 4 of the first edition of the *Doctrine of Chances* (de Moivre, 1718), de Moivre states

> . . .if a Fraction expresses the Probability of an Event, and another Fraction the Probability of another Event, and those two Events are independent; the Probability that both Events will Happen will be the Product of those two Fractions.

On p. 7 of the second edition of the *Doctrine of Chances*[†] (de Moivre, 1738), de Moivre writes

> ...the Probability of the happening of two Events dependent, is the product of the Probability of the happening of one of them, by the Probability which the other will have of happening, when the first shall have been consider'd as having happened; and the same Rule will extend to the happening of as many Events as may be assign'd.

In modern notation, the conditional probability of an event A, given that an event B has already occurred, is given by

$$\Pr\{A|B\} \equiv \frac{\Pr\{A\cap B\}}{\Pr\{B\}},$$

where $\Pr\{B\} > 0$. If A and B are independent, then $\Pr\{A|B\} = \Pr\{A\}$ so that $\Pr\{A\cap B\} = \Pr\{A\}\Pr\{B\}$.

Let us now make some remarks on the method of probability generating functions. This method turns out to be useful in at least three commonly encountered situations. The first has to do when one is dealing with the sum of independent discrete[‡] random variables, as was the case in **Problem 9**. The second occurs when one wishes to calculate slightly more complex probabilities than those requiring the simple binomial, negative binomial, or hypergeometric distribution. An example of the second situation is the following classic problem, also discussed in slightly modified form by de Moivre in the second edition of the *Doctrine of Chances* (de Moivre, 1738):

Find the probability of obtaining a first run of three heads (i.e., three consecutive heads) in n tosses of a coin for which the probability of a head is p.

Let h_n be the probability above. A standard way of solving the problem is to condition on the first tosses. By looking in this "backward" manner, there are only three ways of obtaining a first run of three heads in n tosses: the first toss could be a tail and then a first run of three heads occurs in the remaining $(n-1)$ tosses, or the first two tosses could be head and tail in that order and then a first run of three heads occurs in the remaining $(n-2)$ tosses, or finally the first three tosses could be two heads and one tail in that order and then a first run of three heads occurs in the remaining $(n-3)$ tosses, that is,

$$h_n = qh_{n-1} + pqh_{n-2} + p^2qh_{n-3}, \tag{9.3}$$

with $q = 1-p$, $h_1 = h_2 \equiv 0$ and $h_3 \equiv p^3$. Although Eq. (9.3) can be solved recursively for any h_n, a better alternative is to define the probability generating function $H_3(s) = \sum_{n=1}^{\infty} h_n s^n$. Equation (9.3) then becomes

---

[†] De Moivre also treated conditional probability in the first edition of the *Doctrine of Chances* (de Moivre, 1718, pp. 6–7), but he did not explicitly state the definition there.

[‡] Note that continuous random variables do not have probability generating functions since $\Pr\{X=x\}=0$ for all x if X is continuous.

$$\sum_{n=1}^{\infty} h_{n+3}s^n = q\sum_{n=1}^{\infty} h_{n+2}s^n + pq\sum_{n=1}^{\infty} h_{n+1}s^n + p^2q\sum_{n=1}^{\infty} h_n s^n$$

$$\frac{1}{s^3}\sum_{n=4}^{\infty} h_n s^n = \frac{q}{s^2}\sum_{n=3}^{\infty} h_n s^n + \frac{pq}{s}\sum_{n=2}^{\infty} h_n s^n + p^2q\sum_{n=1}^{\infty} h_n s^n$$

$$\frac{1}{s^3}\{H_3(s) - p^3 s^3\} = \frac{q}{s^2}H_3(s) + \frac{pq}{s}H_3(s) + p^2q H_3(s)$$

$$H_3(s) = \frac{p^3 s^3}{1 - qs - qps^2 - qp^2 s^3}. \tag{9.4}$$

To obtain an exact expression for h_n, one can factorize the denominator in the above, then expand $H_3(s)$ into partial fractions, and finally extract the coefficient of s^n. Alternatively, a standard mathematical package can be used to expand $H_3(s)$ up to any powers of s. For example, the first ten terms of $H_3(s)$ are listed in Table 9.1.

We can proceed in a similar manner to find the generating function $H_r(s)$ of N_r, the number of tosses until the first run of r heads occurs. It can be shown that (e.g., see DasGupta, 2011, p. 27),

$$H_r(s) = \frac{p^r s^r}{1 - qs\{1 + ps + \cdots + (ps)^{r-1}\}} = \frac{p^r s^r (1 - ps)}{1 - s + qp^r s^{r+1}}.$$

Since $H_r'(1) = \mathcal{E}N_r$ and $H_r''(1) = \mathcal{E}N_r(N_r - 1)$, we also have

$$\mathcal{E}N_r = \sum_{j=1}^{r}\left(\frac{1}{p}\right)^j,$$

$$\text{var } N_r = \frac{1 - p^{1+2r} - qp^r(1 + 2r)}{q^2 p^{2r}}.$$

Table 9.1   The Probability h_n of Three Consecutive Heads in n Tosses of a Coin (the Probability of a Head is p)

Value of n	h_n = coefficient of s^n
3	p^3
4	$p^3 - p^4$
5	$p^3 - p^4$
6	$p^3 - p^4$
7	$p^7 - p^6 - p^4 + p^3$
8	$-p^8 + 3p^7 - 2p^6 - p^4 + p^3$
9	$-2p^8 + 5p^7 - 3p^6 - p^4 + p^3$
10	$-3p^8 + 7p^7 - 4p^6 - p^4 + p^3$

The expressions were obtained by expanding $H_3(s)$ in Eq. (9.4) as powers of s; the coefficient of s^n then gives h_n.

A third instance where probability generating functions can be useful is in the calculation of the expectation of rational functions of a random variable. A classic example is the following[†]:

Given that $X \sim B(n, p)$, calculate $\mathcal{E} \frac{1}{1+X}$.

First note that the probability generating function of X is

$$\mathcal{E} s^X = \sum_{x=0}^{n} s^x \Pr\{X = x\}$$

$$= \sum_{x=0}^{n} s^x \binom{n}{x} p^x (1 - p)^{n-x}$$

$$= \sum_{x=0}^{n} \binom{n}{x} (ps)^x (1 - p)^{n-x}$$

$$= (ps + 1 - p)^n.$$

Integrating the above with respect to s,

$$\int_{0}^{t} \mathcal{E} s^X \, ds = \int_{0}^{t} (ps + 1 - p)^n \, ds,$$

$$\therefore \quad \mathcal{E} \frac{t^{X+1}}{X + 1} = \int_{0}^{t} (ps + 1 - p)^n \, ds.$$

Setting $t = 1$, we have,

$$\mathcal{E} \frac{1}{1 + X} = \int_{0}^{1} (ps + 1 - p)^n \, ds$$

$$= \left[\frac{(ps + 1 - p)^{n+1}}{(n + 1)p} \right]_{s=0}^{1}$$

$$= \frac{1 - (1 - p)^{n+1}}{(n + 1)p}.$$

As a final illustration of the usefulness of probability generating functions, we solve the *Generalized Problem of Points*[‡]:

[†] See also Grimmet and Stirzaker (2001, p. 162).

[‡] For a full discussion of the Problem of Points for two players and some discussion for three players, see **Problem 4**.

Players $A_1, A_2, \ldots, A_n$ play a game such that their respective probabilities of winning one round are $p_1, p_2, \ldots, p_n$, where $\sum_i p_i = 1$. At some point in the game, the players are respectively $a_1, a_2, \ldots, a_n$ rounds short of winning the game. What is the probability that player A_1 will win the game?

This problem was first solved through the method of generating functions by Pierre-Simon Laplace (1749–1827) in the *Théorie Analytique des Probabilités* (Laplace, 1812, p. 207). Let $u_{a_1, a_2, \ldots, a_n}$ be the required probability. By conditioning on the next round, we obtain the difference equation

$$u_{a_1, a_2, \ldots, a_n} = p_1 u_{a_1-1, a_2, \ldots, a_n} + p_2 u_{a_1, a_2-1, \ldots, a_n} + \cdots + p_n u_{a_1, a_2, \ldots, a_n-1}.$$

The *boundary conditions* are

$$
\begin{aligned}
u_{0, a_2, \ldots, a_n} &= 1, & a_2, \ldots, a_n &> 0, \\
u_{a_1, 0, \ldots, a_n} &= 0, & a_1, a_3, \ldots, a_n &> 0, \\
&\vdots & & \\
u_{a_1, a_2, \ldots, 0} &= 0, & a_1, \ldots, a_{n-1} &> 0.
\end{aligned}
$$

We define the multivariate probability generating function $G(s_1, s_2, \ldots, s_n) \equiv G$, where

$$
\begin{aligned}
G &= \sum_{a_1=1}^{\infty} \sum_{a_2=1}^{\infty} \cdots \sum_{a_n=1}^{\infty} u_{a_1, a_2, \ldots, a_n} s_1^{a_1} s_2^{a_2} \cdots s_n^{a_n} \\
&= \sum_{a_1=1}^{\infty} \sum_{a_2=1}^{\infty} \cdots \sum_{a_n=1}^{\infty} \left(p_1 u_{a_1-1, a_2, \ldots, a_n} + p_2 u_{a_1, a_2-1, \ldots, a_n} + \cdots + p_n u_{a_1, a_2, \ldots, a_n-1} \right) s_1^{a_1} s_2^{a_2} \cdots s_n^{a_n} \\
&= p_1 \sum_{a_1=1}^{\infty} \sum_{a_2=1}^{\infty} \cdots \sum_{a_n=1}^{\infty} u_{a_1-1, a_2, \ldots, a_n} s_1^{a_1} s_2^{a_2} \cdots s_n^{a_n} + p_2 \sum_{a_1=1}^{\infty} \sum_{a_2=1}^{\infty} \cdots \sum_{a_n=1}^{\infty} u_{a_1, a_2-1, \ldots, a_n} s_1^{a_1} s_2^{a_2} \cdots s_n^{a_n} \\
&\quad + \cdots + p_n \sum_{a_1=1}^{\infty} \sum_{a_2=1}^{\infty} \cdots \sum_{a_n=1}^{\infty} u_{a_1, a_2, \ldots, a_n-1} s_1^{a_1} s_2^{a_2} \cdots s_n^{a_n} \\
&= p_1 \left(s_1 \sum_{a_1=1}^{\infty} \sum_{a_2=1}^{\infty} \cdots \sum_{a_n=1}^{\infty} u_{a_1-1, a_2, \ldots, a_n} s_1^{a_1} s_2^{a_2} \cdots s_n^{a_n} + s_1 \sum_{a_2=1}^{\infty} \cdots \sum_{a_n=1}^{\infty} u_{0, a_2, \ldots, a_n} s_2^{a_2} \cdots s_n^{a_n} \right) \\
&\quad + p_2 \left(s_2 \sum_{a_1=1}^{\infty} \sum_{a_2=1}^{\infty} \cdots \sum_{a_n=1}^{\infty} u_{a_1, a_2-1, \ldots, a_n} s_1^{a_1} s_2^{a_2} \cdots s_n^{a_n} + s_2 \sum_{a_1=1}^{\infty} \sum_{a_3=1}^{\infty} \cdots \sum_{a_n=1}^{\infty} u_{a_1, 0, \ldots, a_n} s_1^{a_1} s_3^{a_3} \cdots s_n^{a_n} \right) \\
&\quad + \cdots + p_n \left(s_n \sum_{a_1=1}^{\infty} \sum_{a_2=1}^{\infty} \cdots \sum_{a_n=1}^{\infty} u_{a_1, a_2, \ldots, a_n-1} s_1^{a_1} s_2^{a_2} \cdots s_n^{a_n} + s_n \sum_{a_1=1}^{\infty} \sum_{a_2=1}^{\infty} \cdots \sum_{a_{n-1}=1}^{\infty} u_{a_1, a_2, \ldots, 0} s_1^{a_1} s_2^{a_2} \cdots s_{n-1}^{a_{n-1}} \right) \\
&= p_1 \left(s_1 G + s_1 \sum_{a_2=1}^{\infty} \cdots \sum_{a_n=1}^{\infty} 1 \cdot s_2^{a_2} \cdots s_n^{a_n} \right) + p_2 \left(s_2 G + s_2 \sum_{a_1=1}^{\infty} \sum_{a_3=1}^{\infty} \cdots \sum_{a_n=1}^{\infty} 0 \cdot s_1^{a_1} s_3^{a_3} \cdots s_n^{a_n} \right) \\
&\quad + \cdots + p_n \left(s_n G + s_n \sum_{a_1=1}^{\infty} \sum_{a_2=1}^{\infty} \cdots \sum_{a_{n-1}=1}^{\infty} 0 \cdot s_1^{a_1} s_2^{a_2} \cdots s_{n-1}^{a_{n-1}} \right) \\
&= p_1 s_1 G + p_1 s_1 \sum_{a_2=1}^{\infty} \cdots \sum_{a_n=1}^{\infty} s_2^{a_2} \cdots s_n^{a_n} + (p_2 s_2 + \cdots + p_n s_n) G.
\end{aligned}
$$

Therefore,

$$
\begin{aligned}
G &= \frac{p_1 s_1 \sum_{a_2=1}^{\infty} \cdots \sum_{a_n=1}^{\infty} s_2^{a_2} \cdots s_n^{a_n}}{1 - p_1 s_1 - p_2 s_2 - \cdots - p_n s_n} \\
&= \frac{p_1 s_1 \sum_{a_2=1}^{\infty} s_2^{a_2} \cdots \sum_{a_n=1}^{\infty} s_n^{a_n}}{1 - p_1 s_1 - p_2 s_2 - \cdots - p_n s_n} \\
&= \frac{p_1 s_1 s_2 \cdots s_n}{(1 - s_2)(1 - s_3) \cdots (1 - s_n)(1 - p_1 s_1 - p_2 s_2 - \cdots - p_n s_n)}.
\end{aligned}
$$

Therefore the probability that player A_1 wins the game is

$$
u_{a_1, a_2, \ldots, a_n} = \text{coeff. of } s_1^{a_1} s_2^{a_2} \cdots s_n^{a_n} \text{ in}
$$

$$
\frac{p_1 s_1 s_2 \cdots s_n}{(1 - s_2)(1 - s_3) \cdots (1 - s_n)(1 - p_1 s_1 - p_2 s_2 - \cdots - p_n s_n)}.
$$

Similarly, the probabilities that players $A_2, A_3, \ldots, A_n$ win the game are, respectively,

coeff. of $s_1^{a_1} s_2^{a_2} \cdots s_n^{a_n}$ in $\dfrac{p_2 s_1 s_2 \cdots s_n}{(1 - s_1)(1 - s_3) \cdots (1 - s_n)(1 - p_1 s_1 - p_2 s_2 - \cdots - p_n s_n)}$,

coeff. of $s_1^{a_1} s_2^{a_2} \cdots s_n^{a_n}$ in $\dfrac{p_3 s_1 s_2 \cdots s_n}{(1 - s_1)(1 - s_2)(1 - s_4) \cdots (1 - s_n)(1 - p_1 s_1 - p_2 s_2 - \cdots - p_n s_n)}$,

$\vdots$

coeff. of $s_1^{a_1} s_2^{a_2} \cdots s_n^{a_n}$ in $\dfrac{p_n s_1 s_2 \cdots s_n}{(1 - s_1)(1 - s_2) \cdots (1 - s_{n-1})(1 - p_1 s_1 - p_2 s_2 - \cdots - p_n s_n)}$.

Thus, in **Problem 4,** we had for two players $p_1 = p_2 = 1/2$ and $a_1 = 1, a_2 = 3$. Therefore, the probability that A wins is[†]

$$
\text{coeff. of } s_1 s_2^3 \text{ in } \frac{s_1 s_2/2}{(1 - s_2)(1 - s_1/2 - s_2/2)} = \frac{7}{8},
$$

giving a division ratio of 7:1 between A and B. This is the same answer we obtained in **Problem 4.** On p. 30, we also considered an example with three players such that $p_1 = p_2 = p_3 = 1/3$ and $a_1 = 1, a_2 = 2$, and $a_3 = 2$. Therefore, the probability that A wins is

$$
\text{coeff. of } s_1 s_2^2 s_3^2 \text{ in } \frac{s_1 s_2 s_3/3}{(1 - s_2)(1 - s_3)(1 - s_1/3 - s_2/3 - s_3/3)} = \frac{17}{27}.
$$

The probability that B wins is

$$
\text{coeff. of } s_1 s_2^2 s_3^2 \text{ in } \frac{s_1 s_2 s_3/3}{(1 - s_1)(1 - s_3)(1 - s_1/3 - s_2/3 - s_3/3)} = \frac{5}{27},
$$

resulting in a division ratio of $17/27 : 5/27 : 5/27 = 17:5:5$ between A, B, and C, again the same answer we obtained previously.

[†] This answer was obtained by doing two successive series expansions of $(s_1 s_2/2)/[(1-s_2)(1-s_1/2-s_2/2)]$ using the software Scientific Workplace 5.50, Build 2890.

Problem 10

De Moivre, Gauss, and the Normal Curve (1730, 1809)

Problem. *Consider the binomial expansion of $(1 + 1)^n$, where n is even and large. Let M be the middle term and Q be the term at a distance d from M. Show that*

$$Q \approx M \, e^{-2d^2/n}. \qquad (10.1)$$

Solution. We let $n = 2m$, where m is a positive integer. Then

$$\frac{Q}{M} = \frac{\binom{2m}{m+d}}{\binom{2m}{m}} = \frac{\binom{2m}{m-d}}{\binom{2m}{m}}$$

$$= \frac{(2m)!/[(m+d)!(m-d)!]}{(2m)!/[(m!)^2]}$$

$$= \frac{(m!)^2}{(m+d)!(m-d)!}.$$

Applying Stirling's formula $N! \sim \sqrt{2\pi}N^{N+1/2}\, e^{-N}$ for large N, we have

$$\frac{Q}{M} \approx \frac{m^{2m+1}\, e^{-2m}}{(m+d)^{m+d+1/2}e^{-m-d} \cdot (m-d)^{m-d+1/2}\, e^{-m+d}}$$

$$= \frac{m^{2m+1}}{(m+d)^{m+d+1/2}(m-d)^{m-d+1/2}}$$

$$\approx \left(\frac{m^2}{m^2 - d^2}\right)^m \left(\frac{m-d}{m+d}\right)^d.$$

Classic Problems of Probability, Prakash Gorroochurn.
© 2012 John Wiley & Sons, Inc. Published 2012 by John Wiley & Sons, Inc.

Taking logarithms on both sides, with d/m small, we have

$$\ln\frac{Q}{M} \approx -m \ln\left(1 - \frac{d^2}{m^2}\right) + d \ln\left(\frac{1 - d/m}{1 + d/m}\right)$$

$$\approx (-m)\left(-\frac{d^2}{m^2}\right) + (d)\left(-\frac{2d}{m}\right)$$

$$= -\frac{d^2}{m}.$$

Writing $m = n/2$, we have $Q \approx M\, e^{-2d^2/n}$, as required.

10.1 Discussion

The above result was first obtained by Abraham de Moivre (1667–1754) (Fig. 10.3) in an attempt to approximate the symmetric binomial distribution. De Moivre started this work in the *Miscellanea analytica* (de Moivre, 1730) and finished it in a November 13, 1733 paper entitled *Approximatio ad summam terminorum binomii* $(a + b)^n$ *in seriem expansi*. De Moivre later himself translated the paper in English and incorporated it in the second edition of his *Doctrine of Chances*[†] (de Moivre, 1738) (see Figs. 10.1 and 10.2). De Moivre's derivation was slightly different from the one presented above and is worth revisiting. From $Q/M = (m!)^2/[(m + d)!(m - d)!]$ (see solution), it follows that

$$\frac{Q}{M} = \frac{m!}{(m + d)(m + d - 1)\cdots(m + 1)m!} \cdot \frac{m(m - 1)\cdots(m - d + 1)(m - d)!}{(m - d)!}$$

$$= \frac{m(m - 1)\cdots(m - d + 1)}{(m + d)(m + d - 1)\cdots(m + 1)}$$

$$= \frac{m}{m + d} \cdot \frac{(m - 1)(m - 2)\cdots(m - d + 1)}{(m + 1)(m + 2)\cdots(m + d - 1)}.$$

[†] The *Doctrine of Chances* appears in Grattan-Guinness' *Landmark Writings in Western Mathematics*, and a full analysis is written by Schneider (2005a). See also Stigler (1986, pp. 70–88).

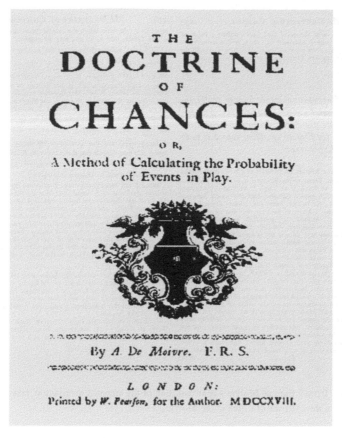

Figure 10.1 Cover of the first edition of the *Doctrine of Chances*.

Taking logarithms on both sides,

$$\ln\left(\frac{Q}{M}\right) = \ln\left\{\left(\frac{m}{m+d}\right) \cdot \frac{(m-1)(m-2)\cdots(m-d+1)}{(m+1)(m+2)\cdots(m+d-1)}\right\}$$

$$= \ln\left(\frac{1}{1+d/m}\right) + \sum_{i=1}^{d-1}\ln\left(\frac{m-i}{m+i}\right)$$

$$= -\ln(1+d/m) + \sum_{i=1}^{d-1}\ln\left(\frac{1-i/m}{1+i/m}\right).$$

Now, $\ln(1+d/m) \approx d/m$ for small d/m, and $\ln\left(\frac{1-i/m}{1+i/m}\right) \approx -2i/m$ for small i/m. Therefore,

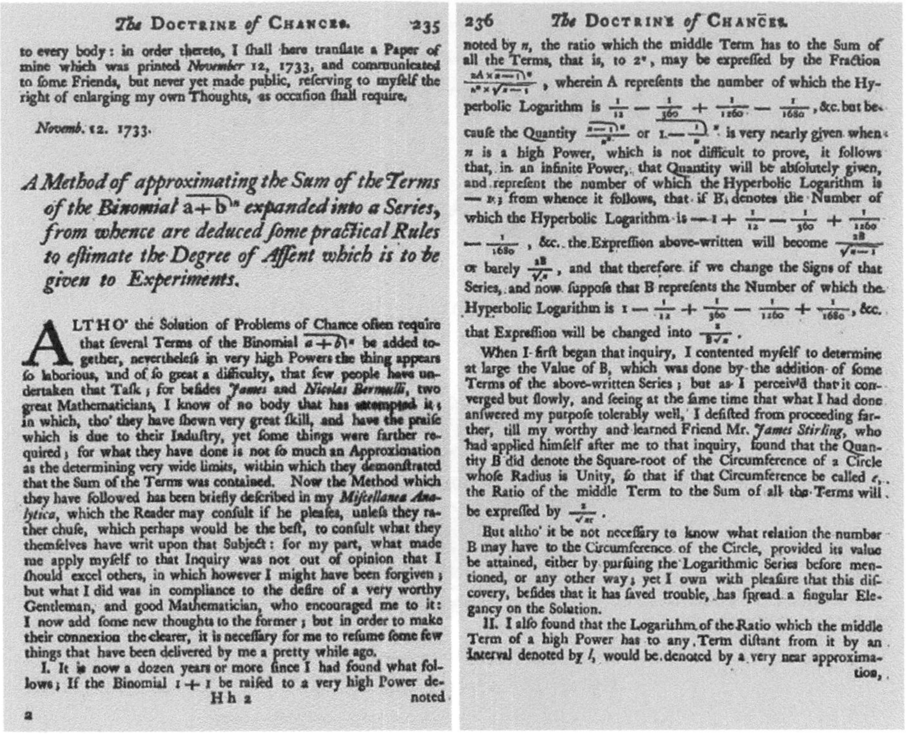

The DOCTRINE *of* CHANCES.   235

to every body: in order thereto, I shall here translate a Paper of mine which was printed *November* 12, 1733, and communicated to some Friends, but never yet made public, reserving to myself the right of enlarging my own Thoughts, as occasion shall require.

Novemb. 12. 1733.

A Method of approximating the Sum of the Terms of the Binomial a + b|ⁿ *expanded into a Series, from whence are deduced some practical Rules to estimate the Degree of Assent which is to be given to Experiments.*

ALTHO' the Solution of Problems of Chance often require that several Terms of the Binomial *a + b*|ⁿ be added together, nevertheless in very high Powers the thing appears so laborious, and of so great a difficulty, that few people have undertaken that Task; for besides *James* and *Nicolas Bernoulli*, two great Mathematicians, I know of no body that has attempted it; in which, tho' they have shewn very great skill, and have the praise which is due to their Industry, yet some things were farther required; for what they have done is not so much an Approximation as the determining very wide limits, within which they demonstrated that the Sum of the Terms was contained. Now the Method which they have followed has been briefly described in my *Miscellanea Analytica*, which the Reader may consult if he pleases, unless they rather chuse, which perhaps would be the best, to consult what they themselves have writ upon that Subject: for my part, what made me apply myself to that Inquiry was not out of opinion that I should excel others, in which however I might have been forgiven; but what I did was in compliance to the desire of a very worthy Gentleman, and good Mathematician, who encouraged me to it: I now add some new thoughts to the former; but in order to make their connexion the clearer, it is necessary for me to resume some few things that have been delivered by me a pretty while ago.

I. It is now a dozen years or more since I had found what follows; If the Binomial 1 + 1 be raised to a very high Power denoted by *n*, the ratio which the middle Term has to the Sum of

H h 2   noted

236   *The* DOCTRINE *of* CHANCES.

noted by *n*, the ratio which the middle Term has to the Sum of all the Terms, that is, to 2ⁿ, may be expressed by the Fraction $\frac{2A \times \overline{n-1}|^n}{n^n \times \sqrt{n-1}}$, wherein A represents the number of which the Hyperbolic Logarithm is $\frac{1}{12} - \frac{1}{360} + \frac{1}{1260} - \frac{1}{1680}$, &c. but because the Quantity $\frac{\overline{n-1}|^n}{n^n}$ or $\overline{1 - \frac{1}{n}}|^n$ is very nearly given when *n* is a high Power, which is not difficult to prove, it follows that, in an infinite Power, that Quantity will be absolutely given, and represent the number of which the Hyperbolic Logarithm is − *n*; from whence it follows, that if B denotes the Number of which the Hyperbolic Logarithm is $-1 + \frac{1}{12} - \frac{1}{360} + \frac{1}{1260} - \frac{1}{1680}$, &c. the Expression above-written will become $\frac{2B}{\sqrt{n-1}}$ or barely $\frac{2B}{\sqrt{n}}$, and that therefore if we change the Signs of that Series, and now suppose that B represents the Number of which the Hyperbolic Logarithm is $1 - \frac{1}{12} + \frac{1}{360} - \frac{1}{1260} + \frac{1}{1680}$, &c. that Expression will be changed into $\frac{2}{B\sqrt{n}}$.

When I first began that inquiry, I contented myself to determine at large the Value of B, which was done by the addition of some Terms of the above-written Series; but as I perceiv'd that it converged but slowly, and seeing at the same time that what I had done answered my purpose tolerably well, I desisted from proceeding farther, till my worthy and learned Friend Mr. *James Stirling*, who had applied himself after me to that inquiry, found that the Quantity B did denote the Square-root of the Circumference of a Circle whose Radius is Unity, so that if that Circumference be called *c*, the Ratio of the middle Term to the Sum of all the Terms will be expressed by $\frac{2}{\sqrt{nc}}$.

But altho' it be not necessary to know what relation the number B may have to the Circumference of the Circle, provided its value be attained, either by pursuing the Logarithmic Series before mentioned, or any other way; yet I own with pleasure that this discovery, besides that it has saved trouble, has spread a singular Elegancy on the Solution.

II. I also found that the Logarithm of the Ratio which the middle Term of a high Power has to any Term distant from it by an Interval denoted by *l*, would be denoted by a very near approximation,

Figure 10.2 Extract from de Moivre's approximation of the binomial distribution, taken from the second edition of the *Doctrine of Chances* (de Moivre 1738).

$$\ln\left(\frac{Q}{M}\right) \approx -\frac{d}{m} - \sum_{i=1}^{d-1}\frac{2i}{m}$$

$$= -\frac{d}{m} - \frac{2}{m}\cdot\frac{(d-1)d}{2}$$

$$= -\frac{d^2}{m}.$$

Since $m = n/2$, de Moivre obtains $Q \approx Me^{-2d^2/n}$. This completes de Moivre's solution to **Problem 10**.

Furthermore, if we divide both sides of de Moivre's result by 2^n and let $X \sim B(n, 1/2)$, in modern notation we obtain

$$\Pr\left\{X = \frac{n}{2} + d\right\} \approx \Pr\left\{X = \frac{n}{2}\right\}e^{-2d^2/n}. \tag{10.2}$$

To be able to approximate the symmetric binomial distribution, de Moivre had first also calculated the ratio of the middle term M to the sum of all the terms in the

Figure 10.3   Abraham de Moivre (1667–1754).

expansion of $(1 + 1)^n$, where n is large and even. This ratio is the same as the following probability, which we derived in **Problem 8**:

$$\Pr\left\{X = \frac{n}{2}\right\} \approx \frac{2}{\sqrt{2\pi n}} \quad (n \text{ large}). \tag{10.3}$$

Again, de Moivre's method was different from the one we used in **Problem 8** and will now be described. With $n = 2m$, we have

$$M = \binom{n}{n/2} = \binom{2m}{m}$$

$$= \frac{(2m)(2m-1)\cdots(m+2)(m+1)}{m(m-1)\cdots(2)(1)}$$

$$= 2 \cdot \frac{(m+1)}{(m-1)} \cdot \frac{(m+2)}{(m-2)} \cdots \frac{m+(m-1)}{m-(m-1)}.$$

Taking logarithms on both sides,

$$\ln M = \ln 2 + \ln\left(\frac{1+1/m}{1-1/m}\right) + \ln\left(\frac{1+2/m}{1-2/m}\right) + \cdots + \ln\left[\frac{1+(m-1)/m}{1-(m-1)/m}\right].$$

Now,

$$\ln\left(\frac{1+1/m}{1-1/m}\right) = 2\left(\frac{1}{m} + \frac{1}{3m^3} + \frac{1}{5m^5} + \cdots\right),$$

$$\ln\left(\frac{1+2/m}{1-2/m}\right) = 2\left(\frac{2}{m} + \frac{2^3}{3m^3} + \frac{2^5}{5m^5} + \cdots\right),$$

$$\cdots$$

$$\ln\left[\frac{1+(m-1)/m}{1-(m-1)/m}\right] = 2\left[\frac{m-1}{m} + \frac{(m-1)^3}{2m^3} + \frac{(m-1)^5}{5m^5} + \cdots\right].$$

By adding the above equations "vertically" (i.e., adding together the first term on the right side of each equation, then the second term, and so on),

$$\ln M = \ln 2 + \frac{2}{m}[1 + 2 + \cdots + (m-1)] + \frac{2}{3m^3}[1^3 + 2^3 + \cdots + (m-1)^3]$$

$$+ \frac{2}{5m^5}[1^5 + 2^5 + \cdots + (m-1)^5] + \cdots$$

Now, each of the sums in the square brackets above multiplied by the corresponding term before the brackets can be written as a polynomial in m (involving the so-called "Bernoulli numbers"). De Moivre adds the highest powers of each polynomial and obtains $(2m-1)\ln(2m-1) - 2m\ln m$. For the sum of the second highest powers, he obtains $(1/2)\ln(2m-1)$. The sum of the other highest powers are more difficult to obtain, but de Moivre shows that, as $m\to\infty$, these become $1/12, -1/360, 1/1260,$ $-1/1680, \ldots$ Therefore, de Moivre is able to write

$$\ln M \approx \ln 2 + \{(2m-1)\ln(2m-1) - 2m\ln m\} + \frac{1}{2}\ln(2m-1) + \ln T,$$

where $\ln T = 1/12 - 1/360 + 1/1260 - 1/1680 + \cdots$, so that

$$\frac{M}{2^n} \approx \frac{2T(n-1)^{n-1}(n/2)^{-n}(n-1)^{1/2}}{2^n} = \frac{2T(n-1)^n}{n^n\sqrt{n-1}}.$$

Now $(n-1)^n/n^n \to e^{-1}$ as $n\to\infty$, and de Moivre lets $T/e = 1/B$, so that the above equation becomes

$$\frac{\binom{n}{n/2}}{2^n} \approx \frac{2}{B\sqrt{n}},$$

where $\ln B = 1 - 1/12 + 1/360 - 1/1260 + 1/1680 - \cdots$ Regarding the constant B, de Moivre (1738, p. 236) says (see Fig. 10.2)

When I first began that enquiry, I contended myself to determine at large the Value of B, which was done by the addition of some Terms of the above-written Series, but as I perceived that it converged but slowly, and seeing at the same time that what I had done answered my purpose tolerably well, I desisted from proceeding farther, till my worthy and learned friend, Mr. *James Stirling*, who applied himself after me to that enquiry, found that the Quantity B did denote the Square-Root of the Circumference of a Circle whose Radius is Unity, so that if that Circumference be called c, the Ratio of the middle Term to the Sum of all the Terms will be expressed by $2/\sqrt{nc}$...

In the above, de Moivre says his friend James Stirling has found that $B = \sqrt{2\pi}$.[†]

Using this and since $\Pr\{X = n/2\} = \binom{n}{n/2} 2^{-n}$ de Moivre is thus able to reach Eq. (10.3).

Combining Eqs. (10.2) and (10.3), de Moivre is finally able to show in his 1733 paper that

$$\Pr\left\{X = \frac{n}{2} + d\right\} \approx \frac{2}{\sqrt{2\pi n}} e^{-2d^2/n}, \tag{10.4}$$

where $X \sim B(n, 1/2)$. This is the normal approximation to the binomial distribution for the case $p = 1/2$. Speaking of the above result, de Moivre (1738, p. 234) says

I'll take the liberty to say, that this is the hardest Problem that can be proposed on the Subject of Chance, for which reason I have reserved it for the last, but I hope to be forgiven if my Solution is not fitted to the capacity of all Readers; however I shall derive from it some Conclusions that may be of use to every body.

To say that the conclusions *may* be of use to everybody is an understatement, for de Moivre's result is the first Central Limit Theorem (CLT).[‡] This suggests that he was actually the first to discover the normal curve (Pearson, 1924), although the credit is usually given to Carl Freidrich Gauss (1777–1855).[§] In all fairness, it must be

---

[†] Stirling had actually shown that $n! \sim \sqrt{2\pi n}\, n^{n+1/2}\, e^{-n}$ in Example 2 of Proposition 28, in his *Methodus differentialis* (Stirling, 1730). This was the same year that de Moivre published his *Miscellanea analytica*. Although de Moivre had first discovered $n! \propto n^{n+1/2}\, e^{-n}$ for large n, the formula later earned the name of only Stirling. Pearson (1924) says "I consider that the fact that Stirling showed that de Moivre's arithmetical constant was $\sqrt{2\pi}$ does not entitle him to claim the theorem, and that it is erroneous to term it Stirling's Theorem." On the other hand, Bressoud (2007, p. 293) writes "Even though it was a joint effort, the formula that we will find is called Stirling's formula. This is primarily de Moivre's own fault. When he published his result he gave Stirling credit for finding the constant, but his language was sufficiently imprecise that the attribution of the constant to Stirling could easily be misread as crediting him with the entire identity."

[‡] The name "Central Limit Theorem" was given by Pólya (1920). Two books entirely devoted to the CLT are Adams (2009) and Fischer (2010).

[§] Some might consider the normal or Gaussian curve to be an example of Stigler's law of eponymy: no scientific discovery is named after its original discover (Stigler, 1980). By self-application, Stigler attributed his law to the sociologist Robert King Merton (1910–2003). De Moivre seems to have been "dealt a double blow" in so far as due credit is concerned, given his substantial work in establishing Stirling's formula (see footnote above).

mentioned that de Moivre recognized neither the universality nor importance of the normal curve he had discovered. In particular, he did not recognize that he had derived the $N(n/2, n/4)$ as an approximation to the $B(n, 1/2)$ distribution, although he did realize the importance of the quantity $\sqrt{n}$. He noted that the accuracy of an estimated proportion was inversely proportional to the quantity $\sqrt{n}$.[†] About 50 years later, Pierre-Simon Laplace (1749–1827) generalized de Moivre's results for general p in Chapter 3 of his *Théorie Analytique des Probabilités* (Laplace, 1812, pp. 275–282), by showing that, for $X \sim B(n, p)$,

$$\lim_{n \to \infty} \Pr\left\{ \frac{X/n - p}{\sqrt{p(1-p)/n}} \le z \right\} = \int_{-\infty}^{z} \frac{1}{\sqrt{2\pi}} \exp\left(-\frac{u^2}{2}\right) du.$$

The above formula is the celebrated de Moivre–Laplace Theorem. It is a case of convergence of distribution, which is an even weaker form of convergence than convergence in probability.

De Moivre's result in Eq. (10.4) is also a refinement of Bernoulli's law of large numbers for the symmetric case ($p = 1/2$). To see why, recall that[‡] Bernoulli's law states, for any natural number c, one can find a sample of size n such that

$$\Pr\left\{ |P_s - p| \le \frac{1}{t} \right\} > \frac{c}{c+1},$$

where

$$n \ge \max\left\{ m_1 t + \frac{st(m_1 - 1)}{r + 1}, m_2 t + \frac{rt(m_2 - 1)}{s + 1} \right\},$$

$$m_1 \ge \frac{\ln\{c(s - 1)\}}{\ln(1 + r^{-1})}, \quad m_2 \ge \frac{\ln\{c(r - 1)\}}{\ln(1 + s^{-1})},$$

$$p = \frac{r}{t}, \quad t = r + s.$$

For example, let us take $r = 30$, $s = 30$, and $c = 1000$. We have $p = 1/2$ and we wish to find how large a sample size is needed so that

$$\Pr\left\{ \left| P_s - \frac{1}{2} \right| \le \frac{1}{60} \right\} > \frac{1000}{1001}.$$

Using the values of r, s, and c, we obtain $m_1 = m_2 = 336$, so that, according to Bernoulli's result, the required sample size is $n = 39{,}612$. Now let us re-estimate the required sample size using de Moivre's result in Eq. (10.4). Letting

---

[†] Today, we would say that the *standard deviation* of an estimated proportion is inversely proportional to $\sqrt{n}$.

[‡] See **Problem 8**, p. 70.

$P_s = X/n$, we have

$$\Pr\left\{\left|P_s - \frac{1}{2}\right| \le \frac{d}{n}\right\} = \Pr\left\{\frac{1}{2} - \frac{d}{n} \le P_s \le \frac{1}{2} + \frac{d}{n}\right\}$$

$$= \Pr\left\{\frac{n}{2} - d \le X \le \frac{n}{2} + d\right\}. \tag{10.5}$$

Using de Moivre's result in Eq. (10.4), Eq. (10.5) becomes

$$\Pr\left\{\left|P_s - \frac{1}{2}\right| \le \frac{d}{n}\right\} \approx \int_{-d}^{d} \frac{2}{\sqrt{2\pi n}} \exp\left(-\frac{2u^2}{n}\right) du = 1 - 2\Phi(-2d/\sqrt{n}), \tag{10.6}$$

where $\Phi(\cdot)$ is the cumulative distribution function of the standard normal distribution. If we substitute $\varepsilon = d/n$, Eq. (10.6) becomes

$$\Pr\left\{\left|P_s - \frac{1}{2}\right| \le \varepsilon\right\} \approx 1 - 2\Phi(-2\varepsilon\sqrt{n}).$$

Let us now take $\varepsilon = 1/60$ and set the probability above to $1000/1001$. Solving $1 - 2\Phi(-\sqrt{n}/30) = 1000/1001$ using standard normal tables, we obtain, according to de Moivre's result, a sample size of $n = 9742$. This is much less than Bernoulli's sample size of $n = 39,612$.

De Moivre also considered the nonsymmetrical binomial case ($p \ne 1/2$). He sketched a generalization of his method and showed how the terms in the general binomial expression $(a + b)^n$, $a \ne b$, could be approximated, but his investigations here did not go very far.

We now describe some of the studies leading to Gauss' own derivation of the normal distribution, which occurred in the context of putting the *method of least squares* on a theoretical foundation. The method of least squares was first published by Adrien-Marie Legendre (1752–1833) (Legendre, 1805, pp. 72–80) (Fig. 10.4).[†] Legendre introduces the method in a purely algebraic way as one of finding the n unknown quantities $x_1, x_2, \ldots, x_n$ such the following set of m equations are satisfied as "nearly" as possible (see Fig. 10.5):

$$a_{11}x_1 + a_{12}x_2 + \cdots + a_{1n}x_n = y_1,$$
$$a_{21}x_1 + a_{22}x_2 + \cdots + a_{2n}x_n = y_2,$$
$$\cdots$$
$$a_{m1}x_1 + a_{m2}x_2 + \cdots + a_{mn}x_n = y_m.$$

Legendre explains that, if $m = n$, it is easy to uniquely determine the unknowns.[‡] However, in many physical and astronomical problems we have $m > n$. In these cases,

---

[†] For a comprehensive survey of the early history of the method of least squares, see the book by Merriman (1877).

[‡] Provided the set of equations is consistent.

Figure 10.4   Adrien-Marie Legendre (1752–1833).

one wishes to minimize the errors $\varepsilon_1, \varepsilon_2, \ldots, \varepsilon_m$, where

$$\varepsilon_1 = a_{11}x_1 + a_{12}x_2 + \cdots + a_{1n}x_n - y_1,$$
$$\varepsilon_2 = a_{21}x_1 + a_{22}x_2 + \cdots + a_{2n}x_n - y_2,$$
$$\cdots$$
$$\varepsilon_m = a_{m1}x_1 + a_{m2}x_2 + \cdots + a_{mn}x_n - y_m.$$

In the above, the a_{ij}'s are known constants. Legendre then states the *least squares principle* as follows (Legendre, 1805, p. 72; Smith, 1929, p. 577) (see Fig. 10.5):

> Of all the principles which can be proposed for that purpose, I think there is none more general, more exact, and more easy of application, than that of which we have made use in the preceding researches, and which consists of rendering the sum of the squares of the errors a *minimum*. By this means there is established among the errors a sort of equilibrium which, preventing the extremes from exerting an undue influence, is very well fitted to reveal that state of the system which most nearly approaches the truth.

Thus, one wishes to minimize

$$\sum_{i=1}^{m} \varepsilon_i^2 = \sum_{i=1}^{m} (a_{i1}x_1 + a_{i2}x_2 + \cdots + a_{in}x_n - y_i)^2. \tag{10.7}$$

By differentiating Eq. (10.7) partially with respect to each of $x_1, x_2, \ldots, x_n$ and equating to zero each time, one obtains the following *normal equations*:

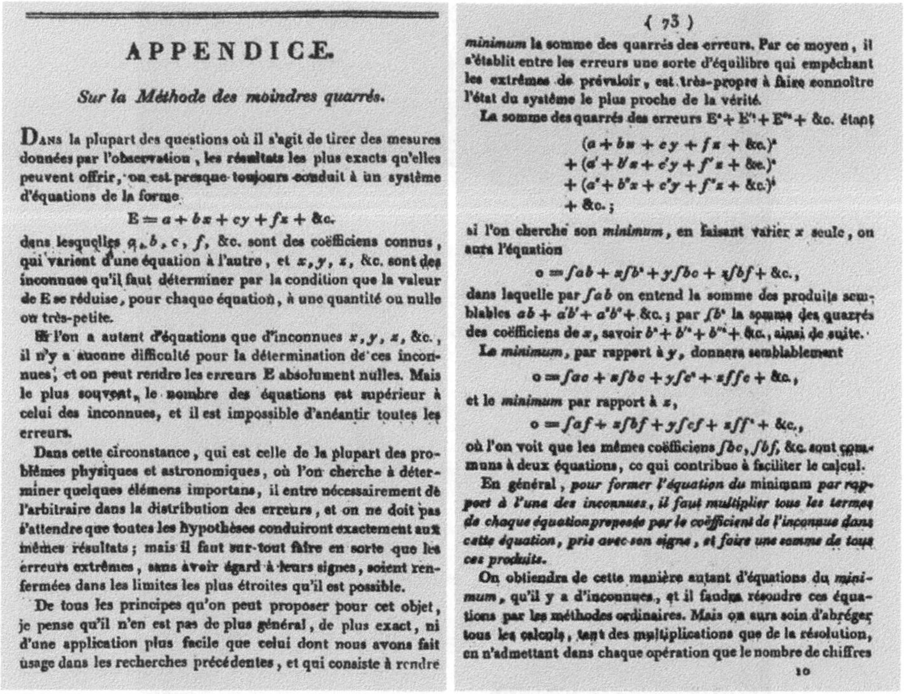

APPENDICE.

Sur la Méthode des moindres quarrés.

Dans la plupart des questions où il s'agit de tirer des mesures données par l'observation, les résultats les plus exacts qu'elles peuvent offrir, on est presque toujours conduit à un système d'équations de la forme.

$$E = a + bx + cy + fz + \&c.$$

dans lesquelles $a, b, c, f,$ &c. sont des coëfficiens connus, qui varient d'une équation à l'autre, et $x, y, z,$ &c. sont des inconnues qu'il faut déterminer par la condition que la valeur de E se réduise, pour chaque équation, à une quantité ou nulle ou très-petite.

Si l'on a autant d'équations que d'inconnues $x, y, z,$ &c., il n'y a aucune difficulté pour la détermination de ces inconnues; et on peut rendre les erreurs E absolument nulles. Mais le plus souvent, le nombre des équations est supérieur à celui des inconnues, et il est impossible d'anéantir toutes les erreurs.

Dans cette circonstance, qui est celle de la plupart des problèmes physiques et astronomiques, où l'on cherche à déterminer quelques élémens importans, il entre nécessairement de l'arbitraire dans la distribution des erreurs, et on ne doit pas s'attendre que toutes les hypothèses conduiront exactement aux mêmes résultats; mais il faut sur-tout faire en sorte que les erreurs extrêmes, sans avoir égard à leurs signes, soient renfermées dans les limites les plus étroites qu'il est possible.

De tous les principes qu'on peut proposer pour cet objet, je pense qu'il n'en est pas de plus général, de plus exact, ni d'une application plus facile que celui dont nous avons fait usage dans les recherches précédentes, et qui consiste à rendre

{ 73 }

minimum la somme des quarrés des erreurs. Par ce moyen, il s'établit entre les erreurs une sorte d'équilibre qui empêchant les extrêmes de prévaloir, est très-propre à faire connoitre l'état du système le plus proche de la vérité.

La somme des quarrés des erreurs E²+ E'²+ E''²+ &c. étant

$$(a + bx + cy + fx + \&c.)^2$$
$$+ (a' + b'x + c'y + f'x + \&c.)^2$$
$$+ (a'' + b''x + c''y + f''x + \&c.)^2$$
$$+ \&c. ;$$

si l'on cherche son *minimum*, en faisant varier x seule, on aura l'équation

$$0 = \int ab + x \int b^2 + y \int bc + x \int bf + \&c.,$$

dans laquelle par $\int ab$ on entend la somme des produits semblables $ab + a'b' + a''b'' +$ &c.; par $\int b^2$ la somme des quarrés des coëfficiens de x, savoir $b^2 + b'^2 + b''^2$ &c., ainsi de suite.

Le *minimum*, par rapport à y, donnera semblablement

$$0 = \int ac + x \int bc + y \int c^2 + x \int cf + \&c.,$$

et le *minimum* par rapport à z,

$$0 = \int af + x \int bf + y \int cf + x \int f^2 + \&c.,$$

où l'on voit que les mêmes coëfficiens $\int bc, \int bf,$ &c. sont communs à deux équations, ce qui contribue à faciliter le calcul.

En général, pour former l'équation du minimum par rapport à l'une des inconnues, il faut multiplier tous les termes de chaque équation proposée par le coëfficient de l'inconnue dans cette équation, pris avec son signe, et faire une somme de tous ces produits.

On obtiendra de cette manière autant d'équations du *minimum*, qu'il y a d'inconnues, et il faudra résoudre ces équations par les méthodes ordinaires. Mais on aura soin d'abréger tous les calculs, tant des multiplications que de la résolution, en n'admettant dans chaque opération que le nombre de chiffres

10

Figure 10.5 Legendre's Method of Least Squares, taken from the *Nouvelles Méthodes* (Legendre, 1805).

$$\left(\sum_{i=1}^m a_{i1}^2\right)x_1 + \left(\sum_{i=1}^m a_{i1}a_{i2}\right)x_2 + \cdots + \left(\sum_{i=1}^m a_{i1}a_{in}\right)x_n = \sum_{i=1}^m a_{i1}y_i,$$

$$\left(\sum_{i=1}^m a_{i2}a_{i1}\right)x_1 + \left(\sum_{i=1}^m a_{i2}^2\right)x_2 + \cdots + \left(\sum_{i=1}^m a_{i2}a_{in}\right)x_n = \sum_{i=1}^m a_{i2}y_i,$$

$$\cdots$$

$$\left(\sum_{i=1}^m a_{in}a_{i1}\right)x_1 + \left(\sum_{i=1}^m a_{in}a_{i2}\right)x_2 + \cdots + \left(\sum_{i=1}^m a_{in}^2\right)x_n = \sum_{i=1}^m a_{in}y_i.$$

The above n equations can now be solved for the n unknowns $x_1, x_2, \ldots, x_n$. As a special case of the above method, Legendre considers the case when n independent observations $a_1, a_2, \ldots, a_n$ are made on an unknown quantity x. The method of least squares then tries to minimize the quantity $(x - a_1)^2 + (x - a_2)^2 + \cdots + (x - a_n)^2$, which happens when the latter's derivative with respect to x is set to zero, that is,

$$(x - a_1) + (x - a_2) + \cdots + (x - a_n) = 0.$$

Figure 10.6   Carl Friedrich Gauss (1777–1855).

$$\therefore \quad x = \frac{1}{n}\sum_{i=1}^{n} a_i = \bar{a},$$

which is the sample mean of the a_i's.[†]

Legendre's clarity of exposition and the usefulness of least squares ensured that his method quickly gained wide acceptance. However, his claim of priority on the method was soon challenged by Gauss. Both Legendre and Gauss were among the greatest mathematicians of their times. In the *Theoria motus corporum coelestium* (Gauss, 1809, English edition, p. 270), Gauss states (Fig. 10.6)

> Our principle, which we have made use of since the year 1795, has lately been published by LEGENDRE in the work *Nouvelles methodes pour la determination des orbites des cometes*, Paris, 1806, where several other properties of this principle have been explained, which, for the sake of brevity, we here omit.

[†] To confirm that the sum of squares is indeed *minimized* at $x = \bar{a}$, one can show that its second derivative, evaluated at $x = \bar{a}$, is *positive*.

Legendre took umbrage at Gauss' claim, especially his use of the term "our principle." In an 1809 letter to Gauss, Legendre writes[†]

> I will therefore not conceal from you, Sir, that I felt some regret to see that in citing my memoir p. 221 you say *principium nostrum quo jam inde ab anno 1795 usi sumus etc.* There is no discovery that one cannot claim for oneself by saying that one had found the same thing some years previously; but if one does not supply the evidence by citing the place where one has published it, this assertion becomes pointless and serves only to do a disservice to the true author of the discovery. In Mathematics it often happens that one discovers the same things that have been discovered by others and which are well known; this has happened to me a number of times, but I have never mentioned it and I have never called *principium nostrum* a principle which someone else had published before me.

It seems that Gauss did indeed use least squares before Legendre's publication[‡] but did not get his work published before 1805. Similarly, the Irish-born mathematician Robert Adrain (1775–1843) independently developed and used least squares around 1806. However, the priority of discovery goes to Legendre for having first published the method.

In the *Theoria motus*, Gauss set out to give a *probabilistic* justification for the principle of least squares (Gauss, 1809, English edition, pp. 257–260) (see Fig. 10.7). Recall that, when n observations $a_1, a_2, \ldots, a_n$ are made on an unknown quantity x, Legendre's principle of least squares states the sum of squares of the errors, that is $(x - a_1)^2 + (x - a_2)^2 + \cdots + (x - a_n)^2$, is minimized when $x = \bar{a}$. However, Legendre provided no theoretical basis as to why the sum of *squares* of the errors ought to be minimized. *Gauss' aim was to find the type of distribution, given a set of independent measurements $a_1, a_2, \ldots, a_n$ for an unknown value x, which leads to the sample mean as the most probable value of x.* In doing so, he formally derived the normal distribution, a feat that enabled him to coin his name on this distribution.[§] Moreover, he was also able to justify the method of least squares. We now describe his derivation.[**] Given the independent observations $a_1, a_2, \ldots, a_n$ on x, let the probability density of the ith error $\varepsilon_i = a_i - x$ be

$$f_\varepsilon(\varepsilon_i | x) = \phi(\varepsilon_i) = \phi(a_i - x).$$

Therefore, the density of each a_i is

$$f_a(a_i | x) = f_\varepsilon(\varepsilon_i | x) = \phi(a_i - x).$$

[†] The following English translation is taken from Plackett (1972).

[‡] See, for example, Adams (2009, p. 28).

[§] A derivation of the normal law of error was actually given one year earlier by Adrain in the paper *Research concerning the probabilities of the errors which happen in making observations* (Adrain, 1808). Both of these occurred after de Moivre's initial derivation in 1733. The paper went largely unnoticed until it was reprinted in 1871. For more details, see Merriman (1877, pp. 163–164).

[**] See also Whittaker and Robinson (1924, pp. 218–220).

258 DETERMINATION OF AN ORBIT FROM [Book II.

this cannot be defined *a priori*, we will, approaching the subject from another point of view, inquire upon what function, tacitly, as it were, assumed as a base, the common principle, the excellence of which is generally acknowledged, depends. It has been customary certainly to regard as an axiom the hypothesis that if any quantity has been determined by several direct observations, made under the same circumstances and with equal care, the arithmetical mean of the observed values affords the most probable value, if not rigorously, yet very nearly at least, so that it is always most safe to adhere to it. By putting, therefore,

$$V = V' = V'' \text{ etc.} = p,$$

we ought to have in general,

$$\varphi'(M-p) + \varphi'(M'-p) + \varphi'(M''-p) + \text{etc.} = 0,$$

if instead of p is substituted the value

$$\frac{1}{\mu}(M + M' + M'' + \text{etc.}),$$

wnatever positive integer μ expresses. By supposing, therefore,

$$M' = M'' = \text{etc.} = M - \mu N,$$

we shall have in general, that is, for any positive integral value of μ,

$$\varphi'(\mu - 1)N = (1-\mu)\varphi'(-N),$$

whence it is readily inferred that $\frac{\varphi'\Delta}{\Delta}$ must be a constant quantity, which we will denote by k. Hence we have

$$\log \varphi\Delta = \tfrac{1}{2}k\Delta\Delta + \text{Constant},$$

$$\varphi\Delta = \varkappa e^{\frac{1}{2}k\Delta\Delta},$$

denoting the base of the hyperbolic logarithms by e and assuming

$$\text{Constant} = \log \varkappa.$$

Moreover, it is readily perceived that k must be negative, in order that Ω may really become a maximum, for which reason we shall put

$$\tfrac{1}{2}k = -hh;$$

and since, by the elegant theorem first discovered by LAPLACE, the integral

$$\int e^{-hh\Delta\Delta}\,d\Delta$$

SECT. 3.] ANY NUMBER OF OBSERVATIONS. 259

from $\Delta = -\infty$ to $\Delta = +\infty$ is $\frac{\sqrt{\pi}}{h}$, (denoting by π the semicircumference of the circle the radius of which is unity), our function becomes

$$\varphi\Delta = \frac{h}{\sqrt{\pi}} e^{-hh\Delta\Delta}.$$

178.

The function just found cannot, it is true, express rigorously the probabilities of the errors: for since the possible errors are in all cases confined within certain limits, the probability of errors exceeding those limits ought always to be zero. while our formula always gives some value. However, this defect, which every analytical function must, from its nature, labor under, is of no importance in practice, because the value of our function decreases so rapidly, when $h\Delta$ has acquired a considerable magnitude, that it can safely be considered as vanishing. Besides, the nature of the subject never admits of assigning with absolute rigor the limits of error.

Finally, the constant h can be considered as the measure of precision of the observations. For if the probability of the error Δ is supposed to be expressed in any one system of observations by

$$\frac{h}{\sqrt{\pi}} e^{-hh\Delta\Delta},$$

and in another system of observations more or less exact by

$$\frac{h'}{\sqrt{\pi}} e^{-h'h'\Delta\Delta},$$

the expectation, that the error of any observation in the former system is contained between the limits $-\delta$ and $+\delta$ will be expressed by the integral

$$\int \frac{h}{\sqrt{\pi}} e^{-hh\Delta\Delta}\,d\Delta$$

taken from $\Delta = -\delta$ to $\Delta = +\delta$; and in the same manner the expectation, that the error of any observation in the latter system does not exceed the limits $-\delta'$ and $+\delta'$ will be expressed by the integral

$$\int \frac{h'}{\sqrt{\pi}} e^{-h'h'\Delta\Delta}\,d\Delta$$

extended from $\Delta = -\delta'$ to $\Delta = +\delta'$: but both integrals manifestly become

Figure 10.7 Gauss' derivation of the normal distribution, taken from the English translation of Gauss' *Theoria motus* (Gauss, 1809).

Then Gauss uses Laplace's form of Bayes Theorem† to obtain the posterior distribution of x, given the observations

$$f(x|a_1, a_2, \ldots, a_n) = \frac{f(a_1, a_2, \ldots, a_n|x)f(x)}{\displaystyle\int f(a_1, a_2, \ldots, a_n|x)f(x)dx}. \tag{10.8}$$

Next, Gauss follows Laplace again by assuming a uniform prior on x, that is, $f(x) = 1$, based on the principle of indifference.‡ Since the observations are independent, Eq. (10.8) becomes

$$\begin{aligned} f(x|a_1, a_2, \ldots, a_n) &= \frac{f_a(a_1|x)f_a(a_2|x)\cdots f_a(a_n|x)}{\displaystyle\int f_a(a_1|x)f_a(a_2|x)\cdots f_a(a_n|x)dx} \\ &= \frac{\prod_{i=1}^{n}\phi(a_i - x)}{\displaystyle\int \prod_{i=1}^{n}\phi(a_i - x)dx}. \end{aligned}$$

Now, $f(x|a_1, a_2, \ldots, a_n)$, hence $\ln f(x|a_1, a_2, \ldots, a_n)$, is maximized when $x = \bar{a}$. This happens when

$$\sum_{i=1}^{n} \frac{d}{dx} \ln \phi(a_i - x) = 0. \tag{10.9}$$

† See **Problem 14**, p. 141.

‡ See **Problem 14**, pp. 135–137.

When $x = \bar{a}$, we also have

$$\sum_{i=1}^{n}(a_i - x) = 0. \tag{10.10}$$

Comparing Eqs. (10.9) and (10.10), we see that

$$\frac{d}{dx}\ln\phi(a_i - x) = k_1(a_i - x),$$

where k_1 is a constant. To solve the above differential equation, we write

$$\int d\ln\phi(a_i - x) = -k_1\int(a_i - x)d(a_i - x)$$

$$\ln\phi(a_i - x) = -\frac{k_1}{2}(a_i - x)^2 + k_2$$

$$\phi(a_i - x) = k_3\, e^{-k_1(a_i-x)^2/2}$$

$$\therefore\ \phi(\varepsilon_i) = k_3\, e^{-k_1\varepsilon_i^2/2},$$

where k_2, k_3 are constants. Since $\int_{-\infty}^{\infty}\phi(\varepsilon_i)d\varepsilon_i = k_3\int_{-\infty}^{\infty}e^{-k_1\varepsilon_i^2/2}\,d\varepsilon_i = 1$, we can make the substitution $u = \varepsilon_i\sqrt{k_1/2}$ and use $\int_{-\infty}^{\infty}e^{-u^2}\,du = \sqrt{\pi}$ to show that $k_3 = \sqrt{k_1/(2\pi)}$. Writing $h \equiv \sqrt{k_1/2}$, Gauss finally obtains the normal density $N(0, \frac{1}{2h^2})$ (see Fig. 10.7):

$$\phi(\varepsilon_i) = \frac{h}{\sqrt{\pi}}e^{-h^2\varepsilon_i^2}. \tag{10.11}$$

Having derived the normal density, it is then easy for Gauss to argue that the joint density

$$\phi(\varepsilon_1)\phi(\varepsilon_2)\cdots\phi(\varepsilon_n) = \left(\frac{h}{\sqrt{\pi}}\right)^n e^{-h^2\sum_{i=1}^{n}\varepsilon_i^2}$$

is maximized when the sum of squares of the errors, that is, $\sum_{i=1}^{n}\varepsilon_i^2$, is minimized, just justifying the principle of least squares. In his own words (Gauss, 1809, English edition, p. 260)

It is evident, in order that the product

$$\Omega = h^{\mu}\pi^{-\mu/2}\,e^{-hh(vv+v'v'+v''v''+\cdots)}$$

may become a maximum, that the sum

$$vv + v'v' + v''v'' + \text{etc.},$$

must become a minimum. *Therefore, that will be the most probable system of values of the unknown quantities p, q, r, s, etc., in which the sum of the squares of the differences between the observed and computed values of V, V', V'', etc. is a minimum...*

This completes Gauss' probabilistic justification of the method of least squares.

Any description of Gauss' derivation of the normal curve would be incomplete without any mention of Laplace, a French mathematician of unrivalled analytic superiority. As early as 1774, on his work on the theory of errors, Laplace had proposed the double exponential function[†]

$$f(x) = \frac{m}{2} e^{-m|x|}, \quad -\infty < x < \infty, \quad m > 0,$$

as a law of error (Laplace, 1774a, pp. 634–644). Now, as we have described, Gauss had proved that the distribution that results in the sample mean as the most likely value of a set of observations is the normal distribution in Eq. (10.11). In the *Mémoire sur les approximations des formules qui sont fonctions de très grands nombres et sur leur application aux probabilités* (Laplace, 1810a), Laplace provided the first formal statement of the CLT after de Moivre's work in the *Miscellanea analytica* (de Moivre, 1730). Laplace considered the sum $S_n = X_1 + X_2, \ldots + X_n$, where the X_i's are independent and identically distributed discrete random variables with probability mass function

$$\Pr\left\{ X_i = \frac{ih}{2m} \right\} = \frac{1}{2m + 1}, \quad \text{for } h > 0, \quad i = 0, 1, \ldots, 2m.$$

By using characteristic functions[‡] for the first time, Laplace showed that, for large n,

$$\Pr\left\{ -u \leq \frac{S_n}{n} - \mu \leq u \right\} \approx \frac{\sqrt{2}}{\sigma\sqrt{\pi}} \int_0^u e^{-x^2/(2\sigma^2)} dx,$$

where $\mu = \mathscr{E}X_i = h/2$ and $\sigma^2 = \text{var } X_i = 1/3$. Equipped with this result, Laplace was able to further reinforce Gauss' result on the normal curve. In the *Supplément au Mémoire sur les approximations des formules qui sont fonctions de très grands nombres* (Laplace, 1810b), Laplace essentially argued that if the errors in Gauss' derivation were the sum of a large number of random variables, then the CLT implies that the errors are approximately normally distributed. This fact gives further validity to Gauss' derivation of the normal curve.

Let us now outline some of the key developments in the CLT after de Moivre's and Laplace's initial works on it. Just like in the case of the law of large numbers,[§] the aim of these investigations was to further relax the conditions under which the CLT can hold. The first rigorous proof of the CLT was provided by Alexander Mikhailovich Lyapunov (1857–1918) (Fig. 10.8) in 1901 for the case of mutually independent random variables $X_1, X_2, \ldots, X_n$ such that $\mathscr{E}X_i = \mu_i$, var $X_i = \sigma_i^2 < \infty$, and $\mathscr{E}|X_i|^{2+\delta} < \infty$ where $\delta > 0$ (Lyapunov, 1901). Let $S_n = X_1 + X_2 + \cdots + X_n$,

---

[†] Now called Laplace's (first) distribution.

[‡] The characteristic function of a random variable X is defined by $\phi_X(t) = \mathscr{E} e^{itX}$, where $i = \sqrt{-1}$.

[§] See **Problem 8**.

Figure 10.8   Alexander Mikhailovich Lyapunov (1857–1918)

$m_n = \mu_1 + \mu_2 + \cdots + \mu_n$, and $B_n^2 = \sigma_1^2 + \sigma_2^2 + \cdots + \sigma_n^2$. Then if

$$\lim_{n \to \infty} \frac{1}{B_n^{2+\delta}} \sum_{j=1}^{n} \mathcal{E} |X_j - \mu_j|^{2+\delta} = 0,$$

we have

$$\lim_{n \to \infty} \Pr\left\{ \frac{S_n - m_n}{B_n} \le z \right\} = \int_{-\infty}^{z} \frac{1}{\sqrt{2\pi}} \exp\left(-\frac{u^2}{2}\right) du.$$

Note that Liapunov's CLT requires the existence of moments beyond the second order. A simpler and more general form of the above is due to the Finnish mathematician Jarl Waldemar Lindeberg (1876–1932) (Lindeberg, 1922) and the

French mathematician Paul Pierre Lévy (1886–1971) (Lévy, 1925): If $X_1, X_2, \ldots, X_n$ are IID random variables with common mean $\mathscr{E}X_i = \mu$ and common variance var $X_i = \sigma^2 < \infty$, then

$$\lim_{n \to \infty} \Pr\left\{ \frac{X_1 + X_2 + \cdots + X_n - n\mu}{\sigma\sqrt{n}} \leq z \right\} = \int_{-\infty}^{z} \frac{1}{\sqrt{2\pi}} \exp\left(-\frac{u^2}{2}\right) du.$$

Sufficient and necessary conditions for the CLT were provided by Lindeberg (1922) and William Feller (1906–1970) (Feller, 1935), respectively: Let $X_1, X_2, \ldots, X_n$ be independent random variables such that $\mathscr{E}X_i = \mu_i$ and var $X_i = \sigma_i^2 < \infty$. Let $S_n = X_1 + X_2 + \cdots + X_n$, $m_n = \mu_1 + \mu_2 + \cdots + \mu_n$, and $B_n^2 = \sigma_1^2 + \sigma_2^2 + \cdots + \sigma_n^2$. Then

$$\lim_{n \to \infty} \Pr\left\{ \frac{S_n - m_n}{B_n} \leq z \right\} = \int_{-\infty}^{z} \frac{1}{\sqrt{2\pi}} \exp\left(-\frac{u^2}{2}\right) du$$

and

$$\lim_{n \to \infty} \max_{1 \leq i \leq n} \frac{\sigma_i}{B_n} = 0$$

if and only if, for any $\varepsilon > 0$,

$$\lim_{n \to \infty} \left\{ \frac{1}{B_n^2} \sum_{i=1}^{n} \int_{|x - \mu_i| > \varepsilon B_n} (x - \mu_i)^2 f_{X_i}(x) dx \right\} = 0. \tag{10.12}$$

In the above, $f_{X_i}(x)$ is the probability density of each X_i.

What are the implications of the CLT? Briefly, it implies that, for the simplest case, *the sum (or sample mean) of IID random variables with finite second moments can be approximated by a normal distribution for "large enough" sample sizes*. It is immaterial whether the original random variables themselves have a normal distribution or not, or whether they are discrete or continuous.[†] This result is of tremendous importance in as much as it enables us to make probability statements about sums or means of random variables. It also leads, under certain conditions, to useful normal approximations for various distributions such as the binomial, the Poisson, the chi-square, and the gamma distributions. Finally, the CLT is extremely important in inferential problems involving the sample mean.

A look at Eq. (10.12) reveals the basic condition for the CLT to hold in large enough sample sizes. First note that[‡]

[†] Note that the theorem does not say that any *random variable* will become approximately normal for large sample sizes, it makes a statement of approximate normality regarding the *sum* (and *sample mean*) of the random variables.

[‡] In the formulas which follow, if the random variable is discrete we replace the density function with the mass function, and the integration with a summation.

$$\frac{1}{B_n^2} \sum_{i=1}^{n} \int_{|x-\mu_i|>\varepsilon B_n} (x-\mu_i)^2 f_{X_i}(x)dx \ge \frac{1}{B_n^2} \sum_{i=1}^{n} \int_{|x-\mu_i|>\varepsilon B_n} \varepsilon^2 B_n^2 f_{X_i}(x)dx$$

$$= \varepsilon^2 \sum_{i=1}^{n} \int_{|x-\mu_i|>\varepsilon B_n} f_{X_i}(x)dx$$

$$= \varepsilon^2 \sum_{i=1}^{n} \Pr\{|X_i - \mu_i| > \varepsilon B_n\}$$

$$\ge \varepsilon^2 \max_{1 \le i \le n} \Pr\{|X_i - \mu_i| > \varepsilon B_n\}.$$

Condition (10.12) therefore implies, as $n \to \infty$,

$$\max_{1 \le i \le n} \Pr\{|X_i - \mu_i| > \varepsilon B_n\} \to 0.$$

Thus for the CLT to hold, the Lindeberg–Feller conditions state that no single random variable should dominate the sum, that is, each $(X_i - \mu_i)/\sigma_i$ is small relative to the sum $(S_n - \mu_n)/B_n$ as n increases.[†]

[†] See Spanos (1986, pp. 174–175).

Problem 11

Daniel Bernoulli and the
St. Petersburg Problem (1738)

Problem. *A player plays a coin-tossing game with a fair coin in a casino. The casino agrees to pay the player 1 dollar if heads appears on the initial throw, 2 dollars if head first appears on the second throw, and in general 2^{n-1} dollars if heads first appears on the nth throw. How much should the player **theoretically** give the casino as an initial down-payment if the game is to be fair (i.e., the expected profit of the casino or player is zero)?*

Solution. The player wins on the nth throw if all previous $(n-1)$ throws are tails and the nth thrown is a head. This occurs with probability $(1/2)^{n-1}(1/2) = 1/2^n$ and the player is then paid 2^{n-1} dollars by the casino. The casino is therefore expected to pay the player the amount

$$\sum_{n=1}^{\infty} \frac{1}{2^n} \times 2^{n-1} = \sum_{n=1}^{\infty} \frac{1}{2} = \infty.$$

Thus it seems that no matter how large an amount the player initially pays the casino, she will always emerge with a profit. Theoretically, this means that only if the player initially pays the casino an infinitely large sum will the game be fair.

11.1 Discussion

This problem has been undoubtedly one of the most discussed in the history of probability and statistics. The problem was first proposed by Nicholas Bernoulli

Classic Problems of Probability, Prakash Gorroochurn.
© 2012 John Wiley & Sons, Inc. Published 2012 by John Wiley & Sons, Inc.

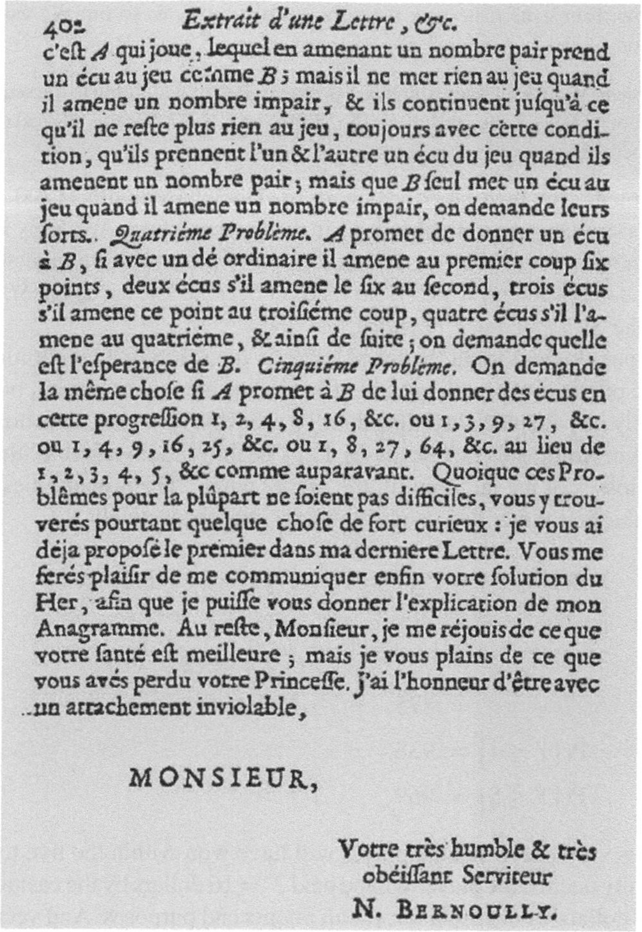

402 *Extrait d'une Lettre, &c.*

c'eſt *A* qui joue, lequel en amenant un nombre pair prend un écu au jeu comme *B*; mais il ne met rien au jeu quand il amene un nombre impair, & ils continuent juſqu'à ce qu'il ne reſte plus rien au jeu, toujours avec cette condition, qu'ils prennent l'un & l'autre un écu du jeu quand ils amenent un nombre pair; mais que *B* ſeul met un écu au jeu quand il amene un nombre impair, on demande leurs ſorts. *Quatriéme Problême.* *A* promet de donner un écu à *B*, ſi avec un dé ordinaire il amene au premier coup ſix points, deux écus s'il amene le ſix au ſecond, trois écus s'il amene ce point au troiſiéme coup, quatre écus s'il l'amene au quatriéme, & ainſi de ſuite; on demande quelle eſt l'eſperance de *B*. *Cinquiéme Problême.* On demande la même choſe ſi *A* promet à *B* de lui donner des écus en cette progreſſion 1, 2, 4, 8, 16, &c. ou 1, 3, 9, 27, &c. ou 1, 4, 9, 16, 25, &c. ou 1, 8, 27, 64, &c. au lieu de 1, 2, 3, 4, 5, &c comme auparavant. Quoique ces Problêmes pour la plûpart ne ſoient pas difficiles, vous y trouverés pourtant quelque choſe de fort curieux : je vous ai déja propoſé le premier dans ma derniere Lettre. Vous me ferés plaiſir de me communiquer enfin votre ſolution du Her, afin que je puiſſe vous donner l'explication de mon Anagramme. Au reſte, Monſieur, je me réjouis de ce que votre ſanté eſt meilleure; mais je vous plains de ce que vous avés perdu votre Princeſſe. J'ai l'honneur d'être avec un attachement inviolable,

MONSIEUR,

Votre très humble & très obéiſſant Serviteur

N. BERNOULLY.

Figure 11.1 Extract from letter sent to Montmort by Nicholas Bernoulli, taken from the second edition of Montmort's *Essay* (Montmort, 1713). The St. Petersburg Problem first appears here (*Quatrième Problème, Cinquième Problème*).

(1687–1759) in a letter to Pierre Rémond de Montmort (1678–1719) in 1713.† It was published in Montmort's second edition of the *Essay d'Analyse sur les Jeux de Hazard* (Montmort, 1713, p. 402) (see Fig. 11.1):

> *Fourth Problem.* A promises to give one ecu to B, if with an ordinary die he scores a six, two ecus if he scores a six on the second throw, three ecus if he scores this number on the

† An even earlier version of the St. Petersburg Problem appears in Cardano's *Practica arithmetice* and is as follows: "A rich and a poor man play for equal stakes. If the poor man wins, on the following day the stakes are doubled and the process continues; if the rich man wins once, the play is ended once and for all." See Coumet (1965a) and Dutka (1988) for more details.

third throw, four ecus if he gets it on the fourth throw, & so on. We ask what the expectation of B is.

Fifth Problem. We ask the same thing if A promises B ecus according to the sequence $1, 2, 4, 8, 16$, etc., or $1, 3, 9, 27$, etc., or $1, 4, 9, 16, 25$, etc., or $1, 8, 27, 64$, etc., instead of $1, 2, 3, 4, 5$, etc. like before.

The problem was then taken up again by Daniel Bernoulli (1700–1782)[†] and published in a similar form to **Problem 11** in the St. Petersburg Academy Proceedings in 1738 (Bernoulli, 1738) (Fig. 11.2). Almost every prominent mathematician of the time discussed the problem, which from then on became known as the *St. Petersburg Problem* or the *St. Petersburg Paradox.*[‡]

What is paradoxical about the problem? It is the solution. Our solution seems to imply that the player is expected to be paid an infinite amount of money, but in fact she will most likely be paid much less than that. To see why, let us calculate how likely it is that she will win within the first few tosses. Let Y be the random variable "the number of independent tosses of the fair coin till the player wins." The probabilities of winning within, say, the first three, four, and five tosses are, respectively,

$$\Pr\{Y \le 3\} = \Pr\{Y = 1\} + \Pr\{Y = 2\} + \Pr\{Y = 3\}$$

$$= \left(\frac{1}{2}\right) + \left(\frac{1}{2}\right)^1 \left(\frac{1}{2}\right) + \left(\frac{1}{2}\right)^2 \left(\frac{1}{2}\right)$$

$$= .875,$$

$$\Pr\{Y \le 4\} = .938,$$

$$\Pr\{Y \le 5\} = .969.$$

Thus, it is very likely that the player will have won within the five tosses. Given that this actually occurs, the player will be paid $2^4 = 16$ dollars by the casino. So paying the casino 16 dollars would seem fair for all intents and purposes. And yet our solution appears to imply that the player should pay the casino an infinite amount to make the game fair. On one hand, the theory seems to predict the player is expected to emerge with a huge profit however large a sum she pays the casino. On the other hand, even paying the casino 1000 dollars seems unfair since most certainly the player will end up being paid at most 16 dollars.

There have been numerous attempts to get around this conundrum. We now outline some of them. First let us consider Jean le Rond d'Alembert's (1717–1783) take on the matter. d'Alembert rejected the solution we presented earlier, namely that the player should pay the casino an infinite amount. In the *Croix ou Pile* article

---

[†] Daniel Bernoulli was the younger cousin of Nicholas Bernoulli.

[‡] The St. Petersburg paradox is also discussed in Clark (2007, p. 196), Samuelson (1977), Shackel (2008), Keynes (1921, pp. 316–323), Nickerson (2004, p. 182), Parmigiani and Inoue (2009, p. 34), Epstein (2009, pp. 111–113), Chernoff and Moses (1986, p. 104), Chatterjee (2003, p. 190), Székely (1986, p. 27), and Everitt (2008, p. 117).

Figure 11.2   Daniel Bernoulli (1700–1782).

of the *Encyclopédie* he states,[†] regarding the infinite solution (d'Alembert, 1754, Vol. IV, p. 513):

> ...& there is some scandal here that deserves to occupy the Algebraists.

Elsewhere, d'Alembert has claimed that there is no paradox at all, based on an argument that we now describe. First, d'Alembert distinguishes between two types of

[†] See also Fig. 12.3.

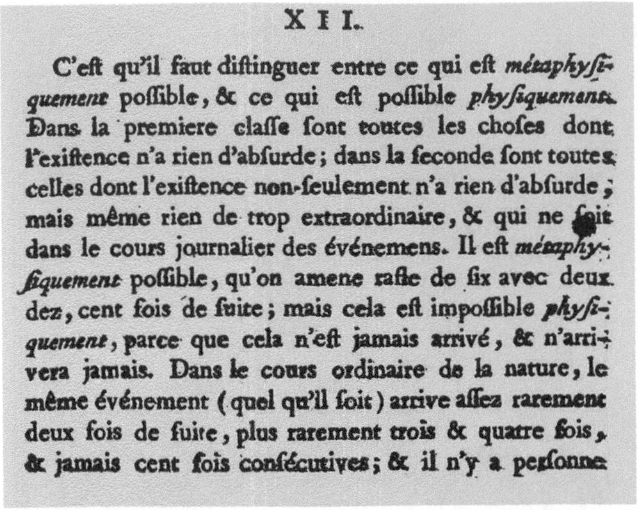

X I I.

C'eft qu'il faut diftinguer entre ce qui eft *métaphyfi-quement* poffible, & ce qui eft poffible *phyfiquement.* Dans la premiere claffe font toutes les chofes dont l'exiftence n'a rien d'abfurde; dans la feconde font toutes celles dont l'exiftence non-feulement n'a rien d'abfurde; mais même rien de trop extraordinaire, & qui ne foit dans le cours journalier des événemens. Il eft *métaphy-fiquement* poffible, qu'on amene rafle de fix avec deux dez, cent fois de fuite; mais cela eft impoffible *phyfi-quement*, parce que cela n'eft jamais arrivé, & n'arri-vera jamais. Dans le cours ordinaire de la nature, le même événement (quel qu'il foit) arrive affez rarement deux fois de fuite, plus rarement trois & quatre fois, & jamais cent fois confécutives; & il n'y a perfonne

Figure 11.3   D'Alembert's definition of metaphysical and physical probability, taken from the *Opuscules (Vol. 2)* (d'Alembert, 1761, p. 10).

probabilities, which he calls "metaphysical" and "physical" probabilities.[†] According to d'Alembert, an event is metaphysically possible if its probability is mathematically greater than zero. On the other hand, an event is physically possible if it is not so rare that its probability is very close to zero. In the *Opuscules Mathématiques* (d'Alembert, 1761, p. 10), d'Alembert thus states (see Fig. 11.3)

> One must distinguish between what is *metaphysically* possible, & what is *physically* possible. In the first class are all things whose existence is not absurd; in the second are all things whose existence not only has nothing that is absurd, but also has nothing that is very extraordinary either, and that is not in the everyday course of events. It is *metaphysically* possible, that we obtain a double-six with two dice, one hundred times in a row; but this is *physically* impossible, because this has never happened, & will never happen.

In the famous article *Doutes et Questions sur le Calcul des Probabilités*, d'Alembert (1767, pp. 282) addresses the *St. Petersburg Problem*:

> Thus I ask if one has to look very far for the reason of this paradox, & if it does not belie the eyes that this supposed *infinite sum* owed by Paul[‡] [the player] at the beginning of the game, is only apparently infinite, only because it is based on a false assumption; namely on the assumption that *head* can never happen, & that the game can last forever?

[†] See also Arne Fisher's *The Mathematical Theory of Probabilities* (Fisher, 1922, pp. 51–52).

[‡] The use of Paul and Pierre (or Peter) as opponents in games of chances has its origins in Montmort's *Essay d'Analyse* (Montmort, 1708). Gordon (1997, p. 185) makes the following amusing point: "Note that *Pierre* Montmort was hardly neutral between *Pierre* and Paul. We still root for Peter and tend to look at things from his point of view. For example, we shall use the letter *p*, which we have used for the probability of success, for the probability that Peter wins a game; *q*, corresponding to failure, is the probability Paul wins."

It is however true, & even clear, that this assumption is mathematically possible. Only *physically speaking* is it false.

Finally, in the *Opuscules,* d'Alembert explains why he believes the "infinite-expectation" solution (which we presented in our solution) is wrong (p. 8):

... when the probability of an event is very small, it must be regarded & treated as zero; & one must not multiply (as it has been prescribed till now) that probability with the gain, so as to obtain the stake or expectation.

Thus d'Alembert believes that very small probabilities ought to be treated as zero. Although d'Alembert does not explicitly give a cutoff probability to decide when a metaphysically possible event becomes physically impossible, he gives an example of a coin that is tossed 100 times to illustrate the notion of physical impossibility (d'Alembert, 1761, p. 9). Let us therefore take a probability $p \leq 1/2^{100}$ to denote physical impossibility, and use d'Alembert's arguments to solve the *St. Petersburg Problem.* In our solution, the casino is now expected to play the player an amount.

$$\sum_{n=1}^{99} \frac{1}{2^n} \times 2^{n-1} + \sum_{n=100}^{\infty} 0 \times 2^{n-1} = 49.5.$$

Thus, according to d'Alembert and using a probability cutoff of 2^{-100}, the player should give the casino an upfront payment of \$49.50 for the game to be fair. However, how should we decide at what point a probability becomes so small[†] that it should be treated as zero? D'Alembert does not provide any convincing argument for a cutoff and his views on the *St. Pertersburg Problem* have largely been superseded today.

The great naturalist, Georges-Louis Leclerc, Comte de Buffon (1707–1788) favored a solution similar to d'Alembert, except that he considered a probability less than $1/10,000 \, (\approx 1/2^{13.29})$ to be "morally" impossible. With this prescription, the casino is expected to pay the player an amount[‡]

$$\sum_{n=1}^{13} \frac{1}{2^n} \times 2^{n-1} + \sum_{n=14}^{\infty} 0 \times 2^{n-1} = 6.5,$$

so that the player should give the casino an upfront payment of \$6.50 for the game to be fair. Buffon was also one of the foremost empiricists of his times. He employed a child to toss a coin until a head was obtained, and the game was repeated 2048 times (Buffon, 1777, p. 84). Buffon reports that there were 1061 games that produced one crown, 494 games that produced two crowns, and so on. The total amount was

---

[†] The issue of small probabilities is also treated in **Problem 24**.

[‡] In his own calculations, Buffon seems to have arbitrarily changed the payment to $(9/5)^{n-1}$ for a first head on the nth head (Coolidge, 1990, p. 174). In this way he is able to obtain an upfront payment of 5 crowns, which matches his empirical results.

Table 11.1   A Modified St. Petersburg's Game Where a Limit 2^N is Placed on the Amount Owned by the Casino

Value of N	Maximum amount owned by casino (2^N)	Amount player must pay upfront for a fair game ($N/2 + 1$)
1	2	1.5
2	4	2.0
3	8	2.5
4	16	3.0
5	32	3.5
10	1,024	6.0
15	32,768	8.5
20	1.03×10^6	11.0

For a given value of N, the player must pay an amount ($N/2 + 1$) for the game to be fair.

10,057 crowns in 2084 games, implying an average upfront payment of about 5 crowns.

On the other hand, mathematicians such as the Marquis de Condorcet (1743–1794), Siméon Denis Poisson (1781–1840), and Eugène Charles Catalan (1814–1894) argued that no casino could pay an infinite amount and that its fortune should be restricted. Let us therefore assume that the casino has a maximum of 2^N dollars, where N is a fixed integer. Furthermore, we assume that the player is paid 2^{n-1} dollars if heads first occurs on the nth toss for $n \leq N + 1$. For $n > N + 1$, the player is always paid the same amount of 2^N dollars. For such a game, the bank is expected to pay the player an amount

$$\frac{1}{2} \cdot 1 + \frac{1}{2^2} \cdot 2 + \cdots + \frac{1}{2^{N+1}} \cdot 2^N + 2^N \left(\frac{1}{2^{N+2}} + \frac{1}{2^{N+3}} + \cdots \right) = \frac{1}{2}(N+1) + \left(\frac{1}{4} + \frac{1}{8} + \cdots \right)$$
$$= \frac{N}{2} + 1.$$

For the game to be fair, the player must therefore pay ($N/2 + 1$) dollars upfront. Table 11.1 lists several values of N, 2^N, and $N/2 + 1$. An upfront payment of ($N/2 + 1$) seems much more reasonable than an "infinite" one.

However, Condorcet et al.'s position of restricting the casino's fortune is criticized by Joseph Bertrand (1822–1900), who argues that this does not solve the problem. In the *Caculs des Probabilités*, he writes (Bertrand, 1889, p. 61)

> If we play for cents instead of francs, for grains of sand instead of cents, for molecules of hydrogen instead of grains of sand, the fear of insolvency can be diminished without limit. This should not make a difference to the theory. It does not also assume that the stakes be paid before every throw of the coin. However much Peter [the casino] may owe, the pen can write it, one can write the calculations on paper; the theory will triumph if the calculations confirm its prescriptions. Chance will very probably, one can even say certainly, end by

favoring Paul [the player]. However much he pays for Peter's promise, the game is to his advantage, and if Paul is persistent it will enrich him without limit. Peter, whether he is solvent or not, will owe him a vast sum.

Gabriel Cramer (1704–1752) and Daniel Bernoulli (1700–1782) both favored a solution that involved the concept of *mean utility* (or *moral expectation*).[†] This is one of the more common solutions to the *St. Petersburg Problem* and needs some explaining. Suppose a player has an initial fortune x_0, and has a probability p_j (for $j = 1, \ldots, K$) of winning an amount a_j, where $\sum_j p_j = 1$. If at any stage the player's fortune changes from x to $(x + dx)$, then Bernoulli suggests that the change in the player's *utility* U for that fortune is $dU \propto dx/x$. This means that, for a given change in the player's fortune, the more money the player has the less will the change in utility be. From $dU \propto dx/x$, we get $U = k \ln x + C$, where k and C are both constants, *that is, the utility of the fortune x is a function of the log of x*. Bernoulli then denotes the mean utility of the player's final fortune by

$$\sum_{j=1}^{K} p_j \ln \left(\frac{x_0 + a_j}{x_0} \right)$$

(Dutka, 1988). Compare this with the mathematical expectation of the player's final fortune: $\sum_j p_j(x_0 + a_j)$. For the *St. Petersburg Problem*, suppose the player has an initial fortune x_0 and pays an initial amount ξ for the game. Then, for a fair game, Bernoulli sets the player's mean utility for her final fortune to be zero, obtaining

$$\sum_{j=1}^{\infty} \frac{1}{2^j} \ln(x_0 + 2^j - \xi) = \ln x_0.$$

This equation can be solved for ξ when the value of x_0 is specified. For example, if the player's initial fortune is $x_0 = 10$ dollars, then she should pay an upfront amount of $\xi \approx 3$ dollars. On the other hand if $x_0 = 1000$ dollars, then $\xi \approx 6$ dollars (Shafer, 2004, p. 8320). Again, these are much more reasonable upfront payments than that suggested in our solution. Note that this version of the *St. Petersburg Problem* does not restrict the casino's fortune, but instead conditions on the player's initial fortune.

A fourth solution due to the eminent Croatian-American mathematician William Feller (1906–1970) (Feller, 1937, 1968) (Fig. 11.5) ought to be mentioned. Feller reasoned in terms of a *variable entrance fee*, that is, an entrance fee that is a function of the number n of tosses. He defined a fair game to be one such that, for any small $\varepsilon > 0$,

$$\lim_{n \to \infty} \Pr\left\{ \left| \frac{S_n}{e_n} - 1 \right| < \varepsilon \right\} = 1,$$

[†] For more details, see Bernstein (1996, p. 105). See also **Problems 4** and **25**.

Table 11.2 Feller's Solution to the St. Petersburg Game, Whereby a Variable Amount is Paid by the Player at the Start of the nth Toss

Value of n	Cumulative amount paid by player on nth toss ($n \log_2 n$)	Actual amount paid by player on nth toss $[n \log_2 n - (n-1)\log_2 n(n-1), n > 1]$
1	.00	.00
2	2.00	2.00
3	4.75	2.75
4	8.00	3.25
5	11.61	3.61
10	33.22	4.69
15	58.60	5.30
20	86.44	5.73

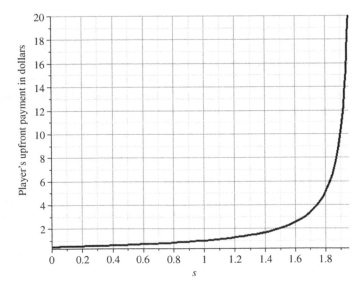

Figure 11.4 Plot of player's upfront payment, $1/(2 - s)$, against s for the modified St. Petersburg game where the player is paid s^{n-1} (where $0 < s < 2$) dollars if head occurs on the nth toss.

where S_n and e_n are the *cumulative* amounts paid to and by the player, respectively. The last formula is analogous to the weak law of large numbers.[†] Feller then proved that the game becomes fair for

$$e_n = n \log_2 n.$$

[†] See **Problem 8**.

Figure 11.5 William Feller (1906–1970).

Table 11.2 shows values of the cumulative and actual amounts the player should play on the nth toss.

Finally, it is worthwhile reconsidering the original St. Petersburg game with one modification: if a head appears first on the nth toss, then assume the casino pays the player an amount of s^{n-1} dollars, where $0 < s < 2$. Then the casino is expected to pay the player an amount

$$\sum_{n=1}^{\infty} \frac{1}{2^n} \times s^{n-1} = \frac{1}{s} \sum_{n=1}^{\infty} \left(\frac{s}{2}\right)^n$$
$$= \frac{s/2}{s(1 - s/2)}$$
$$= \frac{1}{2-s}, \quad 0 < s < 2.$$

Therefore, for the game to be fair, the player must pay the *finite* amount of $(2 - s)^{-1}$ dollars upfront. The graph in Fig. 11.4 shows the variation of the player's upfront payment with the value of s.

Of all the possible resolutions we have presented (including our own in the *Solution*), Daniel Bernoulli's one based on the concept of utility seems the most appealing. A little "tweaking" of the original concept is needed, through, to avoid a "Super St. Petersburg Paradox" (Menger, 1934). The latter will occur with a log-utility formulation (i.e., $u(x) = \ln x$) if the casino pays $\exp(2^{n-1})$ dollars when a head first appears on the nth toss. However, by bounding the utility function from above, the Super St. Petersburg Paradox can be avoided.[†]

[†] For more details, see Ingersoll's *Theory of Financial Decision Making* (Ingersoll, 1987, p. 19).

Problem 12

d'Alembert and the "Croix ou Pile" Article (1754)

Problem. *In two tosses of a fair coin, what is the probability that heads will appear at least once?*

Solution. Let H stand for heads and T for tails. Then the sample space when the coin is tossed is $\Omega = \{HH, HT, TH, TT\}$. Each of these sample points is equally likely, so applying the classical definition of probability

$$\text{Pr}\{\text{heads at least once}\} \equiv \frac{\text{number of favorable outcomes}}{\text{total number of outcomes}}$$

$$= \frac{3}{4}.$$

12.1 Discussion

Jean le Rond d'Alembert (1717–1783) was one of the foremost intellectuals of the 18th century and coedited the monumental *Encyclopédie ou Dictionnaire Raisonné des Sciences, des Arts et des Métiers* (Figs. 12.1 and 12.2).[†] However, d'Alembert was often at odds (no pun intended) with his contemporaries on questions of probability. On the above problem, he denied that 3/4 could be the correct answer. He reasoned as follows: once a head occurs, there is no need for a second throw; the possible outcomes are thus *H, TH, TT*, and the required probability is 2/3. Of course, d'Alembert's reasoning is wrong because he failed to realize that each of *H, TH, TT* is *not* equally likely. The erroneous answer was even included in his article *Croix ou Pile*[‡] of the *Encyclopédie* (d'Alembert, 1754, Vol. IV, pp. 512–513). Let us examine the first part of his article (see Fig. 12.3):

[†] Encyclopedia of the Sciences, Arts, and Crafts.

[‡] Heads or tails.

Classic Problems of Probability, Prakash Gorroochurn.
© 2012 John Wiley & Sons, Inc. Published 2012 by John Wiley & Sons, Inc.

Figure 12.1   Jean le Rond d'Alembert (1717–1783).

Heads or tails (*analysis of chances*). This game which is well known, & which has no need of definition, will provide us the following reflections. One asks how much are the odds that one will bring heads in playing two successive tosses. The answer that one will find in all authors, & following the ordinary principles, is this one here: There are four combinations,

First toss. Second toss.

Heads. Heads.

Tails. Heads.

Heads. Tails.

Tails. Tails.

Of these four combinations one alone makes a loss, & three make wins; the odds are therefore 3 against 1 in favor of the player who casts the piece. If he wagered on three tosses, one will find eight combinations of which one alone makes a loss, & seven make wins; thus the odds will be 7 against 1. See *Combinaison & Avantage*. However is this quite correct? For in order to take here only the case of two tosses, is it not necessary to reduce to one the two combinations which give heads on the first toss? For as soon as heads comes one time,

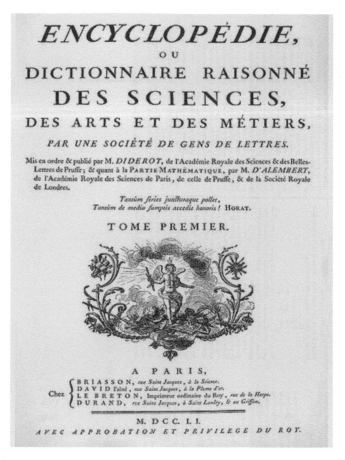

Figure 12.2 Cover of the *Encyclopédie*.

the game is finished, & the second toss counts for nothing. So there are properly only three possible combinations:

Heads, first toss.

Tails, heads, first & second toss.

Tails, tails, first & second toss.

Therefore the odds are 2 against 1. Likewise in the case of three tosses, one will find

Heads.

Tails, heads.

Tails, tails, heads.

Tails, tails, tails.

numens de piété que sur la route de S. Denys, où se remarquent quantité de ces monumens dans le goût gothique.

On appelle aussi *croix*, les amortissemens placés au-dessus des portails & des faîtes des monumens sacrés. Enfin on appelle *croix greque* ou *latine* dans une église, la partie qui traverse l'église entre le chœur & la nef. *Voyez* ÉGLISE. (*P*)

CROIX, (*Marine.*) On dit, *il y a une croix sur les cables*; ce qui signifie que les cables qui sont mouillés, sont passés l'un sur l'autre. (*Z*)

CROIX DE S. ANDRÉ, (*Charpenterie.*) servent a remplir & à entretenir les combles & pans de bois où ils sont employés. *Voyez Pl. du Charpentier, fig.* 17.

* CROIX, (*Manufact. en drap.*) morceau de bois dont le nom désigne assez la figure, sur lequel sont montées les têtes de chardon qui servent au lainage des étoffes.

* CROIX, (*Manuf. en drap.*) petite courroie de cuir qui appartient à la manicle des Tondeurs de draps. *Voyez* MANICLE.

* CROIX, (*Manuf. de fer-blanc.*) marque que ces Manufacturiers placent sur le fond des barrils qu'ils remplissent de fer-blanc : elle désigne que ce fer est de la sorte la plus forte : elle s'imprime avec un fer chaud : elle donne au fer-blanc le nom de *fer à la croix*, qui se vend plus cher que l'autre.

CROIX, *en terme de Fourbisseur*, sont deux sortes de bras recourbés en-dessous, qui passent au haut du corps de la garde, l'un dessous la branche, & l'autre vis-à-vis; ce qui avec le corps représente effectivement une *croix*. Voy. *la fig. Pl. du Ciseleur-Damasquineur.*

Henri II. par édit de 1548, ordonna que sa pourtraiture, d'après son pourtrait, seroit gravée & empreinte sur les monnoies d'or, d'argent . . . *&c.* ce qui a été continué jusqu'à présent.

CROIX DE S. ANDRÉ, *terme de Riviere*, charpente qui porte en décharge la lisse d'un pont.

CROIX DE CERF, (*Venerie.*) c'est un os que l'on trouve dans le cœur de cet animal : il a à-peu-près la forme d'une *croix*. On croit que mis en poudre dans du vin, c'est un remede pour les femmes en travail; & que pendu au cou en amulette, il soulage dans les palpitations de cœur.

CROIX OU PILE, (*analyse des hasards.*) Ce jeu qui est très-connu, & qui n'a pas besoin de définition, nous fournira les réflexions suivantes. On demande combien il y a à parier qu'on amenera *croix* en jouant deux coups consécutifs. La réponse qu'on trouvera dans tous les auteurs, & suivant les principes ordinaires, est celle-ci : Il y a quatre combinaisons,

Premier coup.	Second coup.
Croix.	*Croix.*
Pile.	*Croix.*
Croix.	*Pile.*
Pile.	*Pile.*

De ces quatre combinaisons une seule fait perdre, & trois font gagner; il y a donc 3 contre 1 à parier en faveur du joüeur qui jette la piece. S'il parioit en trois coups, on trouveroit huit combinaisons dont une seule fait perdre, & sept font gagner; ainsi il y auroit 7 contre 1 à parier. *Voyez* COMBINAISON & AVANTAGE. Cependant cela est-il bien exact? Car pour ne prendre ici que le cas de deux coups, ne faut-il

CRO

faut-il pas réduire à une les deux combinaisons qui donnent *croix* au premier coup? Car dès qu'une fois *croix* est venu, le jeu est fini, & le second coup est compté pour rien. Ainsi il n'y a proprement que trois combinaisons de possibles :

Croix, premier coup.
Pile, *croix*, premier & second coup.
Pile, *pile*, premier & second coup.

Donc il n'y a que 2 contre 1 à parier. De même dans le cas de trois coups, on trouvera

Croix.
Pile, *croix.*
Pile, *pile*, *croix.*
Pile, *pile*, *pile.*

Donc il n'y a que 3 contre 1 à parier : ceci est digne, ce me semble, de l'attention des Calculateurs, & iroit à réformer bien des regles unanimement reçûes sur les jeux de hasard.

Autre question. Pierre joüe contre Paul à cette condition, que si Pierre amene *croix* du premier coup, il payera un écu à Paul; s'il n'amene *croix* qu'au second coup, deux écus; si au troisieme coup, quatre, & ainsi de suite. On trouve par les regles ordinaires (en suivant le principe que nous venons de poser), que l'espérance de Paul, & par conséquent ce qu'il doit mettre au jeu est $\frac{1+2+4+\&c.}{1+1+1\ \&c.}$ quantité qui se trouve infinie. Cependant il n'y a personne qui voulût mettre à ce jeu une somme un peu considérable. On peut voir dans les *mémoires de l'académie de Petersbourg*, *tome V.* quelques tentatives pour résoudre cette difficulté; mais nous ne savons si on en sera satisfait; & il y a ici quelque scandale qui mérite bien d'occuper les Algébristes. Ce qui paroît surprenant dans la solution de ce problème, c'est la

CRO 513

minute des sommes immenses. Or, pour joüer avec un fou à jeu égal, il faut se faire fou comme lui. Si Pierre joüant en un seul coup, parioit un million qu'il amenera *pile*, il faudroit que chacun mît au jeu un demi-million : cela est incontestable. Il n'y a pourtant que deux insensés qui pussent joüer un pareil jeu.

Nous remarquerons à cette occasion, que pour rendre plus complettes, & pour ainsi dire plus usuelles, les solutions de problèmes concernans les jeux, il seroit à souhaiter qu'on pût y faire entrer les considérations morales, relatives, soit à la fortune des joüeurs, soit à leur état, soit à leur situation, à leur force même (quand il s'agit des jeux de commerce), & ainsi du reste. Il est certain, par exemple, que de deux hommes inégalement riches qui joüent à jeu égal suivant les regles ordinaires, celui qui est le moins riche risque plus que l'autre. Mais toutes ces considérations étant presque impossibles à soûmettre au calcul à cause de la diversité des circonstances, on est obligé d'en faire abstraction, & de résoudre les problèmes mathématiquement, en supposant d'ailleurs les circonstances morales parfaitement égales de part & d'autre, ou en les négligeant totalement. Ce sont ensuite ces circonstances, quand on vient à y faire attention, qui font croire le calcul en faute, quoiqu'il n'y soit pas. *Voyez* AVANTAGE, JEU, PARI, *&c.* (*O*)

CROIX, (*Sainte-*) *Géog.* île de l'Amérique septentrionale, l'une des Antilles.

CROIX, (*Sainte-*) *Géog.* petite ville de France dans la haute Alsace.

CROKETHORN, (*Géog.*) petite ville d'Angleterre dans la province de Sommerset, sur la riviere de Perd.

CROLER, (*Fauconn.*) il se dit du bruit que font

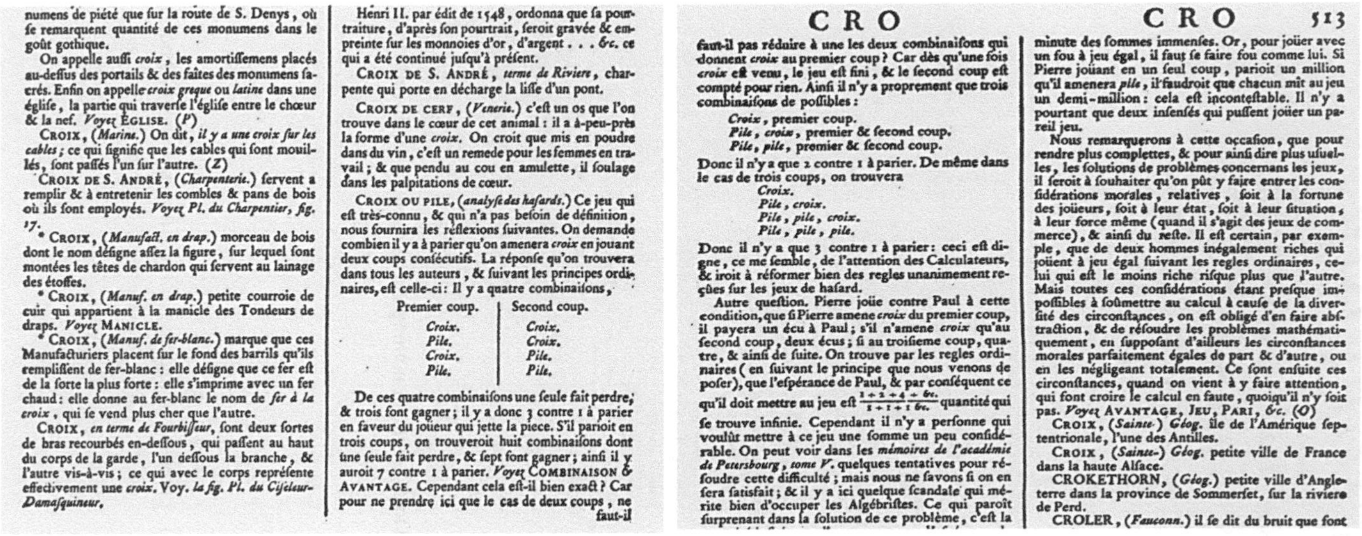

Figure 12.3 d'Alembert article "Croix ou Pile," taken from the *Encyclopédie* (Vol. IV) (d'Alembert, 1754).

Therefore the odds are only 3 against 1: this is worthy, it seems me, of the attention of the Calculators, & would go to reform well some unanimously received rules on the games of chance.

d'Alembert comes back to the issue again, in his article *Gageure*[†] (d'Alembert 1757, pp. 420–421), after having received objections to his *Croix ou Pile* article from a certain Mr. Necker, who was a Professor of Mathematics. Necker gives the correct reasoning, but the end of the article shows a still unconvinced d'Alembert.

d'Alembert notoriously also refuted the solution to the *St. Petersburg's Problem* (see **Problem 11**) and fell in the trap of the *Gambler's Fallacy* (see **Problem 13**). Bertrand (1889, pp. ix–x) did not mince his words about d'Alembert's various *faux pas* in the games of chance:

When it comes to the calculus of probability, d'Alembert's astute mind slips completely.

Similarly, in his *History of Statistics*, Karl Pearson writes (Pearson, 1978, p. 535)

What then did d'Alembert contribute to our subject? I think the answer to that question must be that he contributed absolutely nothing.

In spite of Bertrand's and Pearson's somewhat harsh words, it would be misleading for us to think that d'Alembert, a man of immense mathematical prowess, was so naïve that he would have no strong basis for his probabilistic reasoning. In the *Croix ou Pile* article, a sample space of {*HH, HT, TH, TT*} made no sense to d'Alembert because it did not correspond to reality. In real life, no person would ever observe *HH*, because once an initial *H* was observed the game would end. By proposing an alternative model for the calculus of probabilities, namely that of equiprobability on *observable* events, d'Alembert was effectively asking why his model could not be right, given the absence of an existing theoretical framework for the calculus of probabilities. d'Alembert's skepticism was partly responsible for later mathematicians to seek a solid theoretical foundation for probability, culminating in its axiomatization by Kolmogorov in 1933 (Kolmogorov, 1933).[‡]

[†] Wager.

[‡] See **Problem 23**.

Problem 13

d'Alembert and the Gambler's Fallacy (1761)

Problem. When a fair coin is tossed, given that heads have occurred three times in a row, what is the probability that the next toss is a tail?

Solution. Since the coin is fair, the probability of tails (or heads) on one toss is 1/2. Because of independence, this probability stays at 1/2, irrespective of the results of previous tosses.

13.1 Discussion

When presented with the problem, d'Alembert insisted that the probability of a tail must "obviously" be greater than 1/2, thus rejecting the concept of independence between the tosses. The claim was made in d'Alembert's *Opuscule Mathématiques (Vol. 2)* (d'Alembert, 1761, pp. 13–14) (see Fig. 13.1). In his own words

> Let's look at other examples which I promised in the previous Article, which show the lack of exactitude in the ordinary calculus of probabilities.
>
> In this calculus, by combining all possible events, we make two assumptions which can, it seems to me, be contested.
>
> The first of these assumptions is that, if an event has occurred several times successively, for example, if in the game of heads and tails, heads has occurred three times in a row, it is equally likely that head or tail will occur on the fourth time? However I ask if this assumption is really true, & if the number of times that heads has already successively occurred by the hypothesis, does not make it more likely the occurrence of tails on the fourth time? Because after all it is not possible, it is even physically impossible that tails never occurs. Therefore the more heads occurs successively, the more it is likely tail will occur the next time. If this is the case, as it seems to me one will not disagree, the rule of combination of possible events is thus still deficient in this respect.

Classic Problems of Probability, Prakash Gorroochurn.
© 2012 John Wiley & Sons, Inc. Published 2012 by John Wiley & Sons, Inc.

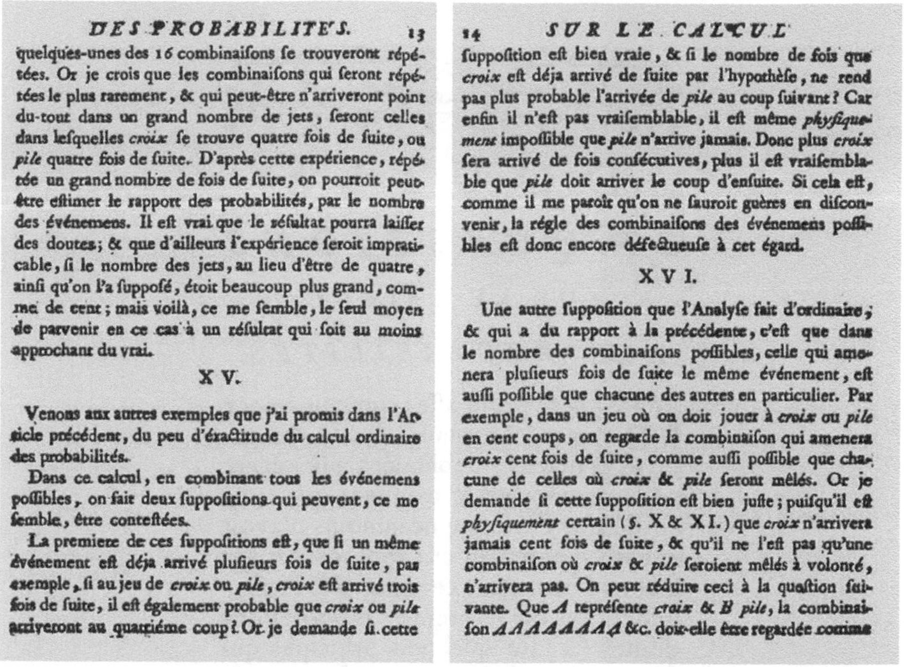

DES PROBABILITÉS. 13

quelques-unes des 16 combinaiſons ſe trouveront répétées. Or je crois que les combinaiſons qui ſeront répétées le plus rarement, & qui peut-être n'arriveront point du-tout dans un grand nombre de jets, ſeront celles dans leſquelles *croix* ſe trouve quatre fois de ſuite, ou *pile* quatre fois de ſuite. D'après cette expérience, répétée un grand nombre de fois de ſuite, on pourroit peut-être eſtimer le rapport des probabilités, par le nombre des événemens. Il eſt vrai que le réſultat pourra laiſſer des doutes; & que d'ailleurs l'expérience ſeroit impraticable, ſi le nombre des jets, au lieu d'être de quatre, ainſi qu'on l'a ſuppoſé, étoit beaucoup plus grand, comme de cent; mais voilà, ce me ſemble, le ſeul moyen de parvenir en ce cas à un réſultat qui ſoit au moins approchant du vrai.

X V.

Venons aux autres exemples que j'ai promis dans l'Article précédent, du peu d'éxactitude du calcul ordinaire des probabilités.

Dans ce calcul, en combinant tous les événemens poſſibles, on fait deux ſuppoſitions qui peuvent, ce me ſemble, être conteſtées.

La premiere de ces ſuppoſitions eſt, que ſi un même événement eſt déja arrivé pluſieurs fois de ſuite, par exemple, ſi au jeu de *croix* ou *pile*, *croix* eſt arrivé trois fois de ſuite, il eſt également probable que *croix* ou *pile* arriveront au quatriéme coup? Or je demande ſi cette

14 *SUR LE CALCUL*

ſuppoſition eſt bien vraie, & ſi le nombre de fois que *croix* eſt déja arrivé de ſuite par l'hypothèſe, ne rend pas plus probable l'arrivée de *pile* au coup ſuivant? Car enfin il n'eſt pas vraiſemblable, il eſt même *phyſiquement* impoſſible que *pile* n'arrive jamais. Donc plus *croix* ſera arrivé de fois conſécutives, plus il eſt vraiſemblable que *pile* doit arriver le coup d'enſuite. Si cela eſt, comme il me paroît qu'on ne ſauroit guères en diſconvenir, la régle des combinaiſons des événemens poſſibles eſt donc encore défectueuſe à cet égard.

X V I.

Une autre ſuppoſition que l'Analyſe fait d'ordinaire, & qui a du rapport à la précédente, c'eſt que dans le nombre des combinaiſons poſſibles, celle qui amenera pluſieurs fois de ſuite le même événement, eſt auſſi poſſible que chacune des autres en particulier. Par exemple, dans un jeu où on doit jouer à *croix* ou *pile* en cent coups, on regarde la combinaiſon qui amenera *croix* cent fois de ſuite, comme auſſi poſſible que chacune de celles où *croix* & *pile* ſeront mêlés. Or je demande ſi cette ſuppoſition eſt bien juſte; puiſqu'il eſt *phyſiquement* certain (§. X & XI.) que *croix* n'arrivera jamais cent fois de ſuite, & qu'il ne l'eſt pas qu'une combinaiſon où *croix* & *pile* ſeroient mêlés à volonté, n'arrivera pas. On peut réduire ceci à la queſtion ſuivante. Que *A* repréſente *croix* & *B* *pile*, la combinaiſon *AAAAAAAA* &c. doit-elle être regardée comme

Figure 13.1 Extract of d'Alembert's article, taken from the *Opuscules*, Vol. 2 (d'Alembert, 1761). Here d'Alembert contends that if a fair coin has come up heads three times in a row, then the next throw is more likely to be a tail.

d'Alembert states that it is *physically* impossible for tails never to occur in a long series of tosses of a coin, and thus uses his concepts of physical and metaphysical probabilities† to support his erroneous argument.

In his later article *Doutes et Questions sur le Calcul des Probabilités*‡ d'Alembert (1767, pp. 275–304) (see Fig. 13.2) further elaborated his objections to the applications of the calculus of probabilities, but his arguments were essentially the same

> The calculus of Probabilities is based in this assumption, that all the various combinations of the same effect are equally likely. For example, if we throw a coin in the air 100 consecutive times, one assumes that it is equally likely that heads occurs one hundred consecutive times or that heads & tails are mixed, by moreover following whatever particular sequence one wishes, for example, heads in the first toss, tails on the next two tosses, heads in the fourth, tails on the fifth, heads on the sixth & seventh, &c.
>
> These two cases are no doubt mathematically equally possible; this is not the difficult point, & mediocre Mathematicians of whom I spoke before have made useless efforts to

† See **Problem 11**, p. 112. Recall that, according to d'Alembert, an event is *metaphysically* possible if its probability is greater than zero, and is *physically* possible if it is not so rare that its probability is very close to zero.

‡ Doubts and questions on the calculus of probabilities.

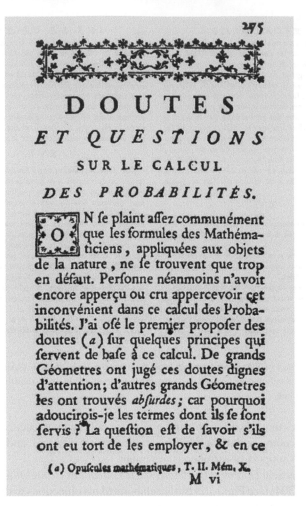

2ʮ5

D O U T E S

ET QUESTIONS

SUR LE CALCUL

DES PROBABILITÉS.

O N ſe plaint aſſez communément que les formules des Mathématiciens, appliquées aux objets de la nature, ne ſe trouvent que trop en défaut. Perſonne néanmoins n'avoit encore apperçu ou cru appercevoir cet inconvénient dans ce calcul des Probabilités. J'ai oſé le premier propoſer des doutes (*a*) ſur quelques principes qui ſervent de baſe à ce calcul. De grands Géometres ont jugé ces doutes dignes d'attention; d'autres grands Géometres les ont trouvés *abſurdes;* car pourquoi adoucirois-je les termes dont ils ſe ſont ſervis ? La queſtion eſt de ſavoir s'ils ont eu tort de les employer, & en ce

(*a*) Opuſcules mathématiques, T. II. Mém. X.

M vi

Figure 13.2 Extract of d'Alembert's article "Doutes et Questions sur le Calcul des Probabilités," taken from the *Mélanges de Littérature* (d'Alembert, 1767).

write lengthy dissertations to prove this equal possibility. But one must ask whether these two cases, which are mathematically equally likely, are also physically as such, taken the order of things into account.

The above remarks made by d'Alembert need some clarification, because the misconceptions are still widely believed. Consider the following two sequences when a fair coin is tossed four times:

Sequence 1: *HHHH*

Sequence 2: *HHHT*

Many would believe that the first sequence is less likely than the second one. After all, it seems highly improbable to obtain four heads in a row. However, the truth of the matter is that it is equally unlikely to obtain the second sequence *in that specific order*. While it is less likely to obtain four heads than to obtain a total of three heads and one tail,[†] *HHHT* is as likely as any other of the same length, even if it contains all heads or tails. To give another example, common lore probably suggests a sequence such as *HHHTTHHTTH* is much more likely than one such as *HHHHHHHHHH*. Yet they both have the same probability,[‡] because it is as unlikely to observe the specific first sequence as the second.

In general, "the belief that if an event has not happened for a long time it is bound to occur soon" is known as the *Gambler's Fallacy* (e.g., Everitt, 2006, p. 168). The mistaken belief is the result of the erroneous application of the so-called "law of averages," which is a colloquialism for the more formal *weak law of large numbers* (WLLN). For a fair coin, the WLLN states that as the coin is tossed again and again, the probability that the actual proportion of heads deviates from 1/2 by more than a certain small amount gets closer to zero.[§] Thus, if the coin gives three heads in a row, believers in the *Gambler's Fallacy* think that the coin will try to compensate for this by being more likely to give tails so as to get closer to 1/2 for the proportion of heads. However, coins have no memory nor do they work through compensation: *the WLLN operates through swamping rather than compensation*. That is to say, once an initial string of heads occurs there is no tendency for the coin to give more tails in further tosses to make up for the initial excess. Rather, the ratio of 1/2 is approached as the effect of the initial excess of heads is overwhelmed by the sheer number of future tosses.

We should mention that, if d'Alembert was merely skeptical of the ways probability calculations were being applied in real-life situations,[**] the philosopher Auguste Comte (1798–1857) was much more acerbic regarding the merits of this doctrine. The following extract clearly shows his thoughts on probability (Comte, 1833, p. 371):

> The calculus of probabilities has seemed to me to be really, by its illustrious inventors, only a convenient text for ingenious and difficult numerical problems, which nonetheless preserves all of its abstract value, like the analytical theories which it has then occasioned, or, if we wish, been the origin for. As to the philosophical conception on which this doctrine rests, I consider it radically false and susceptible to lead to the most absurd consequences.

Commenting on Comte's criticism, Coumet (2003) rightly states

> The theory of chances – I could use, I stress, the expression *probability theory, calculus of probabilities*, which is also in the writings of Auguste Comte in the period 1820–1840 – is not a tired theory that the scientific world can neglect, it is not an emerging theory on which various judgments are allowed – optimistic or pessimistic – but it is a theory with a

---

[†] Remember the specific sequence *HHHT* is one of four possible ways of obtaining a total of three heads and one tail.

[‡] In d'Alembert's language, both sequences have the same metaphysical (mathematical) probability, but not the same physical probability.

[§] The WLLN is fully discussed in **Problem 8**.

[**] d'Alembert's critique of probability theory is further discussed in Daston (1979), Brian (1996), Paty (1988), Henry (2004, pp. 219–223), Samueli and Boudenot (2009, pp. 169–181), and Gorroochurn (2011).

prestigious past, not least by the personalities who devoted themselves to it, Pascal, Fermat, Huygens, the Bernoullis, De Moivre, Condorcet; and especially, since recently, as Auguste Comte is very young, the theory has its monument, the *Théorie Analytique des Probabilités* of Laplace.

Other well-known critics of the theory of probability include the Scottish philosopher David Hume[†] (1711–1776), and the Prussian historian and statesman Johann Peter Friedrich Ancillon[‡] (1767–1837).

[†] See **Problem 14**.

[‡] See Todhunter (1865, p. 453) and Keynes (1921, p. 82).

Problem 14

Bayes, Laplace, and Philosophies of Probability (1764, 1774)

Problem. *Assume that the probability p of an event A is equally likely to be any number between 0 and 1. Show that, conditional on A having previously occurred in a out of n independent and identical trials, the probability that p lies between p_1 and p_2 is*

$$\frac{(n+1)!}{a!(n-a)!} \int_{p_1}^{p_2} p^a (1-p)^{n-a} dp.$$

Solution. Let X_n be the number of times that the event A occurs in n trials. Since p is equally likely to be between 0 and 1, we let p have a probability density function that is uniform on [0,1], that is,

$$f(p) = 1, \qquad 0 \le p \le 1$$

To find $\Pr\{p_1 < p < p_2 | X_n = a\}$, let us first obtain the conditional density $f(p|X_n = a)$ by using Bayes' Theorem[†]:

$$f(p|X_n = a) = \frac{\Pr\{X_n = a|p\}f(p)}{\int_0^1 \Pr\{X_n = a|p\}f(p)dp}$$

$$= \frac{\binom{n}{a} p^a (1-p)^{n-a} \cdot 1}{\int_0^1 \binom{n}{a} p^a (1-p)^{n-a} \cdot 1 \, dp}$$

[†] Which will be explained in the Discussion section.

Classic Problems of Probability, Prakash Gorroochurn.
© 2012 John Wiley & Sons, Inc. Published 2012 by John Wiley & Sons, Inc.

$$= \frac{p^a(1-p)^{n-a}}{\int_0^1 p^a(1-p)^{n-a}\,dp}.$$

Now, we have[†]

$$\int_0^1 p^a(1-p)^{n-a}\,dp = \frac{\Gamma(a+1)\Gamma(n-a+1)}{\Gamma(n+2)} = \frac{a!(n-a)!}{(n+1)!}.$$

Therefore,

$$f(p\,|\,X_n = a) = \frac{(n+1)!p^a(1-p)^{n-a}}{a!(n-a)!}.$$

Hence,

$$\Pr\{p_1 < p < p_2\,|\,X_n = a\} = \int_{p_1}^{p_2} f(p\,|\,X_n = a)dp$$

$$= \frac{(n+1)!}{a!(n-a)!}\int_{p_1}^{p_2} p^a(1-p)^{n-a}\,dp,$$

as required.

14.1   Discussion

This is essentially the problem that the English clergyman, Reverend Thomas Bayes (1702–1761), solved in his famous essay entitled *An Essay towards Solving a Problem in the Doctrine of Chances* (Bayes, 1764) (Fig. 14.2).[‡] The density $f(p)$ is called the *prior density* of p while the density $f(p\,|\,X_n = a)$ is called its *posterior density*.

In actual fact, Bayes framed his problem as follows (see Fig. 14.1): a ball W is thrown on a unit square table on which one of the sides is the x-axis. Suppose W stops at a point with x-coordinate p. Another ball O is then thrown n times and if the x-coordinate of the point where it stops each time is less than or equal to p, then the event A is said to have occurred. The problem then is to find the probability that p lies between p_1 and p_2, given that A has occurred a times and failed $n - a$ times. Bayes gave the answer as the ratio of the two areas that were then evaluated using infinite series expansions. In modern notation, his solution can be written as

---

[†] In the following equation, $\Gamma(\alpha)$ is the gamma function defined by $\Gamma(\alpha) = \int_0^\infty e^{-y}y^{\alpha-1}\,dy$. When n is a positive integer, $\Gamma(n) = (n-1)!$

[‡] Bayes' *Essay* also appears in Grattan-Guinness's *Landmark Writings in Western Mathematics*; see the article in it by Dale (2005).

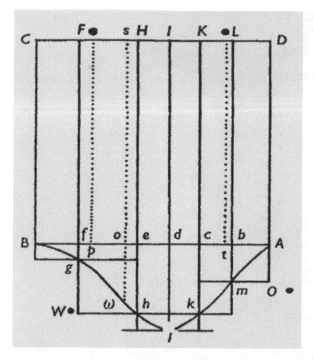

Figure 14.1 Bayes' table as it appears in his paper (Bayes, 1764).

$$\Pr\{p_1 < p < p_2 \,|\, X_n = a\} = \frac{\displaystyle\int_{p_1}^{p_2} p^a (1-p)^{n-a}\, dp}{\displaystyle\int_0^1 p^a (1-p)^{n-a}\, dp}, \tag{14.1}$$

which when simplified gives the same expression as in our solution. Equation (14.1) can be viewed as Thomas Bayes' version of what is known today as "Bayes' Theorem" and assumes a uniform prior for p on [0,1].

The publication of Bayes' paper has an interesting history. During his lifetime, Bayes' published only two anonymous papers.[†] Upon Bayes' death, his friend Richard Price (1723–1791) decided to publish some of his papers to the Royal Society. Bayes' *Essay* was the second posthumous paper and was augmented by an introduction and an appendix written by Price.[‡] This publication is now widely regarded as a major breakthrough not only in probability and statistics, but one that

---

[†] Although Bellhouse has recently discovered more manuscripts by Bayes in the Royal Society and in the Equitable Life Assurance Society in London (Bellhouse, 2002).

[‡] Several reasons have been advanced for the posthumous publication. William Morgan who was Price's nephew claimed Bayes did not publish out of modesty. Good (1988) gives three other reasons, the most serious of which is that Bayes was aware there was a "gap in his mathematics": he had tacitly assumed that a discrete uniform distribution for the number of times an event A occurs *in a given number of trials* implies a uniform (0, 1) distribution for the probability of its occurrence, and this is not true in general.

Figure 14.2   David Hume (1711–1776).

also touches the very core of the philosophy of science. One author, Dupont (1979) has called it

> ...a gem in the history of the Sciences...

In his *Statistical Theory: The Relationship of Probability Credibility and Error*, Lancelot Hogben (1957, p. 111) says

> As matters stand, most subsequent writers on statistical theory up to and including the present time recognize the *Essay towards solving a Problem in the Doctrine of Chance* as a landmark, and its contents as a challenge or programme, according to taste.

Bayes' *Essay* can be regarded as the third major breakthrough after Pascal and Fermat's initial work, and Jacob Bernoulli's[†] *Ars Conjectandi* (Bernoulli, 1713). Recall that while Pascal and Fermat's analyses were restricted to symmetric situations in games of chance, Bernoulli extended the concept of probability to asymmetric situations applicable to diverse real-life problems. In Bernoulli's framework, the probability of an event was a *known* quantity and the problem he tackled was how

---

[†] Whose theorem was then later refined by de Moivre, as we noted in **Problem 8**.

Figure 14.3   Thomas Bayes (1702–1761).

many times an experiment ought to be repeated before one can be "morally" certain that the relative frequency of the event is close enough to the probability. Bayes attacked the inverse problem. Here the numbers of times an event occurred and failed to occur are known quantities, but the probability of that event is *unknown*. The task is then to determine *how likely it is for the unknown probability to lie between any two given probabilities, based on the known number of times the event occurred and did not occur, and some a priori notion of the likelihood of the event.* Bayes' *Essay* was thus concerned with *inverse probability.*[†] As Price puts it in his introduction to Bayes' paper

> . . .the problem enquired after in this essay is no less important than it is curious. It may be safely added, I fancy, that it is also a problem that has never before been solved. Mr. De Moivre, indeed, the great improver of this part of mathematics, has in his *Laws of Chance*, after Bernoulli, and to a greater degree of exactness, given rules to find the probability there is, that if a very great number of trials be made concerning any event, the proportion of the number of times it will happen, to the number of times it will fail in those trials, should differ less than by small assigned limits from the proportion of the probability of its happening to the probability of its failing in one single trial. But I know of no person who has shewn how to deduce the solution of the converse problem to this; namely, 'the number of times an unknown event has happened and failed being given, to find the chance that the

---

[†] Not to be confused with the inverse use of Bernoulli's Theorem, which we explain later in this Discussion.

probability of its happening should lie somewhere between any two named degrees of probability.' What Mr. De Moivre has done therefore cannot be thought sufficient to make the consideration of this point unnecessary: especially, as the rules he has given are not pretended to be rigorously exact, except on supposition that the number of trials made are infinite; from whence it is not obvious how large the number of trials must be in order to make them exact enough to be depended on in practice.

Bayes revolutionary idea was to transition from the old task of finding how probable a particular event is to the new task of finding how probable the probability assigned to the event is. The new task can thus be viewed as attempting to find the "probability of probabilities" or the probability of causes, as it was often called in the literature at one time. It is of fundamental importance because it could potentially provide a method for making probability statements about population parameters (such as population proportions, means, and variances), this being one of the prime objectives of statistical inference. In the words of Hacking (1980a, p. 532)

If Bayes' conclusion is correct, we have a basis for the whole of statistical inference. . .

However, it is also no exaggeration to say that Bayes' work has been as controversial as it has been revolutionary.

Let us now consider two key aspects of Bayes' *Essay* and discuss the extent to which his conclusion was correct. First, we should note Bayes' definition of probability as it appears in his paper

The probability of any event is the ratio between the value at which an expectation depending on the happening of the event ought to be computed, and the value of the thing expected upon its happening.

This definition is dramatically different from the definition used or given by Bayes' predecessors, namely that of the ratio of favorable cases to the total number of possible cases when all of the latter are equally likely. However, Bayes' definition is at the core of the *epistemic or Bayesian* view of probability: if a contract pays $A when it occurs and $0 otherwise, and we are willing to spend up to $B on the contract, then our subjective probability that the contract occurs is B/A. The latter is called the *subjective Bayesian* interpretation of probability. It treats probability as a personal measure of belief and has been championed by the likes of Bruno de Finetti (1906–1985) and Leonard J. Savage (1917–1971). Now a closer look at Bayes' definition shows that he uses the qualifier "ought to be," which suggests that he was more in line with the *objective Bayesian* interpretation (Earman, 1992, p. 8). The latter treats probability as a rational or justified measure of belief, and has been supported by the economist John Maynard Keynes (1883–1946) and the physicist Sir Harrold Jeffreys (1891–1989).

The second key aspect of Bayes' *Essay* is the care[†] with which he set up his ball-and-table example so that p would logically be equally likely to take any value in [0,1]. Thus Bayes writes (Bayes, 1764)

[†] Gillies thus calls Bayes a "cautious" Bayesian (Gillies, 1987).

I suppose the square table or plane ABCD to be so made and levelled, that if either of the balls O or W be thrown upon it, there shall be the same probability that it rests upon any one equal part of the plane as another, and that it must necessarily rest somewhere upon it.

Therefore, under the conditions put forward in the body of his *Essay,* Bayes' use of a uniform prior for the density of p is justified. Consequently, his main result is correct. However, it is in the scholium to his paper that the controversy is to be found:

From the preceding proposition it is plain, that in the case of such an event as I there call *M*, from the number of times it happens and fails in a certain number of trials, without knowing anything more concerning it, one may give a guess whereabouts it's probability is, and, by the usual methods computing the magnitudes of the areas there mentioned, see the chance that the guess is right. And that the same rule is the proper one to be used in the case of an event concerning the probability of which we absolutely know nothing antecedently to any trials made concerning it, seems to appear from the following consideration; viz. that concerning such an event I have no reason to think that, in a certain number of trials, it should rather happen any one possible number of times than another. For, on this account, I may justly reason concerning it as if its probability had been at first unfixed, and then determined in such a manner as to give me no reason to think that, in a certain number of trials, it should rather happen any one possible number of times than another... In what follows therefore I shall take for granted that the rule given concerning the event *M* in prop.9 is also the rule to be used in relation to any event concerning the probability of which nothing at all is known antecedently to any trials made or observed concerning it. And such an event I shall call an unknown event.

Bayes here suggests that, in general, when we have no prior evidence about the probability of an event, our ignorance would place us in a position tantamount to the one in his example where the prior is known to be uniform. Ignorance about the probability of an event is thus equated with an equally likely distribution. This is an example of the *principle of indifference*[†] and a major reason why Bayes' paper overall is controversial.[‡]

Because the principle of indifference is so important in the calculus of probability, let us make some comments on it. The principle can be traced to Jacob Bernoulli's *Ars Conjectandi* (Bernoulli, 1713, p. 219), where we read

---

[†] The principle of indifference is also discussed in Zabell (2005, pp. 22–27), Porter (1986, pp. 82–83), Van Fraassen (1989, Chapter 12), Chatterjee (2003, p. 220), and Howson and Urbach (2006, pp. 266–272).

[‡] The "Bayesian controversy" thus has nothing to do with Bayes' theorem itself. The latter is perfectly valid. The controversy arises from the choice of a suitable prior to be used in the theorem.

All cases are equally possible, that is to say, each can come about as easily as any other.

Later the principle was explicitly used by Laplace to define classical probability.[†] The principle of indifference had at first been called the "principle of insufficient reason" (possibly as a word play on Leibniz's principle of sufficient reason[‡]) by the German physiological psychologist Johannes Adolf von Kries (1853–1928) in the *Die Principien der Wahrscheinlichkeitsrechnung* (von Kries, 1886). However, in 1921, Keynes stated in his *Treatise on Probability* (Keynes, 1921, p. 41)

> A rule, adequate to the purpose, introduced by James [Jacob] Bernoulli, who was the real founder of mathematical probability, has been widely adopted, generally under the title of *The Principle of Non-Sufficient Reason*, down to the present time. This description is clumsy and unsatisfactory, and, if it is justifiable to break away from tradition, I prefer to call it *The Principle of Indifference*.

Later in his book, Keynes provided the following clear definition (Keynes, 1921, p. 42):

> The principle of indifference asserts that if there is no known reason for predicating of our subject one rather than another of several alternatives, then relatively to such knowledge the assertions of each of these alternatives have an equal probability.

Let us now explain how the application of the principle of indifference can easily lead to contradictions,[§] which explains why Bayes' paper is controversial. Consider, for example, the classic example given by von Mises (1981, p. 77):

> We are given a glass containing a mixture of water and wine. All that is known about the proportions of the liquids is that the mixture contains at least as much water as wine, and at most, twice as much water as wine. The range for our assumptions concerning the ratio of water to wine is thus the interval 1 to 2. Assuming that nothing more is known about the mixture, the indifference or symmetry principle or any other similar form of the classical theory tells us to assume that equal parts of this interval have equal probabilities. The probability of the ratio lying between 1 and 1.5 is thus 50%, and the other 50% corresponds to the probability of the range 1.5 to 2.
>
> But there is an alternative method of treating the same problem. Instead of the ratio water/wine, we consider the inverse ratio, wine/water; this we know lies between 1/2 and 1. We are again told to assume that the two halves of the total interval, i.e., the intervals 1/2 to 3/4 and 3/4 to 1, have equal probabilities (50% each); yet, the wine/water ratio 3/4 is equal to the water/wine ratio 4/3. Thus, according to our second calculation, 50% probability corresponds to the water/wine range 1 to 4/3 and the remaining 50% to the range 4/3 to 2.

[†] See later in this Discussion, p. 141.

[‡] In contrast, the principle of sufficient reason states: "Everything that is the case must have a reason why it is the case. Necessarily, every true or at least every contingent true proposition has an explanation. Every event has a cause" (Pruss, 2006, p. 3).

[§] According to Keynes (1921, p. 85), the first person to object to the principle of indifference was the British mathematician Robert Leslie Ellis (1817–1859): "Mere ignorance is no ground for any inference whatever. *Ex nihilo nihil*" (Ellis, 1850).

According to the first calculation, the corresponding intervals were 1 to 3/2 and 3/2 to 2. The two results are obviously incompatible.

A second, more technical, example is due to Fisher (1956, p. 16), one of the key critics of Bayesianism. Suppose nothing is known about the distribution of p and we use the principle of indifference to assign it a uniform distribution on [0, 1]. Now let $p = \sin^2 \phi$, where $0 \leq \phi \leq \pi/2$. Since nothing is known about p and the latter is a one-to-one function of ϕ, we can equally say that nothing is known about ϕ. Applying the principle of indifference to ϕ this time, we assign a uniform distribution to it on $[0, \pi/2]$, that is,

$$f(\phi) = \frac{2}{\pi}, \quad 0 \leq \phi \leq \frac{\pi}{2}.$$

But then

$$f(p) = f(\phi)\left|\frac{dp}{d\phi}\right|^{-1}$$
$$= \frac{2}{\pi} \cdot \frac{1}{2 \sin \phi \cos \phi}$$
$$= \frac{1}{\pi\sqrt{p(1-p)}}, \quad 0 < p < 1.$$

This contradicts our earlier statement that, by the principle of indifference, p should have a uniform distribution on [0,1]. This and other similar examples show why the principle of indifference is usually regarded as deficient.[†] However, we should note that mathematicians such as Edwin T. Jaynes (1922–1998) have defended this principle strongly by using maximum entropy and invariance arguments (Jaynes, 1973).

Coming back to Bayes, what was the philosophical impact of his *Essay*? Fifteen years before its publication, the eminent Scottish philosopher David Hume (1711–1776) (Fig. 14.2) wrote the groundbreaking book *An Enquiry Concerning Human Understanding* (Hume, 1748). In this work, Hume formulated his famous *problem of induction*, which we now explain. Suppose out of a large number n of occurrences of an event A, an event B occurs m times. Based on these observations, an inductive inference would lead us to believe that approximately m/n of all events of type A is also of type B, that is, the probability of B given A is approximately m/n. Hume's problem of induction states that such an inference has no rational justification, but arises only as a consequence of custom and habit. In Hume's own words (Hume, 1748, p. 25)

> That there are no demonstrative arguments in the case [of inductive arguments], seems evident; since it implies no contradiction, that the course of nature may change, and that an object, seemingly like those which we have experienced, may be attended with

---

[†] Other well-known paradoxes arising from the principle of indifference can be found in von Kries (1886), Bertrand (1889), and Keynes (1921).

different or contrary effects. May I not clearly and distinctly conceive, that a body, falling from the clouds, and which, in all other respects, resembles snow, has yet the taste of salt or feeling of fire? Is there any more intelligible proposition than to affirm, that all the trees will flourish in December and January, and decay in May and June? Now whatever is intelligible, and can be distinctly conceived, implies no contradiction, and can never be proved false by any demonstrative argument or abstract reasoning à priori.

Earlier in his book, Hume gave the famous "rise-of-the-sun" example which was meant to illustrate the shaky ground on which "matters-of-fact" or inductive reasoning rested (Hume, 1748, p. 18):

> Matters of fact, which are the second objects of human reason, are not ascertained in the same manner; nor is our evidence of their truth, however great, of a like nature with the foregoing. The contrary of every matter of fact is still possible; because it can never imply a contradiction, and is conceived by the mind with the same facility and distinctness, as if ever so conformable to reality. That the sun will not rise tomorrow is no less intelligible a proposition, and implies no more contradiction, than the affirmation, that it will rise. We should in vain, therefore, attempt to demonstrate its falsehood. Were it demonstratively false, it would imply a contradiction, and could never be distinctly conceived by the mind.

Hume not only attacked the use of induction in probability, he also expressed skepticism regarding the occurrence of miracles. This might have prompted Reverend Bayes to write the paper.

Philosophically, Bayes' paper is a possible solution to *Hume's Problem of Induction*. It is unclear to what extent Bayes himself was aware of Hume's skepticism, but Price was certainly cognizant of the issue. As Price pointed out in his introductory letter to Bayes' paper (Bayes, 1764):

> Every judicious person will be sensible that the problem now mentioned is by no means merely a curious speculation in the doctrine of chances, but necessary to be solved in order to [provide] a sure foundation of all our reasonings concerning past facts, and what is likely to be hereafter. Common sense is indeed sufficient to shew us that, from the observation of what has in former instances been the consequence of a certain cause or action, one may make a judgment what is likely to be the consequence of it another time, and that the larger [the] number of experiments we have to support a conclusion, so much the more reason we have to take it for granted. But it is certain that we cannot determine, at least not to any nicety, in what degree repeated experiments confirm a conclusion, without the particular discussion of the beforementioned problem; which, therefore, is necessary to be considered by any one who would give a clear account of the strength of analogical or inductive reasoning.

Price's argument here essentially is that Bayesian reasoning *does* provide a rational justification for making inductive reasoning regarding the future based on past facts. To illustrate his point, Price considered the following problem:

> *Given that an event has occurred n times in n identical and independent repetitions of an experiment, what is probability that the event has more than an even chance of occurring in the next repetition of the experiment?*

Price solved this by simply applying Eq. (14.1) with $a = n$:

$$\Pr\{1/2 < p < 1 | X_n = n\} = \frac{\int_{1/2}^{1} p^n \, dp}{\int_{0}^{1} p^n \, dp} = 1 - \frac{1}{2^{n+1}}. \qquad (14.2)$$

Note that this calculation also assumes that the probability of the event in a single repetition is uniformly distributed in [0, 1]. Table 14.1 shows the values of this probability for different values of n.

We now address the inevitable question of priority: who first discovered Bayes' Theorem? The issue was raised by Stigler in a 1983 paper[†] (Stigler, 1983). Stigler referred to an interesting paragraph in the 1749 book *Observations on Man* by the British scientist David Hartley (1705–1757) (Hartley, 1749, p. 338):

> Mr. de Moivre has shewn, that where the Causes of the Happening of an Event bear a fixed Ratio to those of its Failure, the Happenings must bear nearly the same Ratio to the Failures, if the Number of Trials be sufficient; and that the last ratio approaches to the first indefinitely, as the Number of Trials increases. . ... An ingenious Friend has communicated to me a Solution of the inverse Problem, in which he has shewn what the Expectation is, when an Event has happened p times, and failed q times, that the original Ratio of the Causes for the Happening or Failing of an Event should deviate in any given Degree from that of p to q. And it appears from this Solution, that where the Number of Trials is very great, the Deviation must be inconsiderable: Which shews that we may hope to determine the Proportions, and, by degrees, the whole Nature, of unknown Causes, by a sufficient Observation of their Effects.

Stigler argues that the "ingenious friend" is actually the first to have discovered Bayes' Theorem. After going through several possible contenders, Stigler concludes in a tongue-in-cheek manner that the odds are 3 to 1 in favor of Nicholas Sanderson, a professor of Mathematics at Cambridge, rather than Bayes for being the real

Table 14.1 Probability that an Event has More than an Even Chance of Occurring in the Next Repetition of an Experiment, Given that it has Already Occurred n Times in n Identical and Independent Repetitions of an Experiment

Value of n	$\Pr\{1/2 < p < 1\}$
1	.750
2	.875
3	.938
4	.969
5	.984
10	.999
100	1.00

The probabilities are obtained from Price's formula in Eq. (14.2).

[†] The paper also appears in Stigler's book *Statistics on the Table* (Stigler, 1999, Chapter 15).

discoverer of Bayes' Theorem. However, a few years after Stigler's paper, both Edwards (1986) and Dale (1988) weighed in on the matter, and concluded that Hartley was not actually referring to Bayes' Theorem in his book. They suggest that Hartley was actually alluding to the "inverse use of Bernoulli's law," not to Bayes' inverse probability as Stigler had thought, and that the ingenious friend was none other than de Moivre who seems to have been Bayes' tutor at one time. Let us explain this inverse use of Bernoulli's law. Recall that in Bernoulli's law, the probability of an event was a *known* quantity and the problem he tackled was how many times an experiment ought to be repeated before one can be "morally" certain that the relative frequency of the event is close enough to the true probability. Now, in the case the probability of the event is unknown and one has a given number of successes and failures, one can use Bernoulli's law inversely by taking the relative proportion of successes as an approximation to the true but unknown probability. As Todhunter (1865, p. 73) explains

> Suppose then that we have an urn containing white balls and black balls, and that the ratio of the number of the former to the latter is *known to be* that of 3 to 2. We learn from the preceding result that if we make 25550 drawings of a single ball, replacing each ball after it is drawn, the odds are 1000 to 1 that the white balls drawn lie between 31/50 and 29/50 of the whole number drawn. This is the *direct* use of James [Jacob] Bernoulli's theorem. But he himself proposed to employ it *inversely* in a far more important way. Suppose that in the preceding illustration we do not know anything beforehand of the ratio of the white balls to the black; but that we have made a large number of drawings, and have obtained a white ball R times, and a black ball S times: then according to James [Jacob] Bernoulli we are to infer that the ratio of the white balls to the black balls in the urn is approximately R/S.

On the other hand, Bayes' aim was to determine the probability that an unknown probability lies between any two given probabilities, based on the known number of times the event occurred and did not occur.

The difference between Bernoulli's law, the inverse use of Bernoulli's law and Bayes' Theorem can be summarized as follows. Let p be the probability of the event in one trial of an experiment, and let X_n be the number of events in n identical trials of the experiment. Then

- **Bernoulli's Law**: for given p and $\varepsilon > 0$, we can find an n such that

$$\Pr\{p - \varepsilon < X_n/n < p + \varepsilon \,|\, p\} \text{ is arbitrarily close to 1.}$$

- **Inverse Use of Bernoulli's Law**: when p is unknown, for given X_n and n, where n is large,

$$\Pr\{p - \varepsilon < X_n/n < p + \varepsilon\} \text{ is close to 1, so that } X_n/n \approx p.$$

- **Bayes' Theorem**: when p is unknown, for given p_1, p_2, X_n, if we assume that p is a random variable with some prior distribution, then we can calculate

$$\Pr\{p_1 < p < p_2 | X_n\}.$$

We next turn our attention to Pierre-Simon Laplace (1749–1827), a real giant in mathematics about whom Lancelot Hogben (1957, p. 133) has said

> The *fons et irigo* of inverse probability is Laplace. For good or ill, the ideas commonly identified with the name of Bayes are largely his.

Indeed, although we have identified Eq. (14.1) as a version of Bayes' Theorem, the form it usually appears in textbooks, namely

$$\Pr\{A_j|B\} = \frac{\Pr\{B|A_j\}\Pr\{A_j\}}{\sum_{i=1}^{n}\Pr\{B|A_i\}\Pr\{A_i\}},\tag{14.3}$$

is due to Laplace. In Eq. (14.3) above $A_1, A_2, \ldots, A_n$ is a sequence of mutually exclusive and exhaustive events, $\Pr\{A_j\}$ is the prior probability of A_j, and $\Pr\{A_j|B\}$ is the posterior probability of A_j given B. The continuous version of Eq. (14.3) can be written as

$$f(\theta|\mathbf{x}) = \frac{f(\mathbf{x}|\theta)f(\theta)}{\displaystyle\int_{-\infty}^{\infty} f(\mathbf{x}|\theta)f(\theta)d\theta},$$

where $f(\theta)$ is the prior density of θ, $f(\mathbf{x}|\theta)$ is the likelihood of the data $\mathbf{x}$, and $f(\theta|\mathbf{x})$ is the posterior density of θ.[†]

Before commenting on Laplace's work on inverse probability, let us recall that it is with him that the classical definition of probability is usually associated, for he was the first to have given it in its clearest terms. Indeed, Laplace's classical definition of probability is the one that is still used today. In his very first paper on probability, *Mémoire sur les suites récurro-recurrentes et leurs usages dans la théorie des hazards* (Laplace, 1774b), Laplace writes

> . . . if each case is equally probable, the probability of the event is equal to the number of favorable cases divided by the number of all possible cases.

This definition was repeated both in Laplace's *Théorie Analytique* and *Essai Philosophique*. It is interesting to note that, in spite of Laplace's monumental contributions to probability, he was an absolute determinist. According to Laplace, any event is causally determined, and a probability is solely our measure of ignorance regarding this event. In the *Essai Philosophique*, we thus read (Laplace, 1814a, English edition, p. 6)

> The curve described by a simple molecule of air or vapor is regulated in a manner just as certain as the planetary orbits; the only difference between them is that which comes from our ignorance.

[†] Thus, the posterior density is proportional to the product of the prior density and the likelihood.

Probability is relative, in part to this ignorance, in part to our knowledge.

Laplace appeals to a "vast intelligence,"[†] which has been dubbed *Laplace's Demon*, to explain universal determinism. In his own words (Laplace, 1814a, English edition, p. 4)

> We ought then to regard the present state of the universe as the effect of its anterior state and as the cause of the one which is to follow. Given for one instant an intelligence which could comprehend all the forces by which nature is animated and the respective situation of the beings who compose it—an intelligence sufficiently vast to submit these data to analysis—it would embrace in the same formula the movements of the greatest bodies of the universe and those of the lightest atom; for it, nothing would be uncertain and the future, as the past, would be present to its eyes.

Laplace's views on universal determinism are no longer tenable today. For more discussions, see Roger Hahn's *Pierre-Simon Laplace, 1749–1827: A Determined Scientist* (Hahn, 2005).

Coming back to our Bayesian discussion, Laplace's definition of inverse probability was first enunciated in the 1774 *Mémoire de la Probabilité des Causes par les Evénements* (Laplace, 1774a). This is how Laplace phrases it:

> If an event can be produced by a number *n* of different causes, the probabilities of the existence of these causes, calculated from the event, are to each other as the probabilities of the event, calculated from the causes; and the probability of each cause is equal to the probability of the event, calculated from that cause, divided by the sum of all the probabilities of the event, calculated from each of the causes.

The above definition by Laplace is the same as Eq. (14.3), but with the added discrete uniform prior assumption that $\Pr\{A_k\} = 1/n$ for $k = 1, 2, ..., n$.

We should note that Eqs. (14.1) and (14.3) are both based on the same principle, although the first equation makes the additional assumption of a uniform prior. It is very likely that Laplace was unaware of Bayes' previous work on inverse probability when he enunciated the rule in 1774. However, the 1778 volume of the *Histoire de l'Académie Royale des Sciences,* which appeared in 1781, contains an interesting summary by the Marquis de Condorcet[‡] (1743–1794) of Laplace's article *Sur les Probabilités*, which also appeared in that volume (Laplace, 1781). While Laplace's article itself makes mention of neither Bayes nor Price,[§] Condorcet's summary explicitly acknowledges the two Englishmen (Laplace, 1781, p. 43) (see Fig. 14.4):

[†] In spite of this, Laplace was an atheist. When asked by Napoleon why God did not appear in his *Celestial Mechanics*, Laplace said, "I have no need of that hypothesis," and later remarked, "But it is beautiful hypothesis; it explains many things." In contrast to Newton's views, Laplace believed that there was no need for divine intervention to prevent the planets from falling into the sun through universal gravitation.

[‡] Condorcet was assistant secretary in the *Académie des Sciences* and was in charge of editing Laplace's papers for the transactions of the Academy.

[§] Laplace's acknowledgment of Bayes appears in his *Essai Philosophique* (Laplace, 1814a) English edition, p. 189.

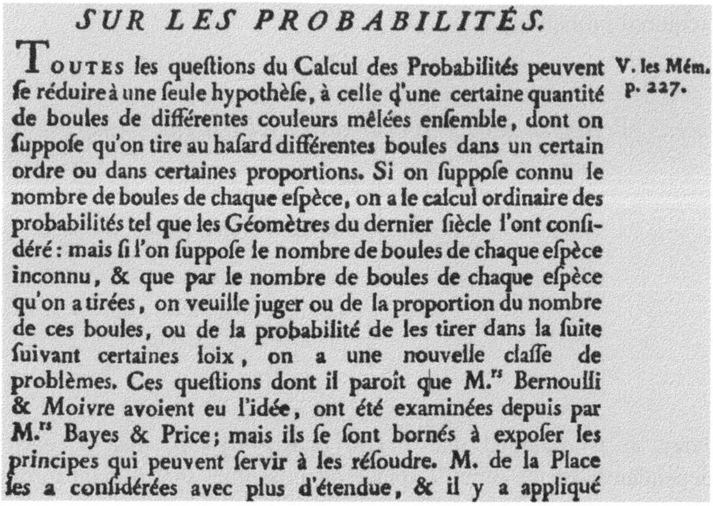

SUR LES PROBABILITÉS.

TOUTES les queſtions du Calcul des Probabilités peuvent ſe réduire à une ſeule hypothèſe, à celle d'une certaine quantité de boules de différentes couleurs mêlées enſemble, dont on ſuppoſe qu'on tire au haſard différentes boules dans un certain ordre ou dans certaines proportions. Si on ſuppoſe connu le nombre de boules de chaque eſpèce, on a le calcul ordinaire des probabilités tel que les Géomètres du dernier ſiècle l'ont conſidéré : mais ſi l'on ſuppoſe le nombre de boules de chaque eſpèce inconnu, & que par le nombre de boules de chaque eſpèce qu'on a tirées, on veuille juger ou de la proportion du nombre de ces boules, ou de la probabilité de les tirer dans la ſuite ſuivant certaines loix, on a une nouvelle claſſe de problèmes. Ces queſtions dont il paroît que M.rs Bernoulli & Moivre avoient eu l'idée, ont été examinées depuis par M.rs Bayes & Price; mais ils ſe ſont bornés à expoſer les principes qui peuvent ſervir à les réſoudre. M. de la Place les a conſidérées avec plus d'étendue, & il y a appliqué

V. les Mém. p. 227.

Figure 14.4 Summary by Condorcet of Laplace's *Sur les Probabilités* memoir, taken from the 1778 volume of the *Histoire de l'Académie Royale des Sciences* (Laplace, 1781, p. 43). Here, Condorcet explicitly acknowledges the works of Bayes and Price on inverse probability.

> These questions [on inverse probability] about which it seems that Messrs. Bernoulli and Moivre had thought, have been since then examined by Messrs. Bayes and Price; but they have limited themselves to exposing the principles that can be used to solve them. M. de Laplace has expanded on them. . .

Coming back to the 1774 paper, after having enunciated his principle on inverse probability, Laplace is famous for discussing the following classic problem:

> *A box contains a large number of black and white balls. We sample n balls with replacement, of which b turn out to be black and n − b turn out to be white. What is the conditional probability that the next ball drawn will be black?*

To solve the above problem,† let X_n be the number of black balls out of the sample of size n and let the probability that a ball is black be p. Also, let $B*$ be the event that the next ball is black. From Bayes' Theorem, we have

$$f(p \mid X_n = b) = \frac{\Pr\{X_n = b \mid p\} f(p)}{\Pr\{X_n = b\}}$$

$$= \frac{\Pr\{X_n = b \mid p\} f(p)}{\int_0^1 \Pr\{X_n = b \mid p\} f(p) dp}.$$

† The proof presented here treats the total number of balls as being infinitely large. For a proof that treats the total number of balls as finite, see Jeffreys (1961, p. 127) and Zabell (2005, pp. 38–41).

Then the required probability is

$$\Pr\{B*|X_n = b\} = \int_0^1 \Pr\{B*|p, X_n = b\}f(p|X_n = b)dp$$

$$= \frac{\displaystyle\int_0^1 p.\Pr\{X_n = b|p\}f(p)dp}{\displaystyle\int_0^1 \Pr\{X_n = b|p\}f(p)dp}.$$

In the above, it is assumed that $\Pr\{B*|p, X_n = b\} = p$, that is, each ball is drawn independently of the other. Laplace also assumes that p is uniform in $[0,1]$, so that

$$\Pr\{B*|X_n = b\} = \frac{\displaystyle\int_0^1 p \cdot \binom{n}{b} p^b(1-p)^{n-b} \cdot 1dp}{\displaystyle\int_0^1 \binom{n}{b} p^b(1-p)^{n-b} \cdot 1dp}$$

$$= \frac{\displaystyle\int_0^1 p^{b+1}(1-p)^{n-b} dp}{\displaystyle\int_0^1 p^b(1-p)^{n-b} dp}$$

$$= \frac{\Gamma(b+2)\Gamma(n-b+1)}{\Gamma(n+3)} \cdot \frac{\Gamma(n+2)}{\Gamma(b+1)\Gamma(n-b+1)}$$

$$= \frac{(b+1)!(n-b)!}{(n+2)!} \cdot \frac{(n+1)!}{b!(n-b)!}$$

$$= \frac{b+1}{n+2}.$$

In particular, if all of the n balls turn out to be black, then the probability that the next ball is also black is $(n+1)/(n+2)$. The above problem has been much discussed in the literature and is known as *Laplace's rule of succession*.[†] Using the rule of succession, Laplace considered the following question: *given that the sun has risen everyday for the past 5000 years, what is the probability that it will rise tomorrow?* Substituting $n = 5000 \times 365.2426 = 1,826,213$ in the above formula, Laplace obtained the probability $1,826,214/1,826,215$ ($\approx .9999994$). Thus, in his *Essai*

[†] Laplace's rule of succession is also discussed in Pitman (1993, p. 421), Sarkar and Pfeifer (2006, p. 47), Pearson (1900, pp. 140–150), Zabell (2005, Chapter 2), Jackman (2009, p. 57), Keynes (1921, p. 376), Chatterjee (2003, pp. 216–218), Good (1983, p. 67), Gelman et al. (2003, p. 36), Blom et al. (1994, p. 58), Isaac (1995, p. 36), Chung and AitSahlia (2003, p. 129), and Gorroochurn (2011).

Figure 14.5   Pierre-Simon Laplace (1749–1827).

Philosophique sur les Probabilités (Figs. 14.5–14.7),[†] Laplace (1814a, English edition, p. 19) says

> Thus we find that an event having occurred successively any number of times, the probability that it will happen again the next time is equal to this number increased by unity divided by the same number, increased by two units. Placing the most ancient epoch of history at five thousand years ago, or at 1,826,213 days, and the sun having risen constantly in the interval at each revolution of twenty-four hours, it is a bet of 1,826,214 to one that it will rise again tomorrow.

Laplace's calculation was meant to be an answer to *Hume's Problem of Induction*, which we discussed earlier. Laplace, who so often has been called France's Newton, was harshly criticized for his calculation. Zabell (2005, p. 47) says

> Laplace has perhaps received more ridicule for this statement than for any other.

[†] Philosophical essay on probabilities.

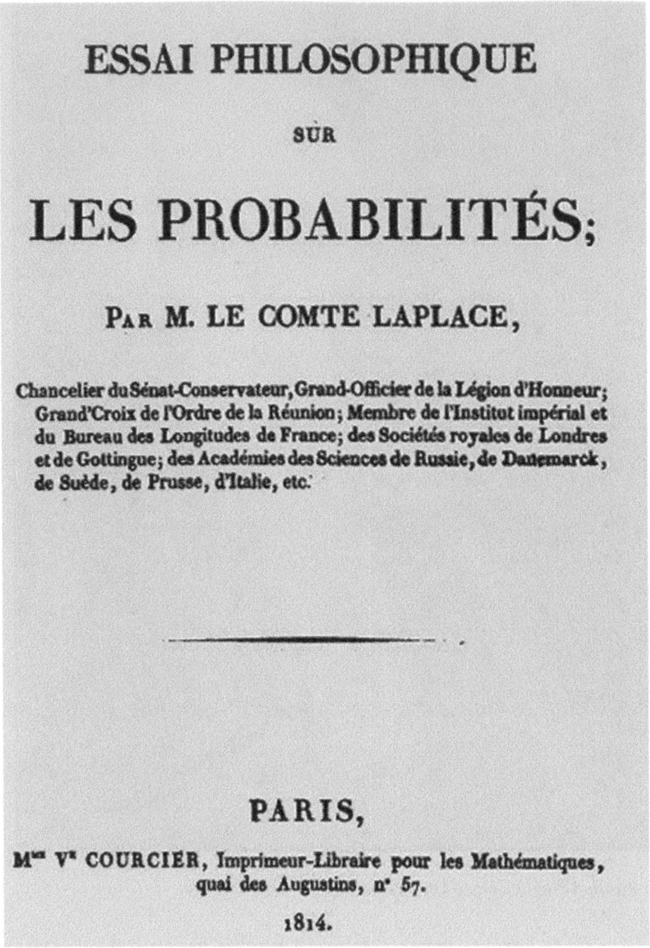

ESSAI PHILOSOPHIQUE

SÙR

LES PROBABILITÉS;

Par M. LE COMTE LAPLACE,

Chancelier du Sénat-Conservateur, Grand-Officier de la Légion d'Honneur;
Grand'Croix de l'Ordre de la Réunion; Membre de l'Institut impérial et
du Bureau des Longitudes de France; des Sociétés royales de Londres
et de Gottingue; des Académies des Sciences de Russie, de Danemarck,
de Suède, de Prusse, d'Italie, etc.

PARIS,

Mᵐᵉ Vᵉ COURCIER, Imprimeur-Libraire pour les Mathématiques,
quai des Augustins, n° 57.

1814.

Figure 14.6   Title Page of the first edition of the *Essai Philosophique*.

Somehow, Laplace must have felt that there was something amiss with his calculation. For his very next sentence reads

> But this number is incomparably greater for him who, recognizing in the totality of phenomena the principal regulator of days and seasons, sees that nothing at the present moment can arrest the course of it.

Laplace here seems to warn the reader that his method is correct when based only on the information from the sample, but his statement is too timid. To understand the criticism leveled against Laplace's calculation, consider the following example given by the famous Austro-British philosopher Karl Popper (1902–1994) (Popper, 1957; Gillies 2000, p. 73): Suppose the sun rises for 1,826,213 days (5000 years), but then suddenly the earth stops rotating on day 1,826,214. Then, for parts of the globe (say

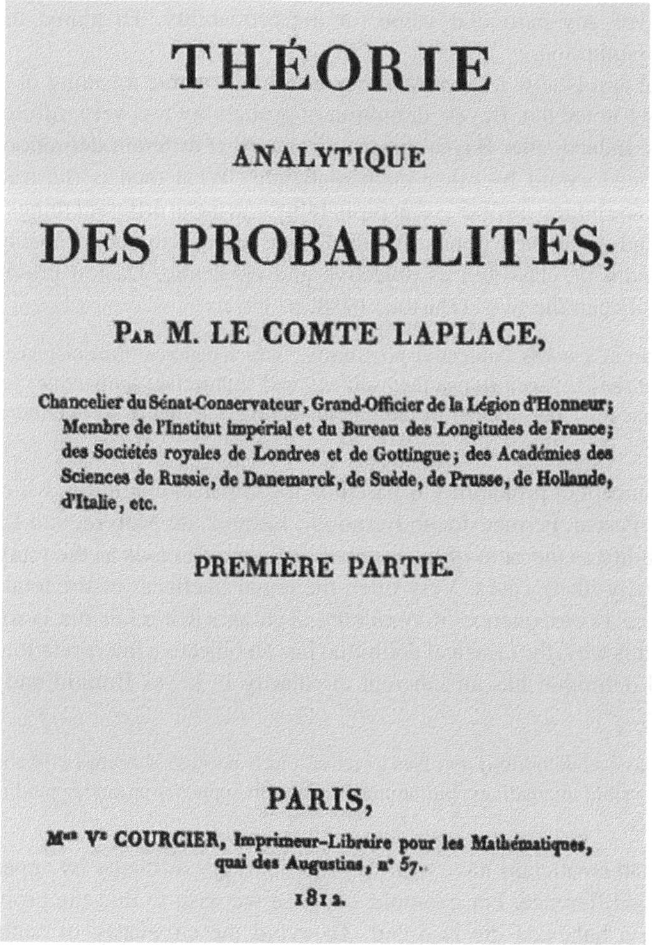

THÉORIE

ANALYTIQUE

DES PROBABILITÉS;

Par M. LE COMTE LAPLACE,

Chancelier du Sénat-Conservateur, Grand-Officier de la Légion d'Honneur;
Membre de l'Institut impérial et du Bureau des Longitudes de France;
des Sociétés royales de Londres et de Gottingue; des Académies des
Sciences de Russie, de Danemarck, de Suède, de Prusse, de Hollande,
d'Italie, etc.

PREMIÈRE PARTIE.

PARIS,

Mᵐᵉ Vᵉ COURCIER, Imprimeur-Libraire pour les Mathématiques,
quai des Augustins, nᵒ 57.

1812.

Figure 14.7   Title Page of the first edition of the *Théorie Analytique*.

Part A), the sun does not rise on that day, whereas other parts (say Part B) the sun will appear fixed in the sky. What then is the probability that the sun will rise again in Part A of the globe? Applying the generalized form of the rule of succession with $n = 1,826,214$ and $b = 1,826,213$ gives a probability of .9999989, which is almost as high as the original probability of .9999994! The answer is preposterous.

The rule of succession is perfectly valid as long as the assumptions it makes are all tenable. Applying the rule of succession to the rising of the sun, however, should be viewed with skepticism for several reasons (see, for example, Schay, 2007, p. 65). First, it is dubious if the rising of the sun on a given day can be considered a random event at all. Second, the assumption of independence is questionable. A third criticism lies in the reliance of the solution on the principle of indifference: the probability of the sun rising is equally likely to take any of the values in [0, 1] because there is no reason to favor any particular value for the probability. To many, this is not a reasonable assumption.

Our final aim is now to address the question of the true meaning of probability. Previously, we noted that Bayes' definition of probability was very different from his predecessors. Indeed, after Bayes' times, several other different definitions of probability were put forward by other mathematicians. What then is the true nature of probability? The honest answer is that probability can mean different things to different people, depending on their points of view. There are two major competing theories[†] that can broadly be classified as objective and epistemic. Daston provides a clear distinction between the two[‡] (Daston, 1988, p. 197):

> . . .in Cournot's words, "objective possibility," which denotes "the existence of a relation which subsists between things themselves," and "subjective [epistemic] probability," which concerns "our manner of judging or feeling, varying from one individual to the next".

The oldest concept of probability is based on the classical definition, which was used by Cardano, Pascal, Fermat, Jacob Bernoulli, Leibniz, de Moivre, and Laplace, and treats probability as the ratio of the number of favorable cases to the total number of possible equally likely cases. Very often the equal likeliness of the total number of cases arises as a consequence of *symmetry*, such as when a fair die is rolled. When looked at in this way, the classical definition has an objective interpretation. However, the classical definition has an inherent circularity in it. As Bunnin and Yu (2004, p. 563) note:

> . . .[the classical definition] involves a vicious circle because it defines probability in terms of equipossible alternatives, but equipossibility presupposes an understanding of probability.

Some mathematicians have tried to get around this difficulty by appealing to the principle of indifference. For example, suppose we wish to find the probability of a "four" when a balanced die is rolled. To avoid the circularity of stating that the probability of a "four" is 1/6 because all six outcomes {1, 2, 3, 4, 5, 6} are equally probable, an appeal to the principle of indifference would result in the same probability of 1/6 to each of the outcomes (and in particular to a "four") because there is no reason to assign different probabilities to them. This line of reasoning was the one used by Laplace in his definition of classical probability (1776):

[†] The various theories and philosophies of probability are also discussed in Gillies (2000), Fine (1973), Lyon (2010), Galavotti (2005), Fitelson et al. (2006), Keuzenkamp (2004), Barnett (1999, Chapter 3), Chatterjee (2003, Chapter 3), Thompson (2007), Chuaqui (1991), and Mellor (2005).

[‡] Daston refers to the epistemic classification of probability as "subjective probability", as is sometimes done. However, we will reserve the word "subjective" for a specific theory that together with the logical theory of probability form part of the epistemic classification. See later in this problem.

The probability of an event is the ratio of the number of cases favorable to it, to the number of possible cases, when there is nothing to make us believe that one case should occur rather than any other, so that these cases are, for us, equally possible.

Laplace's definition appeals to the principle of indifference and relates to our *beliefs*. It is thus more in accordance with a subjective interpretation of the classical definition of probability. However, we have shown how the principle of indifference can easily lead to contradictions. Moreover, the classical definition cannot be applied to nonsymmetric situations. These factors therefore limit its usefulness.

The second concept of probability is perhaps the most popular one and is based on relative frequency, which is an objective concept. The frequency theory can be applied even if the possible events in an experiment are *not* equally likely. According to Keynes (Keynes, 1921, p. 92), the first logical investigation of the frequency theory of probability was done by the British mathematician Robert Leslie Ellis (1817–1859) (Ellis, 1844) and the French mathematician and philosopher Antoine Augustin Cournot (1801–1877) (Cournot, 1843, p. iii). Major proponents of the frequency approach include John Venn (1834–1923) (Fig. 14.8), Richard von Mises (1883–1953), (Fig. 14.9) and Sir Ronald Aylmer Fisher (1890–1962). According to Venn (1866, p. 163)

Figure 14.8   John Venn (1834–1923).

Figure 14.9   Richard von Mises (1883–1953). [Picture licensed under the Creative Commons Attribution-Share Alike 2.0 Germany license.]

> ...within the indefinitely numerous class which composes this series [i.e. 'a large number or succession of objects'] a smaller class is distinguished by the presence or absence of some attribute or attributes... These larger and smaller classes respectively are commonly spoken of as instances of the 'event' and of 'its happening in a given particular way' ...we may define the probability or chance (the terms are here regarded as synonymous) of the event happening in that particular way as the numerical fraction which represents the proportion between the two different classes in the long run.

Thus, the relative frequency definition of the probability of an event is the relative frequency of occurrence of the event in infinite repeated trials of an experiment. The frequency theory is an objective theory insomuch as it refers to the physical world. A key problem with this theory is that it cannot be applied to a one-off event. For example, consider the statement, "Bill is a smoking American with congenital heart disease, is 55 years old, and has a probability of .001 of living to 60." According to the frequency theory, the probability of .001 means that, out of all people like Bill, one in a thousand will live to 60. But what does "all people like Bill mean"? Should we include only all 55 years old American with congenital heart disease? Surely there can be other factors that influence a person's probability of living up to 60. Could we not also include all 55 years old American with congenital heart disease with the same diet and amount of exercise as Bill? What about including also those who live in the same geographical location as Bill? This is the so-called *reference class problem* and makes the frequency interpretation of the probability of a one-off event indeterminate.

To get around the reference class problem, von Mises put forward a *hypothetical* frequency theory of probability (von Mises, 1928). Whereas the concept of a "series" is central to Venn's definition, von Mises' frequency definition of probability is based on the notion of a "collective," defined as (von Mises, 1981, p. 12)

> . . .a sequence of uniform events or processes which differ by certain observable attributes, say colours, numbers, or anything else.

A collective is thus a hypothetical infinite sequence of objects such that the probability of an event is the relative frequency of that event in the collective. A collective has two main properties:

- convergence, which ensures that the relative frequencies possess limiting values;
- randomness,[†] which ensures that the limiting values are the same in all arbitrary subsequences.[‡]

Although von Mises' frequency theory based on a collective was a positive step toward an axiomatization of probability, it did not gain wide acceptance partly because it could not be empirically verified. Shafer and Vovk (Shafer and Vovk, 2001, p. 48) remark

> In spite of von Mises's efforts. . .the derivation of the rules of probability from the concept of a collective remained messy, and the idea of an infinite sequence seemed to contribute nothing to actual uses of probability.

The third concept of probability is Bayesian or epistemic probability, and can be further divided into subjective and logical probability. In the subjective interpretation, probability is regarded as the subject's degree of belief in a proposition. It was first informally put forward by Bernoulli in his *Ars Conjectandi* (Bernoulli, 1713, pp. 210–211)[§]:

[†] The randomness condition is also known as the impossibility of a gambling system principle. Thus, if an event has a certain odds, the condition implies that it is impossible to choose a particular subsequence to improve the odds.

[‡] As a simple example, denoting a head by a "1" and tail by a "0", the infinite sequence of heads and tails 1, 1, 0, 1, 1, 0, 0, 0, 0, 1, 0, 1, 1, 1, 1, 0, . . . would qualify as a collective if the frequency of 1's approaches 1/2 as the number of tosses increases and the same limiting frequency of 1/2 is achieved for any subsequence of the above sequence. On the other hand, the sequence 1, 0, 1, 0, 1, 0, 1, . . . cannot qualify as a collective because, although the relative frequency of 1's in the sequence is 1/2, this not true for the subsequence of even terms: 0, 0, 0, . . .

[§] The translation is taken from Oscar Sheynin's translations of Chapter 4 of the *Ars Conjectandi* (Sheynin, 2005).

Certainty of some thing is considered either objectively and in itself and means none other than its real existence at present or in the future; or subjectively, depending on us, and consists in the measure of our knowledge of this existence...

Key proponents of the subjective theory include Bruno de Finetti (1906–1985) (Fig 14.10), Frank P. Ramsey (1903–1930), and Leonard J. Savage (1917–1971). The latter is credited for making subjective probability[†] as one of the most influential theories of probability, especially in the *Foundations of Statistics* (Savage, 1972). On the other hand, de Finetti is famous for bluntly saying in the preface of his *Theory of Probability* (de Finetti, 1974, p. x):

My thesis, paradoxically, and a little provocatively, but nonetheless genuinely, is simply this:

PROBABILITY DOES NOT EXIST

de Finetti's point is that there is no such thing as an *objective* measure of probability because all probabilities are *subjective*. The force in de Finetti's statement becomes apparent when we consider the following example. Suppose a coin in thrown in the air and one is asked for the probability that it lands heads. One would undoubtedly say half. Suppose, while the coin is still in the air, one is informed that the coin has two

Figure 14.10   Bruno de Finetti (1906–1985).

[†] Which Savage called "personal" probability.

heads and one is asked the question again, "What is the probability that the coin will land heads?" One would now say one. Why has the answer changed? Nothing has changed in the objective properties of the coin, only one's state of knowledge.

de Finetti also advanced the following modern definition of subjective probability (de Finetti, 1937, p. 101):

> ...the degree of probability attributed by an individual to a given event is revealed by the conditions under which he would be disposed to bet on that event.

Thus, my subjective probability of an event E is p if I am willing to bet up to $\$(px)$ for a payment of $\$x$ when E occurs and a payment of $\$0$ otherwise.

It is important to note that, within the subjective framework, the degree of belief one assigns to an event is not completely arbitrary because it must be coherent, that is, it must satisfy the rules of probability. The justification of coherence is based on the so-called *Dutch-Book Argument*,[†] which was first put forward by Ramsey (1926). Whenever coherence is violated, one suffers a Dutch-book. For example, suppose player P believes that the event E has a probability p_1 and the event $\bar{E}$ has a probability p_2. What should the relationship between p_1 and p_2 be so that P does not incur a Dutch-book? This can be derived as follows. Based on her beliefs, P bets

- $\$ p_1x$ for a payment of $\$x$ when E occurs, and
- $\$ p_2x$ for a payment of $\$x$ when $\bar{E}$ occurs.

The total amount P bets is $\$(p_1 + p_2)x$. If E occurs P gets $\$x$ back. Similarly, if $\bar{E}$ occurs P gets $\$x$ back. Thus, whatever happens, P gets $\$x$. To avoid a Dutch-book, we must have $x \geq (p_1 + p_2)x$, that is,

$$p_1 + p_2 \leq 1. \tag{14.4}$$

Now, based on her beliefs, P can also pay

- $\$x$ to somebody who bets $\$ p_1x$ on E, and
- $\$x$ to somebody else who bets $\$ p_2x$ on $\bar{E}$.

The total amount received by P is $\$(p_1 + p_2)x$. If E occurs P pays $\$x$. Similarly, if $\bar{E}$ occurs P pays $\$x$ as well. Thus, whatever happens, P pays $\$x$. To avoid a Dutch-book, we must have $(p_1 + p_2)x \geq x$, that is,

$$p_1 + p_2 \geq 1. \tag{14.5}$$

By using Eqs. (14.4) and (14.5), we see that the only way for P to avoid a Dutch-book is to have $p_1 + p_2 = 1$, that is, $\Pr\{E\} + \Pr\{\bar{E}\} = 1$, in accordance with the probability axioms.

[†] A Dutch-book is a collection of bets that ensures a monetary loss. Although the idea was first used by Ramsey, the term "Dutch-book" was introduced by Lehman (1955). The Dutch-book argument is also discussed in Resnick (1987, p. 71), Sarkar and Pfeifer (2006, p. 213), Jeffrey (2004, p. 5), Kyburg (1983, p. 82), and Burdzy (2009, p. 152); for a critique see Maher (1993, p. 94).

In spite of the coherence structure, the rationale for a subjective prior which can be used in the application of Bayes' Theorem has been the subject of much criticism. Keuzenkamp (2004, pp. 93–94) writes

> ...personal approaches to epistemological probability have been criticized for a lack of scientific objectivity. If scientific inference depends on personal taste or custom and convention, it loses much of its appeal.
>
>   ... Objectivity is not the real battleground. The real dispute is about pragmatic methods and the robustness with respect to the choice of priors.

We next consider the fourth theory of probability that is based on a logical interpretation. This theory is objective in nature and some of its ideas were first put forward by the British logician William Ernest Johnson (1858–1931). However logical probability is usually associated with the works of Johnson's student, the renowned economist John Maynard Keynes (1883–1946) (Fig. 14.11). In his *A Treatise on Probability*, Keynes gives the fundamental idea behind the logical interpretation (Keynes, 1921, p. 52),

> Inasmuch as it is always assumed that we can sometimes judge directly that a conclusion follows from a premiss, it is no great extension of this assumption to suppose that we can sometimes recognize that a conclusion partially follows from, or stands in a relation of probability to a premiss.

Figure 14.11   John Maynard Keynes (1883–1946).

Keynes thus uses the concept of the degree of partial entailment to denote logical probabilities. In logical terminology, if A entails B then whenever A is true, B is also true. Keynes modifies this idea and uses $\Pr\{B \mid A\}$ to denote the *degree* to which A entails B. When this probability is less than 1, then A only partially entails B. Thus, a logical probability is a statement about the logical relation about a hypothesis (B) and the evidence (A), and it does not make sense to speak of the probability of B all by itself. For example, if A is the statement that a die is fair and B is the statement that a "five" is obtained, then the logical probability $\Pr\{B \mid A\} = 1/6$.

The logical theory probability was further developed by the British physicist Sir Harold Jeffreys (1891–1989) and the German-born philosopher Rudolf Carnap (1891–1970). Jeffreys argued for the utilization of the principle of indifference when no evidence was available. Carnap's key contribution was to allow many systems of logic that could be used to calculate logical probabilities. However, the logical theory cannot verify the ways of constructing a logical relation and cannot calculate probabilities in all situations. It has been largely superseded by other theories such as the subjective theory of probability.

There are still other theories of probability, such as the propensity and the fiducial theories. However, no single theory has proved to be completely satisfactory. Feller (1968, p. 19) thus says

> All possible "definitions" of probability fall far short of the actual practice.

Although there is no single satisfactory theory of probability, one or more of the interpretations we have previously outlined will almost always prove to be useful in any given situation. More importantly, the modern axiomatic theory of probability as developed by Kolmogorov (1933) is valid irrespective of one's interpretations of probability.[†]

[†] See **Problem 23** for more on Kolmogorov's axiomatic theory of probability.

Problem 15

Leibniz's Error (1768)

Problem. *With two balanced dice, is a throw of 12 as likely as a throw of 11?*

Solution. The sample space $\Omega = \{(1, 1), (1, 2), \ldots, (6,6)\}$ is made up of 36 equally likely outcomes, where (a, b) are the numbers on the first and second dice, respectively, for $a, b = 1, 2, \ldots, 6$. A throw of 12 can be obtained in only one way, namely (6, 6). However, a throw of 11 can be obtained in two ways, namely (5, 6) and (6, 5). Therefore, Pr{a throw of 12} = 1/36 but Pr{a throw of 11} = 1/18, and the latter is twice as likely as the former.

15.1   Discussion

The renowned German mathematician and philosopher Gottfried Wilhelm Leibniz (1646–1716) (Fig. 15.1) is usually remembered as the coinventor of the differential calculus with archrival Isaac Newton. However, he was also very much interested in probability.

Regarding the question in **Problem 15** above, Leibniz states in the *Opera Omnia* (Leibniz, 1768, p. 217)

> . . .for example, with two dice, it is equally likely to throw twelve points, than to throw eleven; because one or the other can be done in only one manner.

Thus, Leibniz believed the two throws to be equally likely, arguing that in each case the throw could be obtained in a single way. While it is true that a throw of 11 can be realized only with a five and a six, there are two ways in which it could happen: the first die could be a five and the second a six, or vice versa. On the other hand, a throw of 12 can be realized in only one way: a 6 on each die. Thus the first probability is twice the second.

The solution to the problem would seem quite elementary nowadays. However, it must be borne in mind that in the times of Leibniz and even afterward, notions about

---

Figure 15.1   Gottfried Wilhelm Leibniz (1646–1716).

probability, sample spaces and sample points were quite abstruse. The classical definition of probability would come only more than a century later in 1812 in Pierre-Simon Laplace's majestic *Théorie Analytique des Probabilités* (Laplace, 1812) (although a few have argued that Cardano actually stated this definition 300 years before). Commenting on Leibniz's error, Todhunter states (Todhunter, 1865, p. 48)

> Leibniz however furnishes an example of the liability to error which seems peculiarly characteristic of our subject.

However, this should not in any way undermine some of the contributions Leibniz made to probability theory. For one thing, he should be credited as one of the very first to give an explicit definition of classical probability, phrased in terms of an expectation (Leibniz, 1969, p. 161):

> If a situation can lead to different advantageous results ruling out each other, the estimation of the expectation will be the sum of the possible advantages for the set of all these results, divided into the total number of results.

In spite of being conversant with the classical definition, Leibniz was very interested in establishing a logical theory for different degrees of certainty. He may rightly be regarded as a precursor to later developments in the logical foundations of probability by Keynes, Jeffreys, Carnap, and others. Since Jacob Bernoulli had similar interests, Leibniz started a communication with him in 1703. He undoubtedly had some influence in Bernoulli's *Ars Conjectandi* (Bernoulli, 1713). When Bernoulli communicated to Leibniz about his law of large numbers, the latter reacted critically. As Schneider (2005b, p. 90) explains

> Leibniz's main criticisms were that the probability of contingent events, which he identified with dependence on infinitely many conditions, could not be determined by a finite number of observations and that the appearance of new circumstances could change the probability of an event. Bernoulli agreed that only a finite number of trials can be undertaken; but he differed from Leibniz in being convinced by the urn model that a reasonably great number of trials yielded estimates of the sought-after probabilities that were sufficient for all practical purposes.

Thus, in spite of Leibniz's criticism, Bernoulli was convinced of the authenticity of his theorem. This is fortunate because Bernoulli's law was nothing less than a watershed moment in the history of probability.[†]

[†] See **Problem 8** for more on Bernoulli's law.

Problem 16

The Buffon Needle
Problem (1777)

P*roblem.* *A needle of length l is thrown at random on a plane on which a set of parallel lines separated by a distance d (>l) have been drawn. What is the probability that the needle will intersect one of the lines?*

S*olution.* Consider the diagram in Fig 16.1. Let the distance between the midpoint of the needle to the nearest line be Y ($0 < Y < d/2$), and the acute angle between the needle and the horizontal be Φ ($0 < \Phi < \pi/2$). We assume that $Y \sim U(0, d/2)$ and that $\Phi \sim U(0, \pi/2)$. It is also reasonable to assume that Y and Φ are statistically independent. Therefore, the joint density of Y and Φ is

$$f_{Y\Phi}(y, \varphi) = f_Y(y)f_\Phi(\varphi)$$
$$= \frac{2}{d} \cdot \frac{2}{\pi}$$
$$= \frac{4}{\pi d}, \quad 0 < y < d/2, \ 0 < \varphi < \pi/2.$$

The needle will intersect one of the lines if and only if $Y < (l/2)\sin \Phi$ (see Fig. 16.1). The probability of this event is the double integral of the joint density $f_{Y\Phi}$ over the area A (see Fig. 16.2):

$$\Pr\left\{Y < \frac{l}{2}\sin \Phi\right\} = \iint_A f_{Y\Phi}(y, \varphi)dy \, d\varphi$$
$$= \int_0^{\pi/2}\left\{\int_0^{(l \sin \varphi)/2} \frac{4}{\pi d}dy\right\}d\varphi$$
$$= \frac{2l}{\pi d}. \tag{16.1}$$

Hence the required probability is $(2l)/(\pi d)$.

Classic Problems of Probability, Prakash Gorroochurn.
© 2012 John Wiley & Sons, Inc. Published 2012 by John Wiley & Sons, Inc.

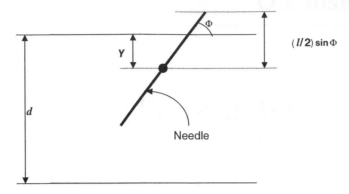

Figure 16.1   The Buffon Needle Problem.

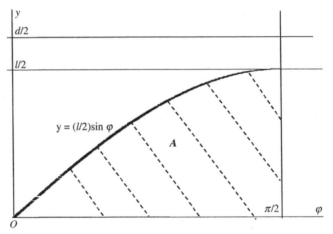

Figure 16.2   Graph of $y = (l/2) \sin \varphi$ vs. φ ($d > l$ case).

16.1 Discussion

A simpler solution than the previous can be obtained by first noting that (Y, Φ) is uniform on $S = \{(y, \varphi) : 0 < y < d/2, 0 < \varphi < \pi/2\}$. Therefore the required probability is a "geometric" probability, namely the ratio of the area of A to that of S (see Fig. 16.2):[†]

$$
\text{Prob. of intersection} = \frac{\int_0^{\pi/2} (l/2)\sin \varphi \, d\varphi}{(\pi/2)(d/2)}
$$

$$
= \frac{2l}{\pi d} \cdot [-\cos \varphi]_{\varphi=0}^{\pi/2}
$$

$$
= \frac{2l}{\pi d},
$$

as we obtained before in Eq. (16.1).

[†] In general, suppose a region G contains a smaller region g. Then if a point is chosen at random in G, the (geometric) probability that the point lies within g is given by the ratio of the measure (e.g. volume, area or length) of g to that of G.

Figure 16.3   Georges-Louis Leclerc, Comte de Buffon (1707–1788).

The *Buffon Needle Problem*[†] is considered to be the first problem in geometric probability, which is an extension of the mathematical definition of probability. It was enunciated in 1733 by Georges-Louis Leclerc, Comte de Buffon (1707–1788), French naturalist and mathematician (Fig. 16.3). The problem was afterward published with the correct solution in Buffon's article *Essai d'Arithmétique Morale* (Buffon 1777, p. 103) (see Fig. 16.4). Laplace later observed that the formula $p = 2l/(\pi d)$ (where p is the probability of intersection) could actually be used to estimate the value of π (Laplace, 1812, p. 360). If a needle is thrown N times and in n cases the needle intersects one of the lines, then n/N is an unbiased estimator of p. An estimator of π is therefore $(2lN)/(nd)$. However, this estimator is *biased*[‡] because $(nd)/(2lN)$ is an *unbiased* estimator of $1/\pi$.

[†] The Buffon needle problem is also discussed in Solomon (1978, p. 2), van Fraassen (1989, p. 303), Aigner and Ziegler (2003, Chapter 21), Higgins (2008, p. 159), Beckmann (1971, p. 159), Mosteller (1987, p. 14), Klain and Rota (1997, Chapter 1), Lange (2010, p. 28), and Deep (2006, p. 74).

[‡] If T is used as an estimator of a parameter t, then the bias of T is defined as $\text{Bias}_t(T) = \mathscr{E}T - t$.

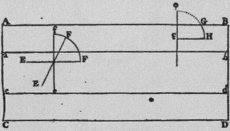

D'ARITHMÉTIQUE MORALE.   101

eft fimplement divifé par des joints parallèles, on jette
en l'air une baguette, & que l'un des joueurs parie que
la baguette ne croifera aucune des parallèles du parquet,
& que l'autre au contraire parie que la baguette croifera
quelques-unes de ces parallèles; on demande le fort de
ces deux joueurs. *On peut jouer ce jeu fur un damier avec
une aiguille à coudre ou une épingle fans tête.*
    Pour le trouver, je tire d'abord entre les deux joints
parallèles *A B* & *C D* du parquet, deux autres lignes

parallèles *a b* & *c d*, éloignées des premières de la moitié
de la longueur de la baguette *E F*, & je vois évidemment
que tant que le milieu de la baguette fera entre ces deux
fecondes parallèles, jamais elle ne pourra croifer les premières dans quelque fituation *E F*, *e f*, qu'elle puiffe fe
trouver; & comme tout ce qui peut arriver au-deffus
de *a b* arrive de même au-deffous de *c d*, il ne s'agit
que de déterminer l'un ou l'autre; pour cela je remarque
que toutes les fituations de la baguette peuvent être

Figure 16.4   Extract from Buffon's Needle Problem, taken from Buffon's *Essai d'Arithmétique Morale* (Buffon, 1777, p. 101).

Table 16.1[†] lists several attempts to estimate the value of π. The one by Lazzarini (1902) is of special interest since it is accurate up to six places of decimal to the true value of π.[‡] Using reasonable assumptions, Coolidge (1925, p. 82) shows that the probability of attaining such an accuracy by chance alone is about .014 and concludes

It is much to be feared that in performing this experiment Lazzarini 'watched his step'.

Indeed, as Kendall and Moran (1963, p. 71) point out, the accurate approximations in Table 16.1 suggest that "optional stopping" was used.[§]

[†] Taken from Kendall and Moran (1963, p. 70).

[‡] And in fact coincides with 355/113, which is a more accurate value than 22/7 to approximate π and which was first discovered by the Chinese mathematician Zu Chongzhi (429–500). See Posamentier and Lehman (2004, p. 61) for more details.

[§] For a fuller analysis of most of these experiments, see Gridgeman (1960) and O'Beirne (1965, pp. 192–197).

Table 16.1   Attempts to Estimate π Using Eq. (16.1)

Experimenter	Needle length	No. of throws	No. of hits	Estimate of π
Wolf (1850)	.8	5000	2532	3.1596
Smith (1855)	.6	3204	1218.5	3.1553
De Morgan (1860)	1.0	600	382.5	3.137
Fox (1884)	.75	1030	489	3.1595
Lazzarini (1902)	.83	3408	1808	3.1415929
Reina (1925)	.5419	2520	859	3.1795

We now present an intriguing alternative solution (e.g., see Uspensky, 1937, p. 253) to the *Buffon Needle Problem*, due to the French mathematician Joseph Emile Barbier (1839–1889) (Barbier, 1860). The peculiarities of the approach are that it does not involve any integration and can be generalized to arbitrary convex shapes. First we write $p \equiv p(l)$ to emphasize that the probability of intersection depends on the length of the needle. Suppose the needle is divided into two parts with lengths l' and l''. Since a line intersects the needle if and only if it intersects either portion, we have $p(l) = p(l') + p(l'')$. The latter equation is satisfied if

$$p(l) = kl, \tag{16.2}$$

where k is a constant of proportionality. Now imagine a polygonal line (not necessarily convex) of total length l made up of n rectilinear segments $a_1, a_2, \ldots, a_n$ where each $a_j < d$. Each segment a_j has a probability $p(j) = ka_j$ of intersecting one of the parallel lines. Let

$$I_j = \begin{cases} 1, & \text{if segment } a_j \text{ intersects the line,} \\ 0, & \text{otherwise.} \end{cases}$$

The total number of intersections is then

$$T = \sum_{j=1}^{n} I_j,$$

and the expectation of T is

$$\mathscr{E}T = \sum_{j=1}^{n} \mathscr{E} I_j$$
$$= \sum_{j=1}^{n} ka_j$$
$$= kl. \tag{16.3}$$

For a circle with diameter d, we have $T = 2$, $\mathscr{E}T = 2$, and $l = \pi d$. Using Eq. (16.3), we solve for k and obtain $k = 2/(\pi d)$. Hence, Eq. (16.2) becomes

$$p(l) = kl = \frac{2l}{\pi d},$$

which is the same formula we obtained using integrals in Eq. (16.1).

Digging deeper, if we consider a sufficiently small convex polygon with circumference C, then we can have only two intersections (with probability, say P) or zero intersection (with probability $1 - P$). Then the expected number of intersections is $2P$. From Eq. (16.3), we have $\mathscr{E}T = kl$, where $k = 2/(\pi d)$ as we showed previously. Substituting $\mathscr{E}T = 2P$, $k = 2/(\pi d)$, and $l = C$ in $\mathscr{E}T = kl$, we have

$$P = \frac{C}{\pi d}. \tag{16.4}$$

This gives a general formula for the probability P of intersection of a small convex polygon (of any shape) with circumference C when thrown on lines separated by a distance d. The formula is interesting because P does not depend either on the number of sides or on the lengths of the sides of the polygon. It therefore also applies to any convex contour. For an alternative derivation of Eq. (16.4), see Gnedenko (1978, p. 39).

Two variations of the classic *Buffon Needle Problem* are also worth considering. First

A needle of length l is thrown at random on a plane on which a set of parallel lines separated by a distance d, where d < l. What is the probability that the needle will intersect one of the lines?

This time, the length of the needle is greater than the separation of the lines but (Y, Φ) is still uniform on $S = \{(y, \varphi) : 0 < y < d/2, \ 0 < \varphi < \pi/2\}$. Therefore the required probability, as a geometric probability, is the ratio of the area of A' to that of S (see Fig 16.5)[†]:

$$
\Pr\left\{ Y < \frac{l}{2}\sin\Phi \right\} = \frac{\displaystyle\int_{0}^{\arcsin(d/l)} \frac{l}{2}\sin\varphi\, d\varphi + \left(\frac{\pi}{2} - \arcsin\frac{d}{l}\right)\frac{d}{2}}{\left(\dfrac{\pi}{2}\right)\left(\dfrac{d}{2}\right)}
$$

$$
= \frac{4}{\pi d}\left\{ -\frac{l}{2}[\cos\varphi]_{\varphi=0}^{\arcsin(d/l)} + \frac{d}{2}\arccos\left(\frac{d}{l}\right) \right\}
$$

$$
= \frac{4}{\pi d}\left[-\frac{l}{2}\left\{\cos\left(\arcsin\frac{d}{l}\right) - 1\right\} \right] + \frac{2}{\pi}\arccos\left(\frac{d}{l}\right)
$$

$$
= \frac{2l}{\pi d}\left(1 - \sqrt{1 - \frac{d^2}{l^2}}\right) + \frac{2}{\pi}\arccos\left(\frac{d}{l}\right). \tag{16.5}
$$

[†] In the following, we will use $\arcsin(x) + \arccos(x) = \pi/2$ and $\cos(\arcsin(x)) = \sqrt{1 - x^2}$.

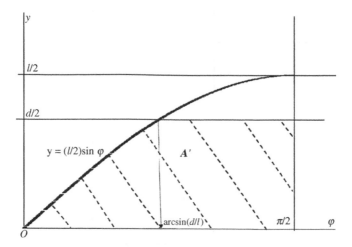

Figure 16.5 Graph of $y = (l/2)\sin \varphi$ vs. φ ($d < l$ case).

The second variation was first considered by Laplace[†] (1812, pp. 360–362) and is as follows (Fig. 16.6):

> *Suppose a needle of length l is thrown on a plane that is made up of congruent rectangles each of dimensions c and d ($l < c, d$). What is the probability that the needle will intersect at least one of the sides of the rectangle?*

Let the distances from the base of the needle to the next vertical line along the horizontal axis be X and to the next horizontal line along the vertical axis be Y. Let the angle made by the length of the needle with the horizontal be Φ. We assume that X, Y, and Φ are uniformly distributed on $(0, c)$, $(0, d)$, and $(0, \pi/2)$, respectively. The needle will cross a vertical line if $X < l \cos \Phi$. The probability of this event is again calculated through geometric probability:

$$p_V = \frac{\int_0^d \left\{ \int_0^{\pi/2} l \cos \varphi \, d\varphi \right\} dy}{\pi c d / 2} = \frac{2l}{\pi c}.$$

Similarly, the needle will cross a horizontal line if $Y < l \sin \Phi$, and this will happen with probability

$$p_H = \frac{\int_0^c \left\{ \int_0^{\pi/2} l \sin \varphi \, d\varphi \right\} dx}{\pi c d / 2} = \frac{2l}{\pi d}.$$

Both horizontal and vertical intersections occur if $X < l \cos \Phi$ and $Y < l \sin \Phi$, and the probability is

$$p_{V \times H} = \frac{\int_0^{(\pi/2)} \int_0^{(l \sin \varphi)} \int_0^{(l \cos \varphi)} dx dy d\varphi}{\pi c d / 2} = \frac{l^2}{\pi c d}.$$

[†] See also Arnow (1994), Solomon (1978, p. 3), and Uspensky (1937, p. 256).

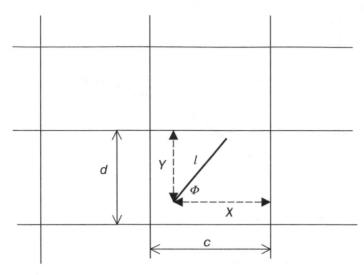

Figure 16.6   Laplace's extension of the Buffon Needle Problem.

Hence, the probability of either a vertical or horizontal intersection is

$$p_{V+H} = p_V + p_H - p_{V \times H}$$
$$= \frac{2l}{\pi c} + \frac{2l}{\pi d} - \frac{l^2}{\pi c d}$$
$$= \frac{2l(c+d) - l^2}{\pi c d}. \qquad (16.6)$$

Note that, as $c \to \infty$, we have

$$\lim_{c \to \infty} p_{V+H} = \lim_{c \to \infty} \frac{2l(c+d) - l^2}{\pi c d} = \frac{2l}{\pi d},$$

which is the same probability as in Eq. (16.1).

We now consider the following issue. Out of the several possible uniform random variables in the *Buffon Needle Problem*, why did we choose the acute angle between the needle and the nearest line (Φ), and the distance between the midpoint of the needle to the nearest line (Y)? Would the probability of intersection have changed if we had instead chosen the vertical rise of the needle from its midpoint and the distance of the midpoint to the nearest line as uniformly distributed each, that is, $\Phi' \sim U(0, l/2)$ and $Y \sim U(0, d/2)$, where $\Phi' = (l \sin \Phi)/2$? The answer is yes. A similar calculation to the one presented in the solution shows that the probability of intersection would then be $l/(2d)$, which is different from Eq. (16.1). This might seem quite disquieting. However, as van Fraassen (1989, p. 312) has shown, $\Phi \sim U(0, \pi/2)$ and $Y \sim U(0, d/2)$ are choices that make the joint density of (Y, Φ) invariant under rigid motions. That is, if the lines were to be rotated or translated before the needle was thrown on them, these choices ensure we get the same probability of intersection.

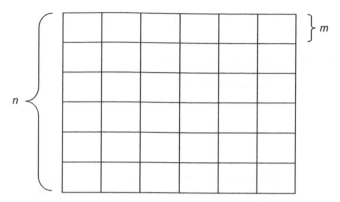

Figure 16.7   Procedure to estimate e geometrically. The big square is of dimension $n \times n$ and each of the small squares is of dimension $m \times m$ (n/m is assumed to be an integer).

The invariance of this probability under such translations seems to be a natural desideratum.[†]

Since we have shown how the value of π could be estimated through geometric probability (see second paragraph in Discussion section), we might ask a related question:

Can we also use a geometric procedure to estimate the exponential number e?

Nahin (2000, p. 30) provides a geometric approach,[‡] which we adapt in the following. Imagine a big square of dimension $n \times n$ made up of identical smaller squares each of dimension $m \times m$, as shown in Fig. 16.7. Let us choose m so that n/m is an integer. The number of small squares is thus $N = n^2/m^2$.

Now, suppose N darts are thrown at random on the big square. Let X be the number of darts that land inside a particular small square. Then $X \sim B(N, 1/N)$ and

$$\Pr\{X = x\} = \binom{N}{x}\left(\frac{1}{N}\right)^x \left(1 - \frac{1}{N}\right)^{N-x}, \quad x = 0, 1, 2, \dots, N.$$

The probability that the small square receives no dart is

$$\Pr\{X = 0\} = \left(1 - \frac{1}{N}\right)^N \to e^{-1} \quad \text{as } N \to \infty.[§]$$

We thus have a method to estimate e. Let us make m small enough so that N is large. Suppose, after the N darts have been thrown, the number of small squares (out of the N) that receive no dart at all is s. Then $s/N \approx 1/e$ so that $e \approx N/s$. Again, N/s is a biased estimator of e since s/N is unbiased for $1/e$.

[†] See **Problem 19** for more on invariance.

[‡] Of course, if one is not interested in a geometric approach one can simply use $e = 1 + 1/1! + 1/2! + 1/3! + \cdots$.

[§] Using the fact that $\lim_{x \to \infty}(1 + a/x)^x = e^a$.

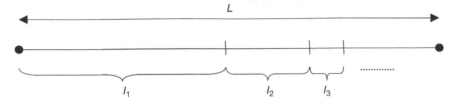

Figure 16.8   Geometric unbiased estimation of e.

There exists a second geometric approach[†] that can give an *unbiased* estimate of e. Suppose a line of length L is divided into successively smaller segments $I_n = \left(\frac{n-1}{n}L, \frac{n}{n+1}L\right)$ for $n = 1, 2, \ldots$ (see Fig. 16.8).

A sharp knife is then thrown at random perpendicular to the line and if it cuts the line across the jth segment, we then assign the random variable X the value $2 + 1/(j-1)!$ We now show that $\mathscr{E}X = e$. We have

$$
\Pr\{X = x\} = \begin{cases} \dfrac{j}{j+1} - \dfrac{j-1}{j} = \dfrac{1}{j(j+1)}, & \text{for } x = 2 + 1/(j-1)!, \ j = 1, 2, \ldots \\ 0 & \text{otherwise} \end{cases}
$$

so that

$$
\begin{aligned}
\mathscr{E}X &= \sum_{j=1}^{\infty} \left\{ 2 + \frac{1}{(j-1)!} \right\} \cdot \frac{1}{j(j+1)} \\
&= \sum_{j=1}^{\infty} \left\{ \frac{2}{j} - \frac{2}{(j+1)} + \frac{1}{(j+1)!} \right\} \\
&= \left(\frac{2}{1} - \frac{2}{2} + \frac{2}{2} - \frac{2}{3} + \frac{2}{3} - \frac{2}{4} + \cdots \right) + \sum_{j=1}^{\infty} \frac{1}{(j+1)!} \\
&= 2 + \sum_{j=1}^{\infty} \frac{1}{(j+1)!} \\
&= \sum_{j=0}^{\infty} \frac{1}{j!} \\
&= e.
\end{aligned}
$$

Thus to obtain an unbiased estimate of e, we throw the knife on the line n times (where n is large), note the segment cut each time and record the corresponding values $x_1, x_2, \ldots, x_n$ of X. Then an unbiased estimate of e is $\bar{x} = (x_1 + x_2 + \cdots + x_n)/n$.

[†] This method was communicated to me by Bruce Levin.

Problem 17

Bertrand's Ballot Problem (1887)

Problem. *In an election, candidate M receives a total of m votes while candidate N receives a total of n votes, where $m > n$. Prove that, throughout the counting, M is ahead of N with probability $(m - n)/(m + n)$.*

Solution. We use mathematical induction to prove that the probability of M being ahead throughout is

$$p_{m,n} = \frac{m - n}{m + n}. \tag{17.1}$$

If $m > 0$ and $n = 0$, A will always be ahead since N receives no votes. The formula is thus true for all $m > 0$ and $n = 0$ because it gives $p_{m,n} = m/m = 1$. Similarly, if $m = n > 0$, M cannot always be ahead of N since they clearly are not at the end. The formula is thus true for all $m = n > 0$ because it gives $p_{m,n} = 0/(2m) = 0$.

Now suppose $p_{m-1,n}$ and $p_{m,n-1}$ are both true. Then M with m votes will always be ahead of N with n votes if and only if either (i) in the penultimate count, M with m votes is ahead of N with $n - 1$ votes, followed by a last vote for N or (ii) in the penultimate count, M with $m - 1$ votes is ahead of N with n votes, followed by a last vote for M. Therefore,[†]

[†] Note that, since M and N receive a total of m and n votes, respectively, the probability that M wins any one vote is $m/(m + n)$.

Classic Problems of Probability, Prakash Gorroochurn.
© 2012 John Wiley & Sons, Inc. Published 2012 by John Wiley & Sons, Inc.

$$p_{m,n} = p_{m,n-1} \frac{n}{m+n} + p_{m-1,n} \frac{m}{m+n} \qquad (17.2)$$

$$= \frac{m-(n-1)}{m+(n-1)} \cdot \frac{n}{m+n} + \frac{(m-1)-n}{(m-1)+n} \cdot \frac{m}{m+n}$$

$$= \frac{m-n}{m+n}.$$

Hence the result is established.

17.1 Discussion

The *Ballot Problem* is usually attributed to Joseph Bertrand (1822–1900) (Fig. 17.1) who proposed it in a half-page article (Bertrand, 1887).[†] In this article, Bertrand gave the correct formula for $p_{m,n}$ but did not provide a proof. Instead, he provided the following recursive equation:

$$f_{m+1,t+1} = f_{m,t} + f_{m+1,t}. \qquad (17.3)$$

In the above equation, $f_{m,t}$ is the number of ways M is always ahead of N when M receives a total of m out of t votes. Bertrand's recursion in Eq. (17.3) is equivalent to Eq. (17.2) above, since

$$p_{m,n} \equiv \frac{f_{m,m+n}}{\dbinom{m+n}{m}}.$$

Thus, although Bertrand did not prove his formula,[‡] he could easily have done so through induction. Bertrand, however, believed that there should be an even simpler proof, for he writes

...it seems probable that such a simple result could be shown in a more direct way.

The "more direct way" was provided only a few weeks later by the French mathematician Désiré André (1887). André reasoned in terms of favorable and unfavorable permutations of the m votes for M and n votes for N, where a favorable permutation is an ordered sequence of votes such that M is always ahead (e.g., MMNM). First, he notes that all permutations starting with an N are

---

[†] The ballot problem is also discussed in Feller (1968, p. 69), Higgins (1998, p. 165), Meester (2008, p. 73), Blom et al. (1994, p. 128), Durrett (2010, p. 202), Gyóari et al. (2008, pp. 9–35), and Shiryaev (1995, p. 107).

[‡] However, Bertrand later did provide a complete demonstration of the result in his *Calculs des Probabilités* (Bertrand, 1889, pp. 19–20), based on André's proof, which we will next describe.

Figure 17.1   Joseph Louis François Bertrand (1822–1900).

necessarily unfavorable, and there are $(m + n - 1)!/[m!(n - 1)!]$ of them. Then André argues that each one of the unfavorable permutations starting with an M can be put into a one-to-one correspondence with a permutation of m M's and $(n - 1)$ N's. Therefore, the total number of favorable permutations is

$$\frac{(m + n)!}{m!n!} - 2\frac{(m + n - 1)!}{m!(n - 1)!} = \frac{(m + n)!}{m!n!} \cdot \frac{m - n}{m + n},$$

whence the formula for $p_{m,n}$ follows. Several authors have stated that André actually used the so-called *reflection principle* in his proof (Doob, 1953, p. 393; Feller, 1968, p. 72; Loehr, 2004), but this is not the case, as we will soon show (see also Renault, 2008; Humphreys, 2010). Let us first explain the reflection principle. Refer to the diagram in Fig. 17.2. The vertical axis X_n denotes the number of votes by which M is ahead of N after a total of n votes on the horizontal axis. The initial point $(0, a)$ above the horizontal axis and the final point (n, b) correspond to candidate M being initially ahead by a votes, and finally ahead by b votes, respectively. *Then the reflection principle states that the number of paths from $(0, a)$ to (n, b) that touch or cross the horizontal axis is equal to the total number of paths from $(0, -a)$ to*

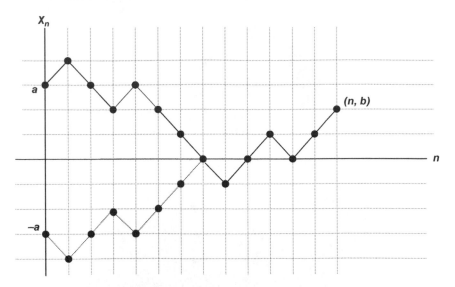

Figure 17.2   The reflection principle. The number of paths from $(0, a)$ to (n, b) which touch or cross the horizontal axis is equal to the total number of paths from $(0, -a)$ to (n, b).

(n, b).[†] The reflection principle has its origin in the works of the physicists Lord Kelvin (1824–1907) and James Clerk Maxwell (1831–1879) (Gray, 1908; Maxwell, 1873). It has since been applied in several fields of probability and statistics, including random walks, Brownian motion, and order statistics.

Using the reflection principle, we can easily solve *Bertrand's Ballot Problem.* As before, we assume that candidates M and N receive a total of m and n votes, where $m > n$. We wish to find the probability $p_{m,n}$ that M is always ahead of N. That is, we want to determine the fraction of the total number of paths from $(0, 0)$ to $(m+n, m-n)$ that always stays above the horizontal axis (except for $(0,0)$). The latter total is the same as the number of paths from $(1, 1)$ to $(m+n, m-n)$ that always stay above the horizontal axis, that is, the total number of paths from $(1, 1)$ to $(m+n, m-n)$ minus the number of paths from $(1, 1)$ to $(m+n, m-n)$ that touch or cross the horizontal axis. Applying the reflection principle, the second quantity in the subtraction is equal to the number of paths from $(1, -1)$ to $(m+n, m-n)$. Note that the number of paths from (i, j) to (k, l) is equal to the number of paths from $(0, 0)$ to $(k - i, l - j)$, which equals $\left(\dfrac{k-i}{\dfrac{k-i+l-j}{2}} \right)$. Therefore, the number of paths above the horizontal axis is

[†] The reflection principle is also discussed in Feller (1968, p. 72), Chuang-Chong and Khee Meng (1992, pp. 91–93), Jacobs (2010, p. 57), Finkelstein and Levin (2001, p. 54), Karlin and Taylor (1975, p. 345), and Khoshnevisan (2007, p. 175).

$$\binom{m+n-1}{m-1} - \binom{m+n-1}{m} = \frac{(m+n)!}{m!n!} \cdot \frac{m-n}{m+n},$$

so that $p_{m,n} = (m-n)/(m+n)$, as required.

We are now in a position to explain why André's method is different from the reflection principle.[†] In the latter, *one* unfavorable path is modified, namely the path that touches or crosses the horizontal axis is modified by reflection. However, in André's method, *two* unfavorable paths are modified, namely one that starts with an M and the other that starts with an N. These are very different procedures, as can be seen when one tries to solve the *Generalized Ballot Problem*:

> *M and N get m and n votes, respectively, such that m > kn, where k is a natural number; find the probability that M is always ahead of N by a factor of at least k.*

Here each vote M gets results in the random walk moving by $(1, 1)$ and each vote N gets results in the random walk moving by $(1, -k)$. A solution using the reflection principle is not possible. This is because an upward movement of one unit is reflected as a downward movement of one unit, but the latter does not make sense here. However, a solution can be obtained by extending André's method, as follows. First, we note that there are $k + 1$ types of "bad" paths, where the bad path of ith ($i = 0, 1, \ldots, k$) type is defined as one for which the first bad step results in the random walk being i units below the x-axis. Now, it can be shown that the total number of bad paths of the ith type is exactly equal to the total number of permutations of m Ms and $(n-1)$ Ns. Therefore, the total number of paths such that M is always ahead of N by a factor of at least k is

$$\binom{m+n}{m} - (k+1)\binom{m+n-1}{m} = \frac{m-kn}{m+n}\binom{m+n}{m}.$$

Hence, we obtain a probability of $(m - kn)/(m + n)$ as the solution to the *Generalized Ballot Problem*.

Finally, a question of priority is in order. Nine years before Bertrand's paper, William Allen Whitworth[‡] (1840–1905) asked the following question (Whitworth, 1878):

> In how many orders can *m* positive units and *n* negative units be arranged so that the sum to any number of terms may never be negative?

Whitworth then gives the solution as $(m + n)!(m - n + 1)/[(m + 1)!n!]$. Whitworth's question is essentially the same as *Bertrand's Ballot Problem*, although the former indirectly asks for the number of ways candidate M is either ahead or on par with candidate N. To see the equivalence between the two problems, let $p_{m,n}^*$ be the

---

[†] See also Renault (2008).

[‡] Author of the popular *Choice and Chance* (Whitworth, 1901).

probability that M with a total of m notes is never behind N with a total of n votes. Then Whitworth's formula gives

$$p^*_{m,n} = \frac{(m+n)!(m-n+1)/[(m+1)!n!]}{(m+n)!/(m!n!)} = \frac{m-n+1}{m+1}.$$

Now the event that M is always ahead of N is equivalent to the event that M gets the first vote and then, *keeping this vote aside*, M is never behind for the remaining $m+n-1$ votes. Therefore,

$$p_{m,n} = \frac{m}{m+n} p^*_{m-1,n}$$

$$= \frac{m}{m+n} \cdot \frac{m-n}{m}$$

$$= \frac{m-n}{m+n},$$

which is the formula given by Bertrand. The *Ballot Theorem* is thus also sometimes rightfully called the *Bertrand–Whitworth Ballot Theorem*.

Problem 18

Bertrand's Strange Three Boxes (1889)

Problem. *There are three boxes, each with two drawers. Box A contains one gold coin in each drawer, box B contains one silver coin in each drawer, and box C contains one gold coin in one drawer and one silver coin in the other.*

(a) *A box is chosen at random. What is the probability that it is box C?*

(b) *A box is chosen at random and a drawer opened at random. The coin is removed. Suppose it is a gold coin. What is the probability that the box chosen is box C?*

Solution. (a) All three boxes have the same chance of being chosen. So, the probability that box C is chosen is 1/3.

(b) Let C be the event "box C is chosen", and similarly for events A and B. Let G be the event "a gold coin is found in one of the drawers from the chosen box". Then

$$\Pr\{A\} = \Pr\{B\} = \Pr\{C\} = \frac{1}{3},$$
$$\Pr\{G|A\} = 1, \quad \Pr\{G|B\} = 0, \quad \Pr\{G|C\} = \frac{1}{2}.$$

Using Bayes' Theorem,[†] we have

$$\Pr\{C|G\} = \frac{\Pr\{G|C\} \cdot \Pr\{C\}}{\Pr\{G|C\} \cdot \Pr\{C\} + \Pr\{G|B\} \cdot \Pr\{B\} + \Pr\{G|A\} \cdot \Pr\{A\}}$$
$$= \frac{(1/2)(1/3)}{(1/2)(1/3) + (0)(1/3) + (1)(1/3)}$$
$$= \frac{1}{3}.$$

[†] See **Problem 14**

18.1   Discussion

Although the name of Joseph Louis François Bertrand (1822–1900) is usually associated with the chord paradox (see **Problem 19**), his treatise *Calculs des Probabilités*[†] (Bertrand, 1889) is a treasure-trove of several interesting probability problems. Of these, the box problem appears as the second problem in the book.[‡] However, the second part of the original question is in a slightly different form from that presented in (b) in the *Problem*. On p. 2 of his book, Bertrand says

> A box is chosen. A drawer is opened. Whatever the coin I find, there are two cases that remain possible. The drawer that was not opened either contains a similar or different coin from the first one. Out of these two possibilities, only one is favorable to the box with two different coins. The probability of having chosen the box with the different coins is thus 1/2.

> Is it possible, however, that the mere opening of a drawer could change the probability from 1/3 to 1/2?

Note that the above problem does not identify the initial coin that is drawn, as opposed to our question in (b) above where it is identified as gold. Nevertheless the answer to Bertrand's question is the same in both cases: the mere opening of a drawer cannot change the probability from 1/3 to 1/2. The (intentional) error in Bertrand's reasoning is this: whatever the coin that is first drawn, there are two cases that remain possible; in accordance with the principle of indifference,[§] each one should have the same probability. However, the two cases are not equally likely. The one that is favorable to the box with two similar coins is twice as likely as the one that is favorable to the box with two different coins. We have shown this in part (b) of the solution for the case when the first coin drawn is gold. We would have obtained the same probability of 1/3 if the first coin drawn was silver.

Bertrand's box problem is important not only because its solution can be counterintuitive but also because it is in fact a precursor to the famous *Monty-Hall Problem*, which we will discuss in **Problem 32.**

We now consider the following classic problem, which is similar to Bertrand's but which has an entirely different solution:

(a) *Mr. Smith has two children and at least one of them is a boy.*

(b) *Mr. Smith meets a friend and tells her, "I have two children and the older is a boy."*

In each case, what is the probability that Mr. Smith has two boys?

To solve the above problem, we first define Y as the event "the younger child is a boy" and O as the event "the older child is a boy." Then $\Pr\{Y\} = \Pr\{O\} = 1/2$

[†] Calculus of Probabilities.

[‡] The Bertrand Box Problem is also discussed in Shafer and Vovk (2006), Shackel (2008), and Greenblatt (1965, p. 107).

[§] See pp. 135–137.

and, since Y and O are independent of each other, $\Pr\{Y \cap O\} = 1/4$. For part (a), we have

$$\Pr\{Y \cap O \mid Y \cup O\} = \frac{\Pr\{(Y \cap O) \cap (Y \cup O)\}}{\Pr\{Y \cup O\}}$$

$$= \frac{\Pr\{Y \cap O\}}{\Pr\{Y\} + \Pr\{O\} - \Pr\{Y \cap O\}}$$

$$= \frac{1/4}{1/2 + 1/2 - 1/4}$$

$$= \frac{1}{3}.$$

For part (b), we have

$$\Pr\{Y \cap O \mid O\} = \frac{\Pr\{(Y \cap O) \cap O\}}{\Pr\{O\}}$$

$$= \frac{\Pr\{Y \cap O)\}}{\Pr\{O\}}$$

$$= \frac{1/4}{1/2}$$

$$= \frac{1}{2}.$$

Intuitively, the sample space for the problem is shown in Table 18.1 (B = boy, G = girl). The sample space is $\Omega = \{\text{BB, BG, GB, GG}\}$, where each sample point is equally likely. Given that Mr. Smith has at least one boy, the new sample space is $\Omega^* = \{\text{BB, BG, GB}\}$, where the first and second letters for each sample point represent the gender of the younger and older children, respectively. The probability of two boys is therefore 1/3. On the other hand, given that we identify the older child to be a boy, the sample space changes to $\Omega^{**} = \{\text{BB, GB}\}$. The probability of two boys is now 1/2.

It might at first seem counterintuitive that the second probability should be different from the first. However, by specifying one of the children to be the oldest, the

Table 18.1   Sample Space for the "Older Child" Problem

Younger child	Older child
B	B
B	G
G	B
G	G

sample space changes. Consequently the probability cannot stay at 1/3. Martin Gardner (1914–2010), the Dean of mathematical puzzles, provides a most interesting illustration of why that should be so (1959, p. 51):

> If this were not the case [i.e. if the second probability did not change from 1/3 to 1/2] we would have a most ingenious way to guess the face of a concealed coin with better than even odds. We would simply flip our own coin. If it came heads we would reason: "There are two coins here and one of them (mine) is heads. The probability the other is heads is therefore 1/3, so I will bet that it is tails." The fallacy of course is that we are specifying which coin is heads. This is the same as identifying the oldest child as the boy, and it changes the odds in a similar fashion.

Problem 19

Bertrand's Chords (1889)

***P**roblem.* *A chord is randomly chosen on a circle. Calculate the probability that the chord is longer than the side of an equilateral triangle inscribed inside the circle.*

***S**olution.* There are at least three methods a chord can be randomly chosen on the circle, as shown in Fig. 19.1.

Solution 1: Random Endpoints Approach. Consider the first figure. We randomly choose two endpoints on the circumference of the circle and join them to obtain the chord. Because of symmetry, we assume one of the points coincides with one of the vertices of the inscribed triangle (at A). The remaining two vertices (B and C) of the triangle then divide the circumference of the circle into three equal arcs (AC, CB, and BA). The chord is longer than the side of the triangle if and only if it intersects the triangle. The required probability is thus 1/3.

Solution 2: Random Radius Approach. Consider the second figure. We randomly choose a radius of the circle and then select a point on it. The chord is obtained by drawing a perpendicular to the radius across the point. Because of symmetry, we assume the radius is perpendicular to one of the edges of the triangle. The chord is longer than the side of the triangle if and only if the selected point lies on the half of the radius that is closer to the center of the circle. The required probability is thus 1/2.

Solution 3: Random Midpoint Approach. Consider the third figure. We randomly choose a point inside the circle and obtain the chord so that the chosen point is its midpoint. The chord is longer than the side of the triangle if and only if the chosen point lies inside the inscribed circle of the triangle. Since the inner inscribed circle has radius half that of the outside circle, the required probability is $(1/2)^2 = 1/4$.

Classic Problems of Probability, Prakash Gorroochurn.
© 2012 John Wiley & Sons, Inc. Published 2012 by John Wiley & Sons, Inc.

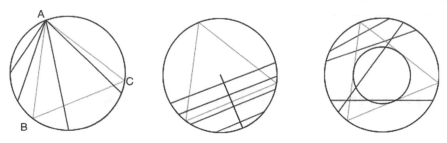

Figure 19.1   Three of the ways in which a chord can be randomly chosen on a circle.

19.1   Discussion

This is the famous problem that comes to mind when the name of Joseph Bertrand is mentioned in probability.[†] It features as the fifth problem in the *Calculs des Probabilités* (Bertrand, 1889, p. 4) (see Fig. 19.2) and has gained nearly immortal status because it apparently shows the limitations of classical probability when applied to problems where the number of possible outcomes is infinite (see, for example, Papoullis, 1991, p. 9). Shafer and Vovk (2006) explain that Bertrand's paradoxes were one of the reasons for the dissatisfaction with classical probability at the turn of the twentieth century,[‡] and such dissatisfaction provided an impetus for the later axiomatization of probability by Kolmogorov in 1933.[§] Indeed, the eminent German mathematician David Hilbert (1862–1943) gave a keynote talk, "Mathematical Problems," in 1900 at the Second International Congress of Mathematicians in Paris. In his talk, he outlined some of the most important unsolved problems of mathematics at that time.[**] Hilbert recognized the need for a rigorous and coherent theory of probability, for his sixth problem reads (Yandell, 2002, pp. 159–160)

> The investigations on the foundations of geometry suggest the problem: To treat in the same manner [as Hilbert had treated geometry], by means of axioms, those physical sciences in which mathematics play an important part; in the first rank are the theory of probabilities and mechanics.

Coming back to our original *Chord Problem*, Bertrand gave an argument which has become the standard explanation of the paradox. In his own words (Bertrand, 1889, p. 4):

[†] Bertrand's paradox is also discussed in Clark (2007, p. 24), Shackel (2008), Ross (1997, pp. 203–204), Erickson and Fossa (1998, p. 21), von Mises (1981, p. 77), Yaglom and Yaglom (1987, pp. 34–35), Papoullis (1991, p. 9), Székely (1986, p. 43), Grinstead and Snell (1997, p. 49), Andel (2001, p. 8), Stirzaker (2003, p. 324), Kaplan and Kaplan (2006, p. 56), and Stapleton (2008, p. 55).

[‡] See also the discussion in Tanton (2005, p. 417).

[§] See **Problem 23**.

[**] These later became part of Hilbert's "program" of 23 canonical problems.

4   CALCUL DES PROBABILITÉS.

Pierre est diminuée. Parmi les quatre cas possibles, ceux dans lesquels Pierre vaincu dans un coup est vainqueur dans l'autre sont moins vraisemblables que ceux dans lesquels ses deux boules ont le même sort.

4. Une remarque encore est nécessaire : l'infini n'est pas un nombre; on ne doit pas, sans explication, l'introduire dans les raisonnements. La précision illusoire des mots pourrait faire naître des contradictions. Choisir *au hasard*, entre un nombre infini de cas possibles, n'est pas une indication suffisante.

On demande, par exemple, la probabilité pour qu'un nombre, entier ou fractionnaire, commensurable ou incommensurable, choisi *au hasard* entre o et 100, soit plus grand que 5o. La réponse semble évidente : le nombre des cas favorables est la moitié de celui des cas possibles. La probabilité est $\frac{1}{2}$.

Au lieu du nombre, cependant, on peut choisir son carré. Si le nombre est compris entre 5o et 100, le carré le sera entre 2500 et 10000.

La probabilité pour qu'un nombre choisi *au hasard* entre o et 10000 surpasse 2500 semble évidente : le nombre des cas favorables est les trois quarts du nombre des cas possibles. La probabilité est $\frac{3}{4}$.

Les deux problèmes sont identiques. D'où vient la différence des réponses? Les énoncés manquent de précision.

Les contradictions de ce genre peuvent être multipliées à l'infini.

5. On trace *au hasard* une corde dans un cercle. Quelle est la probabilité pour qu'elle soit plus petite que le côté du triangle équilatéral inscrit?

On peut dire : si l'une des extrémités de la corde est connue, ce renseignement ne change pas la probabilité; la symétrie du cercle ne permet d'y attacher aucune influence, favorable ou défavorable à l'arrivée de l'événement demandé.

L'une des extrémités de la corde étant connue, la direction doit être réglée par le hasard. Si l'on trace les deux côtés du triangle équilatéral ayant pour sommet le point donné, ils forment entre eux et avec la tangente trois angles de 6o°. La corde, pour être

CHAP. I. — ÉNUMÉRATION DES CHANCES.   5

plus grande que le côté du triangle équilatéral, doit se trouver dans celui des trois angles qui est compris entre les deux autres. La probabilité pour que le hasard entre trois angles égaux qui peuvent le recevoir le dirige dans celui-là semble, par définition, égale à $\frac{1}{3}$.

On peut dire aussi : si l'on connaît la direction de la corde, ce renseignement ne change pas la probabilité. La symétrie du cercle ne permet d'y attacher aucune influence, favorable ou défavorable à l'arrivée de l'événement demandé.

La direction de la corde étant donnée, elle doit, pour être plus grande que le côté du triangle équilatéral, couper l'un ou l'autre des rayons qui composent le diamètre perpendiculaire, dans la moitié la plus voisine du centre. La probabilité pour qu'il en soit ainsi semble, par définition, égale à $\frac{1}{2}$.

On peut dire encore : choisir une corde au hasard, c'est en choisir au hasard le point milieu. Pour que la corde soit plus grande que le côté du triangle équilatéral, il faut et il suffit que le point milieu soit à une distance du centre plus petite que la moitié du rayon, c'est-à-dire à l'intérieur d'un cercle quatre fois plus petit en surface. Le nombre des points situés dans l'intérieur d'une surface quatre fois moindre est quatre fois moindre. La probabilité pour que la corde dont le milieu est choisi au hasard soit plus grande que le côté du triangle équilatéral semble, par définition, égale à $\frac{1}{4}$.

Entre ces trois réponses, quelle est la véritable? Aucune des trois n'est fausse, aucune n'est exacte, la question est mal posée.

6. On choisit au hasard un plan dans l'espace; quelle est la probabilité pour qu'il fasse avec l'horizon un angle plus petit que $\frac{\pi}{4}$?

On peut dire : tous les angles sont possibles entre o et $\frac{\pi}{2}$, la probabilité pour que le choix tombe sur un angle inférieur à $\frac{\pi}{4}$ est $\frac{1}{2}$.

On peut dire aussi : par le centre d'une sphère, menons un rayon perpendiculaire au plan en question. Choisir le plan au

Figure 19.2   Bertrand's Chord Problem, taken from the *Calculs des Probabilités* (Bertrand, 1889).

A further remark is necessary: the infinite is not a number; one should not, without any explanation, introduce it into one's reasoning. The illusion of precision brought about by words can result in contradictions. Choosing at random from infinitely many possibilities is not sufficiently defined.

The paradox thus stems from the fact that there are many ways to choose from an infinite sample space. The problem consequently appears ill-posed: it does not specify *how* the random chord should be chosen. Thus, each of the solutions provided is valid *for the given method the chord was chosen*.

The three different answers also seem to seriously undermine the *principle of indifference.*[†] This principle was first used by Jacob Bernoulli (1654–1705) and Pierre-Simon Laplace (1749–1827). It states that if we have no *a priori* knowledge of how an event can occur, then we must assign equal probabilities to each possible way it can occur. The principle is contradicted here, for we have shown three different ways of applying it, and each results in a different answer. Hunter and Madachy (1975, pp. 100–102), give a fourth possible solution to the Bertrand *Chord Problem*, with a probability of approximately 14/23 that the chord is longer than the side of the triangle. Weaver (1982, p. 355) provides two other possible answers (2/3 and 3/4).

[†] See **Problem 14**, pp. 135–137.

Berloquin (1996, p. 90) mentions a seventh possible answer $(1 - \frac{1}{2}\sqrt{3})$, and provides a proof that the problem possesses infinitely many solutions.

After Bertrand had adduced the *Chord Problem*, geometric probability was seemingly dealt a serious blow. However, the problem was taken up again a few years later by the eminent French mathematician Henri Poincaré (1854–1912) in his own *Calculs des Probabilités*[†] (Poincaré, 1912, p. 118). The invaluable contribution made by Poincaré (1912, p. 130) was to suggest that the arbitrariness in the interpretation of randomness could be removed by associating probability densities that remain invariant under Euclidean transformations. That is to say, the problem could become well-posed if we could find probability densities for random variables such that we would get the same answer if the circle and inscribed triangle were to be rotated, translated, or dilated. Poincaré (1912, p. 130), and later Jaynes (1973), concluded that such probability densities exist only for the random radius method (Figs. 19.3 and 19.4). Let us now explain Jaynes' approach and show how invariance arguments lead to a probability of 1/2.[‡] Let the circle C have radius R and let the randomly drawn chord have its midpoint at (X, Y), relative to the center O of the circle. Suppose the probability density of (X, Y) is $f(x, y)$, where $(x, y) \in C$. We next draw a smaller concentric circle C' with radius aR, where $0 < a \leq 1$. Let the density of (X, Y) be $h(x, y)$ for $(x, y) \in C'$. The two densities f and h should be the same except for a constant factor, because their supports are different. For $(x, y) \in C'$, we have

$$h(x, y) = \frac{f(x, y)}{\iint_{C'} f(x, y) dx \, dy}.$$

Rotational invariance implies that both $f(x, y)$ and $h(x, y)$ depend on (x, y) only through the radial distance $r = \sqrt{x^2 + y^2}$. Therefore, in polar coordinates (r, θ), the above equation becomes

$$h(r) = \frac{f(r)}{\int_0^{2\pi} \int_0^{aR} rf(r) dr \, d\theta}$$

$$= \frac{f(r)}{2\pi \int_0^{aR} rf(r) dr}$$

$$\therefore f(r) = 2\pi h(r) \int_0^{aR} f(r) r \, dr. \tag{19.1}$$

[†] Calculus of Probabilities. Strangely, Poincaré discussed only the first two of the three solutions Bertrand had given.

[‡] See Jaynes (2003, pp. 388–393), and also Howson and Urbach (2006, pp. 282–286), and Székely (1986, pp. 44–47).

Figure 19.3   Henri Poincaré (1854–1912).

We now invoke scale invariance. Imagine the larger circle is shrunk by a factor a, $0 < a \leq 1$, to give the smaller circle. Then scale invariance implies

$$f(r)r \, dr \, d\theta = h(ar)ar \, d(ar)d\theta$$

$$f(r)r \, dr \, d\theta = a^2 h(ar)r \, dr \, d\theta$$

$$\therefore f(r) = a^2 h(ar). \tag{19.2}$$

From Eqs. (19.1) and (19.2), we have for $0 < a \leq 1$ and $0 < r \leq R$,

$$a^2 f(ar) = 2\pi f(r) \int_0^{aR} rf(r)dr.$$

Figure 19.4   Edwin T Jaynes (1922–1998).

If we differentiate the last equation with respect to a, then set $a = 1$, we obtain the differential equation

$$2f(r) + rf'(r) = 2\pi R^2 f(r)f(R)$$

with general solution

$$f(r) = \frac{qr^{q-2}}{2\pi R^q},$$

where q is a positive constant. It can be shown that translational invariance implies $q = 1$. Hence we finally obtain

$$f(r) = \frac{1}{2\pi Rr}, \quad 0 < r \le R, 0 \le \theta \le 2\pi. \tag{19.3}$$

Assuming the above invariance arguments, the probability that the chord is longer than the side of the inscribed triangle can be calculated as

$$\int_{0}^{2\pi R/2}\int_{0}^{} f(r)r\,dr\,d\theta = \int_{0}^{2\pi R/2}\int_{0}^{} \frac{1}{2\pi Rr}r\,dr\,d\theta$$

$$= (2\pi)\left(\frac{R}{2}\right)\cdot\frac{1}{2\pi R}$$

$$= \frac{1}{2}.$$

Thus, adherence to the invariance principle results in an answer of half. Moreover, this is the answer that Jaynes obtained when he carried out experiments of actually throwing straws into a circle (Jaynes, 1973). However, we hasten to add that not all situations where the principle of indifference breaks down are amenable to the kind of invariance treatment we provided above, as Jaynes himself conceded.[†] Additionally, debate on the *Chord Problem* is not over: in 2007, Shackel published a paper casting some doubts on Jaynes' solution and maintained that the *Chord Problem* remains still unsolved (Shackel, 2007). Shackel's main contention is that Jaynes' solution is not to the original Bertrand *Chord Problem* and that it amounts to "substituting a restriction of the problem for the general problem."

[†] For example, von Mises water-and-wine example, mentioned on p. 136, is not resolved by using invariance.

Problem 20

Three Coins and a Puzzle from Galton (1894)

Problem. *Three fair coins are tossed. What is the probability of all three coins turning up alike?*

Solution. The sample space $\Omega = \{HHH, HHT, \ldots, TTT\}$ is made up of eight equally likely outcomes. Of these, all three coins turn up alike with either *HHH* or *TTT*. The required probability is therefore 2/8 = 1/4.

20.1 Discussion

Sir Francis Galton (1822–1911) is considered to be the father of eugenics and one of the most renowned statisticians of his time (Fig. 20.1). His name is usually associated with the notions of correlation and the regression effect.[†] In the February 15, 1894 issue of *Nature*, Galton presented both the correct solution and an intentionally wrong solution to the above problem (Galton, 1894). Concerning the wrong solution, he writes

> I lately heard it urged, in perfect good faith, that as at least two of the coins must turn up alike, and it is an even chance whether a third coin is heads or tails; therefore the chance of being all-alike is as 1 to 2, and not as 1 to 4. Where does the fallacy lie?

[†] Also known as regression to the mean. Galton noted that tall parents tend to produce tall offspring but who, on average, are shorter than their parents. Similarly, tall offspring tend to have tall parents but who, on average, are shorter than their offspring. The regression effect thus describes the natural phenomenon of a variable that is selected as extreme on its first measurement and tends to be closer to the center of the distribution for a later measurement. For more details, see, for example, Campbell and Kenny (1999) who devote a whole book to the subject.

Classic Problems of Probability, Prakash Gorroochurn.
© 2012 John Wiley & Sons, Inc. Published 2012 by John Wiley & Sons, Inc.

Figure 20.1   Sir Francis Galton (1822–1911).

The argument does *seem* to contain an infallible logic. The fallacy, however, lies in the assumption that, given that at least two coins turn up alike, then there is an even chance for the remaining coin to be alike too. We can show this as follows. To fix ideas, let us consider the situation for heads.

Let U be the event "at least two coins are heads," and V be the event "the last coin is a head." Then,

Table 20.1   Illustration of Why the Probability of Obtaining Three Heads, Given that at Least Two Heads have been Obtained, is 1/4 Instead of 1/2

First coin	H	H	H	H	T	T	T	T
Second coin	H	H	T	T	H	H	T	T
Third coin	H	T	H	T	H	T	H	T
	Case 1	Case 2	Case 3	Case 4	Case 5	Case 6	Case 7	Case 8

$$
\begin{aligned}
\Pr\{V\,|\,U\} &\equiv \frac{\Pr\{V \cap U\}}{\Pr\{U\}} \\
&= \frac{\Pr\{\text{three heads}\}}{\Pr\{\text{two heads}\} + \Pr\{\text{three heads}\}} \\
&= \frac{(1/2)^3}{3(1/2)^2(1/2) + (1/2)^3} \\
&= \frac{1}{4}.
\end{aligned}
$$

Similarly for tails. Thus, given that at least two coins turn up alike, it is three times more likely for the remaining coin to be different than to be alike too.

A more visual proof is provided by Northrop (1944, p. 238) (see Table 20.1). Suppose at least two of three coins are heads. As can be seen from Table 20.1, this happens in four of the eight equally likely cases, namely cases 1, 2, 3, and 5. However, only in one of these four cases do we have all three heads, namely case 1. Thus, once we condition on at least two of the coins being the same, the probability of the third coin being the same is 1/4, not 1/2.

Problem 21

Lewis Carroll's Pillow Problem No. 72 (1894)

Problem. *A bag contains two counters that are each equally likely to be either black or white. Can you ascertain their colors without taking them out of the bag?*

Solution. The problem is ill-posed: it is impossible to ascertain the colors of the counters from the given information.

21.1 Discussion

This problem was adduced by the famous English writer, logician, and mathematician Lewis Carroll[†] (1832–1898) in his book *Pillow Problems*[‡] (Dodgson, 1894) (Figs. 21.1 and 21.2). The latter contains a total of 72 problems of which 13 are of a probabilistic nature. Of all these problems, Problem 72[§] has attracted the most attention. It appears under the rather mysterious heading *Transcendental Probabilities*. In the preface of his book, Carroll says (p. xiv)

> If any of my readers should feel inclined to reproach me with having worked too uniformly in the region of Common-place, and with never having ventured to wander out of the beaten tracks, I can proudly point to my one problem in Transcendental Probabilities - a subject on which very little has yet been done by even the most enterprising of mathematical explorers. To the casual reader it may seem abnormal or even paradoxical, but I would have such a reader ask himself candidly the question, 'Is not life itself a paradox'?

[†] Pseudonym used by Charles Lutwidge Dodgson.

[‡] So-called because Carroll apparently conceptualized and solved them at night while in bed.

[§] Carroll's Pillow Problem 72 is also discussed in Weaver (1956), Seneta (1984, 1993), Pedoe (1958, p. 43), Mosteller (1987, p. 33), and Barbeau (2000, p. 81).

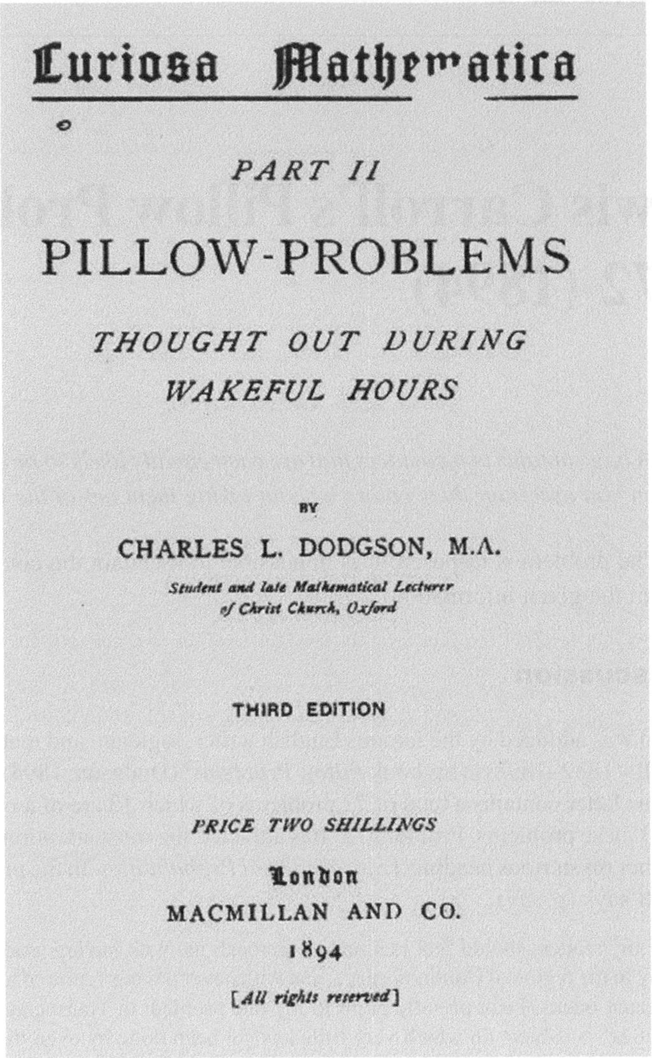

Curiosa Mathematica

○

PART II

PILLOW-PROBLEMS

THOUGHT OUT DURING

WAKEFUL HOURS

BY

CHARLES L. DODGSON, M.A.

*Student and late Mathematical Lecturer
of Christ Church, Oxford*

THIRD EDITION

PRICE TWO SHILLINGS

London

MACMILLAN AND CO.

1894

[All rights reserved]

Figure 21.1   Cover of the Third edition of the *Pillow Problems*.

On p. 27 of his book, Carroll gives the wrong answer to **Problem 21**:

One is black, and the other is white.

Subsequently, Carroll goes on to provide an erroneous argument why the bag *must* contain one black and one white counter. His proof is as follows (Dodgson, 1894, p. 109):

We know that, if a bag contains 3 counters, 2 being black and one white, the chance of drawing a black one would be 2/3; and that of any *other* state would *not* give this chance.

Figure 21.2 Lewis Carroll (1832–1898).

Now the chances, that the given bag contains (α) *BB*, (β) *BW*, (γ) *WW*, are respectively 1/4, 1/2, 1/4.

Add a black counter.

Then the chances, that it contains (α) *BBB*, (β) *BWB*, (γ) *WWB*, are respectively 1/4, 1/2, 1/4.

Hence the chance, of now drawing a black one,

$$= \frac{1}{4} \cdot 1 + \frac{1}{2} \cdot \frac{2}{3} + \frac{1}{4} \cdot \frac{1}{3} = \frac{2}{3}.$$

Hence the bag now contains *BBW* (since any *other* state of things would *not* give this chance).

Hence before the black counter was added, it contained *BW*, i.e. one black counter one white.

Q.E.F.[†]

Carroll has received some flak for his reasoning. Weaver (1956) in particular has been quite harsh:

> Dodgson's solutions of the problems in this collection are generally clever and accurate, but one of them [Problem 72] ludicrously exposes the limitations in his mathematical thinking. . . .

An earlier author, Eperson (1933) writes

> But for this serious tone, one would have suspected the author [Carroll] of indulging in a little leg-pulling at the expense of his readers, since it is curious that so logical a mind as his could have overlooked the fallacy in his solution.

On the other hand, Gardner claims that Carroll intentionally gave the wrong answer to Problem No. 72 as a joke (Gardner, 1996, p. 64):

> The [Carroll's] proof is so obviously false that it is hard to comprehend how several top mathematicians could have taken it seriously and cited it as an example of how little Carroll understood probability theory! There is, however, not the slightest doubt that Carroll intended it as a joke.

Although the latter argument seems less likely, one has to admit that Carroll's reasoning indeed seems very convincing. Through an example, let us see where the error lies. Suppose a bag contains only one counter which is equally likely to be white or black. The probabilities the bag contains B and W are 1/2 and 1/2. We now add one black counter to the bag so that we now have two counters. The probabilities of BB and WB are 1/2 and 1/2. Therefore, the probability of drawing a black counter is $(1)(1/2) + (1/2)(1/2) = 3/4$. Using Carroll's reasoning, this would imply that the bag must contain three black and one white counter, a total of four counters. But this cannot be correct because the bag now contains only two counters! The error in this example lies in the inference that was made from the probability of 3/4. This cannot tell us anything about the actual content of the bag. What the probability of 3/4 tells us is the following. Suppose we repeat an experiment of putting a counter (which half of the time is white and half of the time is black) in a bag, followed by adding a black counter. Then when a counter is taken out of the bag which now contains two counters, it will be black 3/4 of the time.

Similarly, in Carroll's case, the probability of 2/3 of drawing a black counter does not imply that the bag must contain two black and one white counter. Suppose we repeat the following experiment a large number of times: we put two counters (each of which is equally likely to be black or white) in a bag, then we add a third black counter, and finally we count the number of black counters. Then, we can say that in the long run, the average number of back counters we will count is two. But for a given experiment the number will not necessarily be two. Therefore, it is impossible to make the conclusion that, before the black counter was added, there must have been one

---

[†] Q.E.F. means "quod erat faciendum" which is the Latin for "which had to be done". More usually, Q.E.D. is used after a proof and means "quod erat demonstrandum" or "which had to be proved."

white and one black counter in the bag. What we can say is that the *expected* number of white and back counters is each one.

Hunter and Madachy (1975, pp. 98–99) and Northrop (1944, p. 192) give the correct explanation to the Carroll puzzle. However, many authors have either given the wrong solution or have contented themselves with saying that Carroll's reasoning is "obviously" wrong. Eperson, for example, states (1933):

> As I expect you would prefer to enjoy that feeling of intellectual superiority by discovering the fallacy for yourself, I will refrain from pointing it out, but will content myself with remarking that if one applies a similar argument to the case of a bag containing 3 unknown counters, black or white, one reaches the still more paradoxical conclusion that there cannot be 3 counters in the bag!

Similarly, Seneta (1984; 1993) and Székely (1986, p. 68) do not give a definite resolution of the issue. Finally, Vakhania (2009) claims:

> Lewis Carroll's proof of his problem No.72 is really ambiguous, and it has been tacitly accepted that the proof was incorrect. However, it seems that no clear and convincing proof of the incorrectness has been given, and the question "where is the fallacy" continues to be of interest and even challenging for many years already.

The fallacy, however, ceases to be challenging once one realizes that the calculated probability of 2/3 can only give the expected, not the actual, number of counters.

Problem 22

Borel and a Different Kind of Normality (1909)

Problem. *A real number in [0,1] is said to be normal[†] in a given base if any finite pattern of digits occurs with the same expected frequency when the number is expanded in that particular base. Thus a number in [0,1] is normal in base 10 if, in its decimal expansion, each of the digits {0,1, . . ., 9} occurs with frequency 1/10, each of the pairs {00, 01, . . ., 99} occurs with frequency 1/100, and so on. Prove that almost every real number in [0,1] is normal for all bases.*

Solution. Let the random variable ξ_j ($\xi_j = 0, 1, \ldots, 9; j = 1, 2, \ldots$) be the jth digit in the decimal expansion of $x \in [0, 1]$, that is

$$x = \frac{\xi_1}{10} + \frac{\xi_2}{10^2} + \cdots + \frac{\xi_j}{10^j} + \cdots$$

Then, for the digit $b = 0, 1, \ldots, 9$, we define the binary random variable

$$\delta_{b\xi_j} = \begin{cases} 1, & b = \xi_j \text{ with prob. } 1/10, \\ 0, & b \neq \xi_j \text{ with prob. } 9/10. \end{cases}$$

It is reasonable to assume that the $\delta_{b\xi_j}$s are independent (across the j 's)[‡] and identically distributed. Applying the Strong Law of Large Numbers (SLLN),[§] we have

---

[†] Not to be confused with the usual sense in which "normal" is used in statistics to mean following a Gaussian distribution.

[‡] Although this might seem a strong assumption, see Chung (2001, p. 60) for a theoretical justification.

[§] See pp. 72–75.

Classic Problems of Probability, Prakash Gorroochurn.
© 2012 John Wiley & Sons, Inc. Published 2012 by John Wiley & Sons, Inc.

$$\Pr\left\{\lim_{n\to\infty}\frac{\delta_{b\xi_1}+\delta_{b\xi_2}+\cdots+\delta_{b\xi_n}}{n}=\frac{1}{10}\right\}=1.$$

Thus, $(\delta_{b\xi_1}+\delta_{b\xi_2}+\cdots+\delta_{b\xi_n})/n$ converges almost surely to 1/10 (i.e., except for some set with probability zero), that is, the relative frequency of the digit b ($=0$, 1, ..., 9) of almost all $x\in[0,1]$ is 1/10.

We next consider blocks of two digits in the decimal expansion of x. Consider the independent random variables $\delta_{b_1\xi_1}\delta_{b_2\xi_1},\delta_{b_1\xi_2}\delta_{b_2\xi_2},\ldots$, where $b_1,b_2,\xi_1,\xi_2=0,1,\ldots,$ 9. By applying the SLLN again, we see that the relative frequency of the pair $b_1b_2(=00,$ 01, ..., 99) is 1/100 for almost all $x\in[0,1]$. A similar proof applies for all blocks of larger digits. Likewise, similar results follow if we expand x in all base systems other than decimal. Hence, we have proved that almost every real number in [0,1] is normal.

22.1 Discussion

Emile Borel (1871–1956) is regarded as the father of measure theory and one of the key figures in the development of the modern theory of probability (Fig. 22.1). In 1909, he published a landmark paper that contained several key results in probability theory (Borel, 1909b). These included the first version of the SLLN,[†] a theorem on continued fractions, and finally Borel's normal law. This states that almost every real number in [0,1] is normal.[‡] Borel's normal law turns out to be a corollary to the SLLN and will be the topic of interest here.

Two comments should be made here. First, although Borel's results were groundbreaking, his proofs of these results had some slight deficiencies. Writing on the historical development of rigor in probability, the eminent mathematician Joseph Leo Doob (1910–2004) remarked (Doob, 1996) (Fig. 22.2)

> ...but a stronger mathematical version of the law of large numbers was the fact deduced by Borel - in an unmendably faulty proof[§] - that this sequence of averages converges to 1/2 for (Lebesgue measure) almost every value of x.

Borel seemed to have been aware of these limitations, and later both Faber (1910) and Hausdorff (1914) filled in the gaps in his proofs.

Second, although almost every real number in [0,1] is normal so that there are infinitely many such numbers, finding them has been hard.[**] This is indeed surprising.

[†] See **Problem 8**.

[‡] Borel's normal numbers are also discussed in Niven (1956, Chapter 8), Kac (1959, p. 15), Khoshnevisan (2006), Marques de Sá (2007, pp. 117–120), Richards (1982, p. 161), Rényi (2007, p. 202), Billingsley (1995, pp. 1–5), and Dajani and Kraikamp (2002, p. 159).

[§] In particular, Borel assumed a result based on independence for dependent trails, and used a more precise version of the Central Limit Theorem than was available at that time (Barone and Novikoff, 1978).

[**] Although it is suspected that numbers such as π, e, and $\sqrt{2}$ are normal.

Figure 22.1 Emile Borel (1871–1956).

Consider the countably infinite set of rational numbers. Every rational number ends with an infinite tail of zeros or ones, or with a periodic sequence, and therefore cannot be normal. For instance, 2/3 is .6666... in decimal and .101010... in binary. In decimal, 2/3 cannot be normal because only the digit 6 and the blocks {66, 666, ...} are present. In binary, the bits {0, 1} occur with the same frequency, but none of the blocks {00, 11, 000, ...} is present. Thus, 2/3 is not normal in binary, either. However, 2/3 is said to be *simply normal*[†] in binary because each of the bits {0, 1} occurs with a relative frequency of .5. It is therefore possible for a number to be simply normal in a given base but not normal at all in another base. An example of a normal number was provided by the English economist David Gawen Champernowne (1912–2000)

[†] In general a number is said to be simply normal in base r if each of the digits 0, 1, ..., $r-1$ occurs with a relative frequency $1/r$. Pillai (1940) has shown that a number is normal in base r if and only if it is simply normal in each of the bases $r, r^2, r^3, ...$

Figure 22.2   Joseph Leo Doob (1910–2004). [Picture licensed under the Creative Commons Attribution-Share Alike 2.0 Germany license.]

(Champernowne, 1933). He systematically concatenated numbers and blocks in a given base, giving a normal number in that base. In base 10 and base 2, for example,

$$C_{10} = .123456789 \, 10 \, 11 \, 12 \, 13 \, 14 \, 15 \, 16 \ldots$$
$$C_2 = .01 \, 00 \, 01 \, 10 \, 11 \, 000 \, 001 \, 010 \, 011 \, 101 \ldots$$

The systematic construction of these numbers means that randomness is not necessary for normality, although it is sufficient (Kotz et al., 2004, p. 5673).

Of the several bases into which a real number in [0, 1] can be expanded, the binary system is of particular interest. Borel observed that the binary expansion mimics the tossing of a fair coin.[†] The bits in the expansion can be viewed as independent random variables that represent the outcomes when a fair coin is tossed. This fact turns out to be helpful in the following classic problem[‡]:

[†] More precisely, almost every number in [0, 1] can be expanded in base two to represent the behavior of a fair coin. See also Billingsley (1995, pp. 1–5) and Marques de Sá (2007, pp. 117–120).

[‡] Bruce Levin first alerted me to this type of problem. See also Halmos (1991, p. 57).

Two players A and B use a fair coin to decide who wins a prize. Devise a procedure such that player A wins with probability $1/\pi$.

The problem seems all the more intriguing because one would wonder what the number π has to do with the toss of coins. To solve it, first note that $1/\pi$ is a real number in [0,1] with binary expansion .01010001011... This represents the reference sequence *THTHTTTHTHH...*, if we use "0" to denote a tail (T) and "1" to denote a head (H). Now we can imagine player A as winning with probability $1/\pi$ if she picks a number at random from [0,1] and obtains a number less than $1/\pi^{\dagger}$. This means that she obtains a number less than .01010001011... in binary. We thus have the solution to our problem, as follows. Toss the fair coin and record the actual sequence of heads and tails. Stop as soon as this sequence shows a T but the reference sequence *THTHTTTHTHH...* has an H at the corresponding position (for example, stop on the 4th throw if the actual sequence of tosses is *THTT*). This is equivalent to picking a number from [0,1] and obtaining a number less than .01010001011... in binary. This procedure ensures that A will win with probability exactly $1/\pi$. Of course, $1/\pi$ is incidental here: by a similar binary expansion we could devise a procedure such that the probability of winning is *any* real number in [0,1] (such as $1/e$, $1/\sqrt{2}$, and so on).

Finally, let us use the binary expansion of a real number in [0,1] to explain the connection between the Borel's normal numbers and Borel's strong law of large numbers (SLLN).[‡] Imagine a coin for which the probability of a head is p is tossed repeatedly, resulting in a sequence of heads and tails. Suppose the experiment is repeated several times so that many such sequences are obtained. On p. 74, we stated that, according to Borel's SLLN, the set of infinite sequences for which the proportion of heads is not equal to p has probability zero. The question we then asked was, "What type of sequences would behave in this way, that is, their proportion of heads would not equal p (and the set of such sequences would have probability zero)?" The answer is this: the set of all sequences of heads and tails that correspond to the binary expansion of a real number in [0, 1] such that the proportion of 1's in the expansion is not p will have probability zero. For a fair coin ($p = 1/2$), these "rogue" sequences of heads and tails correspond to the binary expansions of real numbers in [0,1] that are *not* simply normal in the Borel sense. *That is, the set of sequences that do not obey Borel's SLLN when a fair coin is tossed correspond to the binary expansions of real numbers in [0,1] that are not simply normal.*

[†] This is because, if $X \sim U(0,1)$ then $\Pr\{X \le x\} = x$ for $0 \le x \le 1$.

[‡] See **Problem 8**.

Problem 23

Borel's Paradox and Kolmogorov's Axioms (1909, 1933)

Problem. *Suppose the random variables X and Y have the joint density function*

$$f_{XY}(x, y) = \begin{cases} 4xy, & 0 < x < 1, \ 0 < y < 1, \\ 0, & \text{otherwise.} \end{cases}$$

Obtain the conditional density of X, given that Y = X.

Solution. Let $U = X$ and $V = Y - X$. Then we aim to find $f_{U|V}(u|v = 0)$. We have

$$f_{UV}(u, v) = f_{XY}(x, y)|J|^{-1},$$

where J is the Jacobian

$$J = \begin{vmatrix} \partial u/\partial x & \partial u/\partial y \\ \partial v/\partial x & \partial v/\partial y \end{vmatrix}$$

$$= \begin{vmatrix} 1 & 0 \\ -1 & 1 \end{vmatrix}$$

$$= 1.$$

Therefore, $f_{UV}(u, v) = 4xy \cdot 1 = 4u(u + v)$ for $0 < u < 1$, $-u < v < -u + 1$ (see Fig. 23.1). The marginal density of V for $-1 < v < 0$ is

---

Classic Problems of Probability, Prakash Gorroochurn.
© 2012 John Wiley & Sons, Inc. Published 2012 by John Wiley & Sons, Inc.

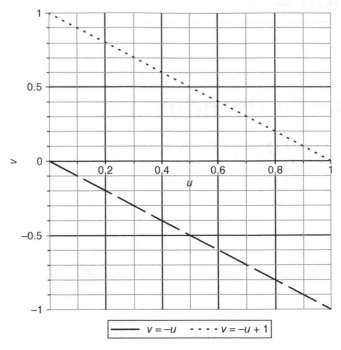

Figure 23.1   The support of (U, V) lies in the region $-u < v < -u + 1$.

$$f_V(v) = 4 \int_{-v}^{1} (u^2 + uv)du$$

$$= \frac{2}{3}(1 + v)^2(2 - v).$$

For $0 < v < 1$, the marginal density of V is

$$f_V(v) = 4 \int_{0}^{1-v} (u^2 + uv)du$$

$$= \frac{2}{3}(v - 1)^2(v + 2).$$

Using either probability density we have $f_V(0) = 4/3$. Therefore,

$$f_{U|V}(u|v = 0) = \frac{f_{UV}(u, 0)}{f_V(0)}$$

$$= \frac{4u^2}{4/3}$$

$$= 3u^2.$$

Hence, the conditional density of X, given that $Y = X$, is $3x^2$ for $0 < x < 1$.

23.1   Discussion

Would the final answer have changed if we had instead used the auxiliary random variables $U = X$ and $W = Y/X$? Let us see. The aim is to find $f_{U|W}(u|w = 1)$. We have $f_{UW}(u, w) = f_{XY}(x, y)|J|^{-1}$, where $J = 1/x$. Therefore, $f_{UW}(u, w) = 4u^3w$ for $0 < u < 1, 0 < w < 1/u$. The marginal density of W is $f_W(w) = w$ for $0 < w < 1$ and is $f_W(w) = 1/w^3$ for $w > 1$. Using either probability density, we have $f_W(1) = 1$. Therefore, $f_{U|W}(u|w = 1) = f_{UW}(u, 1)/f_W(1) = 4u^3$ for $0 < u < 1$. Thus, we seem to get a different answer: the conditional density of X, given that $Y = X$, is $4x^3$ for $0 < x < 1$.

We obtain two different conditional densities depending on how we choose to write the same conditioning event. This might seem troublesome. The paradox occurs because, in both cases, the conditioning event has probability zero, that is, $\Pr\{V = 0\} = \Pr\{W = 1\} = 0$. In view of the definition $\Pr\{A|B\} \equiv \Pr\{A \cap B\}/\Pr\{B\}$, where A and B are two events with $\Pr\{B\} \neq 0$, the reader might even wonder, *is conditional expectation defined when conditioning on events with probability zero?* At the end of this Discussion section, we shall provide a mathematical basis for why the answer to this question is yes and why both conditional densities above are valid.

The historical origin of the paradox can be traced back to Joseph Bertrand's *Calculs des Probabilités* (Bertrand, 1889, pp. 6–7). Bertrand asked the question: "*If we choose at random two points on the surface of a sphere, what is the probability that the distance between them is less than 10′?*" Bertrand then proceeded to give two different solutions to the problem. The first solution was based on the assumption that the probability that a given segment of the surface of the sphere contains the two points is proportional to the area of the segment. The second solution was based on the assumption that the probability that a given arc of a great circle contains the two points is proportional to the length of the arc. Bertrand's conclusion was that both solutions were correct, since the problem is ill-posed. Borel later (1909a, pp. 100–104) re-examined the paradox.[†] He concluded that Bertrand's second assumption (and hence solution) was incorrect. A few years afterward, Kolmogorov gave what is commonly used as an explanation of the paradox[‡] (Kolmogorov, 1933, p. 51):

> ... the concept of a conditional probability with regard to an isolated given hypothesis whose probability equals 0 is inadmissible. For we can obtain a probability distribution for Θ [the latitude] on the meridian circle only if we regard this circle as an element of the decomposition of the entire spherical surface into meridian circles with the given poles.

[†] Which is now also known as the *Paradox of the Great Circle*.

[‡] The Borel paradox is also sometimes known as the Borel–Kolmogorov paradox. For more discussions, see Bartoszynski and Niewiadomska-Bugaj (2008, p. 194), Proschan and Presnell (1998), DeGroot (1984, p. 171), Casella and Berger (2001, pp. 204–205), Jaynes (2003, p. 467), and Singpurwalla (2006, pp. 346–347).

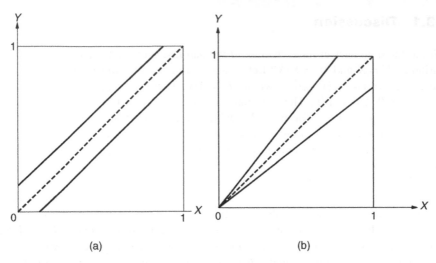

(a)   (b)

Figure 23.2   (a) The region $A = \{(x, y) : |Y - X| < \varepsilon\}$; (b) the region $B = \{(x, y) : |Y/X - 1| < \varepsilon\}$.

This last sentence can help us understand what exactly is going on in **Problem 23** when we use $Y - X = 0$ and $Y/X = 1$. Note that $Y - X = 0$ can be written as $\lim_{\varepsilon \to 0} |Y - X| < \varepsilon$ and that $Y/X = 1$ can be written as $\lim_{\varepsilon \to 0} |Y/X - 1| < \varepsilon$. The region $A = \{(x, y) : |Y - X| < \varepsilon\}$ lies between the two parallel lines in Fig. 23.2a whereas the region $B = \{(x, y) : |Y/X - 1| < \varepsilon\}$ lies between the two diverging lines in Fig. 23.2b (Bartoszynski and Niewiadomska-Bugaj, 2008, p. 194; Proschan and Presnell, 1998). The two conditional densities $f_{U|V}(u|v = 0)$ and $f_{U|W}(u|w = 1)$ are different because the approach to the limiting event $Y = X$ takes place in different ways. Furthermore, Fig. 23.2b shows that the condition $W = 1$ makes values of $x (= u)$ close to one more likely, and values of x close to zero less likely, than the condition $V = 0$. This is reflected in the two respective conditional densities $4x^3$ and $3x^2$ (see Fig. 23.3).

Borel's paradox and several others raised by Bertrand (see **Problems 18** and **19**) were major reasons that lead the Russian mathematician Andrey Nikolaevich Kolmogorov (1903–1987) to formulate an axiomatic theory of probability in his landmark monograph, *Grundbegriffe der Wahrscheinlichkeitsrechnung*[†] (Kolmogorov, 1933). This is how Kolmogorov introduces his axiomatic theory (Kolmogorov, 1956, p. 1):

> The theory of probability, as a mathematical discipline, can and should be developed from axioms in exactly the same way as Geometry and Algebra. This means that after we have defined the elements to be studied and their basic relations, and have stated the axioms by which these relations are to be governed, all further exposition must be based exclusively on these axioms, independent of the usual concrete meaning of these elements and their relations.

[†] Foundations of the Theory of Probability. Kolmogorov's *Grundbegriffe* features as one of the landmark works in Grattan-Guinness' *Landmark Writings in Western Mathematics*. A full analysis of the *Grundbegriffe* in this book is provided by von Plato (2005).

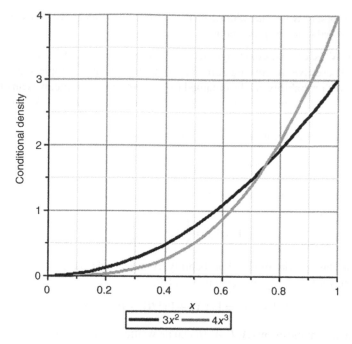

Figure 23.3   The two conditional densities of X given $Y = X$. The first ($3x^2$) writes the conditioning event as $Y - X = 0$, while the second ($4x^3$) writes the conditioning event as $Y/X = 1$

... the concept of a field of probabilities is defined as a system of sets which satisfies certain conditions. What the elements of this set represent is of no importance in the purely mathematical development of the theory of probability (cf. the introduction of basic geometric concepts in the Foundations of Geometry by Hilbert, or the definitions of groups, rings and fields in abstract algebra).

Kolmogorov enunciated six axioms in all. The first five concern a finite set of events. Kolmogorov (1956, p. 2) writes

Let E be a collection of elements $\xi, \eta, \zeta, \ldots$, which we shall call *elementary events*, and $\mathfrak{F}$ a set of subsets of E; the elements of the set $\mathfrak{F}$ will be called *random events*.

(I) $\mathfrak{F}$ *is a field of sets.*

(II) $\mathfrak{F}$ *contains the set E.*

(III)   *To each set A in $\mathfrak{F}$ is assigned a non-negative real number P(A). This number P(A) is called the probability of the event A.*

(IV)   *P(E) equals 1.*

(V)   *If A and B have no element in common, then $P(A + B)^{\dagger} = P(A) + P(B)$*

$^{\dagger} P(A + B)$ is Pr{$A \cup B$} in our notation.

A system of sets, $\mathfrak{F}$, together with a definite assignment of numbers $P(A)$, satisfying Axioms I–V, is called *a field of probability.*

Axiom V is of particular significance and is known as the *Axiom of Finite Additivity.* These five axioms are sufficient to deal with finite sample spaces and the basic results of probability can be obtained from them. For example, we can obtain the classical definition of probability as follows. Suppose Ω is made of N equally likely elementary outcomes $\omega_1, \omega_2, \ldots, \omega_N$. Since elementary outcomes are disjoint by definition, using the Axiom of Finite Additivity implies

$$\Pr\{\omega_j\} = \frac{1}{N}, \quad j = 1, 2, \ldots, N.$$

Now, suppose an event A consists of n_A elementary outcomes. Then the probability of A is

$$\Pr\{A\} = \frac{1}{N} \times n_A = \frac{n_A}{N},$$

which is the classical definition of probability. As a second example, we can prove the following formula for any two events A and B:

$$\Pr\{A \cup B\} = \Pr\{A\} + \Pr\{B\} - \Pr\{A \cap B\}. \tag{23.1}$$

First note that the events $A, B, A \cup B$ can be written as the union of disjoint events:

$$\{A\} = \{A \cap \bar{B}\} \cup \{A \cap B\},$$
$$\{B\} = \{A \cap B\} \cup \{\bar{A} \cap B\},$$
$$\{A \cup B\} = \{A \cap \bar{B}\} \cup \{A \cap B\} \cup \{\bar{A} \cap B\}.$$

Applying the Axiom of Finite Additivity,

$$\Pr\{A\} = \Pr\{A \cap \bar{B}\} + \Pr\{A \cap B\},$$
$$\Pr\{B\} = \Pr\{A \cap B\} + \Pr\{\bar{A} \cap B\},$$
$$\Pr\{A \cup B\} = \Pr\{A \cap \bar{B}\} + \Pr\{A \cap B\} + \Pr\{\bar{A} \cap B\}.$$

Subtracting the sum of the first two above equations from the third, we obtain Eq. (23.1).

Kolmogorov's sixth axiom is much more fundamental, and states (Kolmogorov, 1956, p. 14)

(VI) *For a decreasing sequence of events*

$$A_1 \supset A_2 \supset \cdots \supset A_n \supset \cdots$$

of $\mathfrak{F}$, for which

$$\bigcap_n A_n = 0,$$

the following equation holds:

$$\lim P(A_n) = 0, \quad n \to \infty.$$

Axiom VI is called the *Axiom of Continuity* and can be shown to be equivalent to the *Axiom of Countable Additivity*[†]: if $A_1, A_2, A_3, \ldots$ form a pairwise disjoint sequence of events, then

$$\Pr\left\{\bigcup_{j=0}^{\infty} A_j\right\} = \sum_{j=0}^{\infty} \Pr\{A_j\}.$$

Kolmogorov introduced the Axiom of Countable Additivity to deal with infinite sample spaces, and this was one of the two key fundamental contributions of his axiomatic theory. However, in order to deal with infinite sample spaces, Kolmogorov also needed another important tool, namely measure theory. This is because when the sample space Ω is infinite, such as when choosing a point on $[0,1]$, not all subsets of Ω qualify as events and can therefore be assigned a probability.[‡] By regarding a probability as a measure on $[0,1]$, the problem of indeterminate probabilities can be avoided, as we will soon explain. In measure-theoretic language, for any given set Ω, we define a σ-field $\mathfrak{I}$, which is a collection of measurable subsets of Ω such that

- $\Omega \in \mathfrak{I}$;
- $A \in \mathfrak{I} \Rightarrow \bar{A} \in \mathfrak{I}$;
- $A_1, A_2, \ldots \in \mathfrak{I} \Rightarrow \bigcup_{i=1}^{\infty} A_i \in \mathfrak{I}$.

Thus, when Ω is the sample space, the triple $\{\Omega, \mathfrak{I}, \Pr\}$ is called a probability space. By defining probabilities with respect to measurable subsets $\mathfrak{I}$ of the sample space, Kolmogorov avoids the problem of indeterminate probabilities that arise in infinite sample spaces. Moreover, since in Kolmogorov's framework, the exact probability space must be specified in a given experiment, paradoxes of the type Bertrand[§] raised do not occur.

Having defined the probability space $\{\Omega, \mathfrak{I}, \Pr\}$, Kolmogorov then proceeds to define a random variable, as follows (Kolmogorov, 1956, p. 22):

A real single-valued function $x(\xi)$, defined on the basic set E, is called a *random variable* if for each choice of a real number a the set $\{x < a\}$ for all ξ for which the inequality $x < a$ holds true, belongs to the system of sets $\mathfrak{I}$.

Intuitively, a random variable X is a real-valued *measurable* function with domain the sample space Ω. Furthermore, we can write

$$X : (\Omega, \mathfrak{I}, \Pr) \rightarrow (\mathfrak{R}, \mathfrak{I}_X, P_X),$$

[†] Henceforth, we shall refer to Axiom VI as the Axiom of Countable Additivity.

[‡] Insisting on using all subsets leads to the much-dreaded Banach-Tarski paradox, according to which a unit sphere may be decomposed into a finite number of pieces that can then be reassembled into two unit spheres. For more details, see Kac and Ulham (1968, pp. 48–52), Bewersdorff (2005, pp. 48–49), and Kardaun (2005, pp. 8–9).

[§] See **Problems 18** and **19**.

where $\mathfrak{R}$ is the set of real numbers, $\mathfrak{I}_X$ is the σ-field associated with $\mathfrak{R}$,[†] and P_X is the probability measure induced by X and is known as the probability distribution of X.

We mentioned before that when the sample space Ω is infinite, not all subsets of Ω qualify as events and can be assigned a probability. But can there be nonmeasurable events in *finite* sample spaces? The following example shows why the answer to this question is yes. Consider an experiment that involves tossing a fair coin three times but the result of only the first toss is observed. A requirement for a measurable event is that, once the experiment is performed, we should be able to answer the question of whether the event occurred by a yes or a no. Suppose A is the event "two heads are obtained." In this case, the event A is not measurable and a probability cannot be assigned to it. Thus, for a given $\omega \in \Omega$, where $\Omega = \{H, T\}$, we cannot determine whether $\omega \in A$.

We must note that Kolmogorov's Axiom of Countable Additivity has been somewhat controversial. The eminent subjectivist Bruno de Finetti (1906–1985) was categorically against it and instead favored Finite Additivity (de Finetti, 1972, pp. 86–92). Hájek provides a nice summary of one of de Finetti's key arguments (Hájek, 2008, p. 326)[‡]:

> Suppose a positive integer is selected at random—we might think of this as an infinite lottery with each positive integer appearing on exactly one ticket. We would like to reflect this in a uniform distribution over the positive integers (indeed, proponents of the principle of indifference would seem to be committed to it), but if we assume countable additivity this is not possible. For if we assign probability 0 in turn to each number being picked, then the sum of all these probabilities is again 0; yet the union of all of these events has probability 1 (since it is guaranteed that some number will be picked), and $1 \neq 0$. On the other hand, if we assign some probability $\varepsilon > 0$ to each number being picked, then the sum of these probabilities diverges to ∞, and $1 \neq \infty$. If we drop countable additivity, however, then we may assign 0 to each event and 1 to their union without contradiction.

However, few probabilists have found it convenient to reject Countable Additivity owing to its equivalence to the Axiom of Continuity. The latter is of fundamental importance because it implies among other things that we can take the limit of, and differentiate, probability measures.

Note that Kolmogorov's axioms do not force us to adhere to any particular philosophy of probability, which can be viewed as one of the great merits of his theory. Chaumont et al. (2007, pp. 42–43) write

> The great force of Kolmogorov's treatise is to voluntarily consider a completely abstract framework, without seeking to establish bridges with the applied aspects of the theory of probability, beyond the case of finite probabilities. In general, the search for such bonds inevitably brings to face delicate philosophical questions and is thus likely to darken mathematical modelling. While speaking about the questions of application only in the part devoted to the finite probabilities, Kolmogorov is released from this constraint and can avoid the pitfalls that von Mises had not always circumvented.

[†] In this case, the σ-field is called a *Borel* σ-field.

[‡] The issue of finite versus countable additivity is also thoroughly discussed in Bingham (2010).

However, Kolmogorov seems to have had von Mises' frequency theory in mind when he developed his axiomatic theory, for he writes (Kolmogorov, 1956, p. 3)

> ...Here we limit ourselves to a simple explanation of how the axioms of the theory of probability arose and disregard the deep philosophical dissertations on the concept of probability in the experimental world. In establishing the premises necessary for the applicability of the theory of probability to the world of actual events, the author has used, in large measure, the work of R. V. Mises [1], pp. 21–27.

We previously mentioned that one of the key contributions of Kolmogorov's *Grundbegriffe* was to provide a framework to work with infinite sample spaces. A second fundamental aspect of Kolmogorov's work was the establishment of a rigorous theory for conditional expectation (and conditional probability) when the condition has probability zero. Kolmogorov was able to do this by defining conditional expectation through the *Radon-Nikodym theorem*. The latter states that if $\{\Omega, \mathfrak{I}, \mu\}$ is σ-finite and ν is a measure on $\{\Omega, \mathfrak{I}\}$ that is absolutely continuous with respect to μ (i.e., $\mu = 0 \Rightarrow \nu = 0$), then there exists a real-valued measurable function f on Ω such that

$$\nu(B) = \int_B f \, d\mu.$$

The function f is then called the Radon-Nikodym derivative of ν with respect to μ (i.e., $f = d\nu/d\mu$). *The Radon-Nikodym theorem thus postulates the existence of f.* As a simple application of this theorem, absolutely continuous probability distribution functions $F_X(x)$ *must* possess probability densities $f_X(x)$, where

$$F_X(x) = \int_{-\infty}^{x} f_X(u) \, du$$

and

$$f_X(x) = \frac{d}{dx} F_X(x).$$

Coming back to conditional expectations, by writing

$$EXI_B(Y) = \int_B Z \, dF_Y(y),$$

where I_B is the indicator function on the set B and $F_Y(y)$ is the distribution function of Y, Kolmogorov is able to establish the existence of the random variable Z as the conditional expectation of X with respect to Y, that is, $Z = E(X|Y)$. Furthermore, by writing $X = I_A$, where A is an event, we obtain analogous results regarding the existence of the conditional probability $\Pr\{A|Y\}$. The important point to note is that, as a consequence of the Radon-Nikodym theorem, the conditional expectation $E(X|Y)$ is defined each unique to within a modification on a set of probability zero, that is, for *any* $\Pr\{Y=y\} = 0$, $E(X|Y)$ remains a valid (but not unique) conditional expectation of X given Y.

Problem 24

Of Borel, Monkeys, and the New Creationism (1913)

Problem. *A robot monkey is seated at a typewriter and randomly hits the keys in an infinite sequence of independent trials. Prove that,* ***theoretically,*** *the monkey will eventually type out the complete works of Shakespeare with certainty.*

Solution. Let the number of keys on the keyboard be K, and the complete works of Shakespeare form a sequence of S letters, where both K and S are finite numbers. Divide the sequence of letters typed by the monkey into segments Seg_1, Seg_2, ..., each S letters long. Now, the probability that the ith letter typed by the monkey coincides with the ith letter in the works of Shakespeare is $1/K$. Thus, the probability that segment Seg_1 corresponds to the sequence in the works of Shakespeare, that is, the probability that Seg_1 is a "success" is $1/K^S$. Next consider the first G segments typed by the monkey. The probability that every one of the G segments is a "failure" $(1 - 1/K^S)^G$. Therefore, the probability that at least one of the G segments is a "success" is

$$p = 1 - \Pr\{\text{all } G \text{ segments are "failures"}\}$$

$$= 1 - \left(1 - \frac{1}{K^S}\right)^G.$$

Since the robot monkey types an infinite number of letters, we have

$$\lim_{G \to \infty} p = \lim_{G \to \infty} \left\{ 1 - \left(1 - \frac{1}{K^S}\right)^G \right\}.$$

Now, $(1 - 1/K^S)$ is a real number in (0, 1) so that

$$\lim_{G \to \infty} \left(1 - \frac{1}{K^S}\right)^G = 0$$

Classic Problems of Probability, Prakash Gorroochurn.
© 2012 John Wiley & Sons, Inc. Published 2012 by John Wiley & Sons, Inc.

and

$$\lim_{G \to \infty} p = 1 - 0 = 1.$$

Hence, theoretically, the monkey will eventually type out the complete works of Shakespeare with certainty.

24.1 Discussion

The result we have just proved is the so-called *Infinite Monkey Theorem.*[†] This result, and indeed a stronger form of it, can also be demonstrated by a straightforward application of the second Borel-Cantelli lemma. The latter states that if $A_1, A_2, \ldots$ is a sequence of independent events such that

$$\sum_{j=1}^{\infty} \Pr\{A_j\} = \infty,$$

then for $j = 1, 2, \ldots,$

$$\Pr\{A_j \text{ occurs infinitely often}\} = 1.$$

To apply the second Borel-Cantelli lemma to the *Infinite Monkey Problem*, we let A_j be the event that the segment Seg$_j$ typed by the monkey is a success, that is, coincides with the sequence of letters in the complete works of Shakespeare. Then

$$\Pr\{A_j\} = \frac{1}{K^S} > 0, \quad j = 1, 2, \ldots$$

Therefore,

$$\sum_{j=1}^{\infty} \Pr\{A_j\} = \sum_{j=1}^{\infty} \frac{1}{K^S} = \infty.$$

Thus, the second Borel-Cantelli lemma implies that the monkey will eventually type the complete works of Shakespeare not once, but infinitely many times (i.e., in infinitely many segments).

The *Infinite Monkey Theorem* can be traced to the French mathematician Emile Borel (1871–1956). In a 1913 article, Borel (1913) stated[‡]

Let us imagine that we have assembled a million monkeys each randomly hitting on the keys of a typewriter and that, under the supervision of illiterate foremen, these typing monkeys work hard ten hours a day with a million of various typing machines. The illiterate foremen would gather the blackened sheets and would bind them into volumes.

[†] The Infinite Monkey Theorem is also discussed in Gut (2005, p. 99), Paolella (2007, p. 141), Isaac (1995, p. 48), and Burger and Starbird (2010, p. 609).

[‡] Borel later gave the same example in his book *Le Hasard* (Borel, 1920, p. 164).

And after a year, these volumes would enclose a copy of accurate records of all kinds and of all languages stored in the richest libraries of the world.

The same observation was made later by the British astrophysicist Sir Arthur Eddington[†] (1882–1944). In his *Nature of the Physical World,* Eddington (1929, p. 72) writes

> If I let my fingers wander idly over the keys of a typewriter it might happen that my screed made an intelligible sentence. If an army of monkeys were strumming on typewriters they might write all the books in the British Museum. The chance of their doing so is decidedly more favourable than the chance of the molecules returning to one half of the vessel.

The reason why the *Infinite Monkey Theorem* holds *mathematically* is because the probability, $1/K^S$, that the first S letters typed by the robot monkey corresponds to the complete works of Shakespeare is nonzero. However, this probability is so small that, *practically speaking*, it will take eons before Shakespeare's works are correctly typed out, even for a fast robot monkey.

Let us do some simple calculations to prove our point. For example, assuming $K = 44$ and $S = 5.0 \times 10^6$, we obtain a probability of $1/K^S$ ($\approx 4.1 \times 10^{-8,217,264}$) of correctly typing the works of Shakespeare. Therefore, the expected number of segments the monkey must type before the first successful one is K^S. This corresponds to SK^S characters. Assuming the monkey types 1000 characters per second, the expected time before the first occurrence of the complete works of Shakespeare is about

$$\frac{S \times K^S}{1000 \times 60 \times 60 \times 24 \times 365} \approx 3.8 \times 10^{8,217,259} \text{ years.}$$

To put this number in perspective, the age of the Universe is about 1.4×10^{10} years.

On the other hand, suppose the robot monkey could be programmed to randomly type 1000 characters per second for the next million (10^6) years. What is the probability that it will successfully type Shakespeare's works? One million years corresponds to 3.2×10^{13} seconds or 3.2×10^{16} characters. This means that the monkey will be able to type a total of 6.3×10^9 segments. The probability of one successful segment is $1/K^S$, so the probability of at least one successful segment is

$$= 1 - \Pr\{\text{no successful segment}\}$$

$$= 1 - \left(1 - \frac{1}{K^S}\right)^{6.3 \times 10^9}$$

$$\approx 0.$$

Finally, suppose the robot monkey had a crack at only the first page of Shakespeare's works. Assume as before $K = 44$, but this time $S = 600$. Using a

---

[†] It is interesting to note that both Borel and Eddington gave the infinite monkey example in the context of the kinetic theory of gases.

similar reasoning as before, we obtain a probability of 8.5×10^{-984} of successfully typing the first page, an average of 2.2×10^{978} years before the first occurrence of the first page (assuming a typing rate of 1000 characters per second), and a probability of still zero of at least one successful page if the monkey kept on randomly typing for one million years at a rate of one thousand characters per second.

The nature and meaning of very small probabilities have been topics of interest since the times of Jacob Bernoulli. In the very first chapter of Part IV of his *Ars Conjectandi*, Bernoulli states (Bernoulli, 1713)

> That is morally certain whose probability nearly equals the whole certainty, so that a morally certain event cannot be perceived not to happen: on the other hand, that is morally impossible which has merely as much probability as renders the certainty of failure moral certainty. Thus, if one thing is considered morally certain which has 999/1000 certainty, another thing will be morally impossible which has only 1/1000 certainty.

The great French mathematician Jean le Rond d'Alembert (1717–1783) was also interested in small probabilities. d'Alembert distinguishes between "metaphysical" and "physical" probabilities.[†] According to d'Alembert, an event is metaphysically (or mathematically) possible if its probability is greater than zero. On the other hand, an event is physically possible only if it is not so rare that its probability is almost zero. d'Alembert (1761, p. 10) writes

> One must distinguish between what is *metaphysically* possible, & what is *physically* possible. In the first class are all things whose existence is not absurd; in the second are all things whose existence not only has nothing that is absurd, but also has nothing that is very extraordinary either, and that is not in the everyday course of events. It is *metaphysically* possible, that we obtain a double-six with two dice, one hundred times in a row; but this is *physically* impossible, because this has never happened, & will never happen.

d'Alembert does not give a cutoff for how small a probability ought to be for the corresponding event to be declared physically impossible, although he gives an example of a coin that is tossed 100 times to illustrate the notion of physical impossibility[‡] (this corresponds to a probability of $p \leq 1/2^{100}$ for physical impossibility).

The issue of small probabilities was taken again by Georges-Louis Leclerc, Comte de Buffon (1707–1788), in his celebrated *Essai d'Arithmétique Morale* (Buffon, 1777, p. 58):

> ...in all cases, in one word, when the probability is smaller than 1/10000, it has to be, & it is in fact for us absolutely nil, & by the same reasoning in all cases where this probability is greater than 10000, it makes for us the most complete moral certainty.

An even more comprehensive set of guidelines for interpreting small probabilities was spelled out much later by Borel in his *Probabilities and Life* (Borel 1962, pp. 26–29). Borel defined a probability of

---

[†] See also p. 112.

[‡] See d'Alembert (1761, p. 9).

- 10^{-6} as being "negligible on the human scale." That is, for such a probability "the most prudent and reasonable must act as if this probability is nil."
- 10^{-15} as being "negligible on the terrestrial scale." This kind of probability must be considered "when our attention is turned not to an isolated man but to the mass of living men."
- 10^{-50} as being "negligible on a cosmic scale." Here, "we turn our attention, not to the terrestrial globe, but to the portion of the universe accessible to our astronomical and physical instruments."
- 10^{-500} as being "negligible in a supercosmic scale," that is, when the whole universe is considered.

Borel, who had once been the President of France's pretigious *Académie des Sciences*, hardly knew how his probability guidelines would later be utilized by a very influential group to promote its agendas. The group in question is the Intelligent Design (ID) movement that developed from "scientific" creationism when the latter was barred from being taught in public schools by rulings such as the US Supreme Court's *Edwards v. Aguillard*, 482 U.S. 578 (1987). The ID movement was founded by the American law professor Phillip E Johnson (b. 1940) who popularized the concept in his 1991 book *Darwin on Trial* (Johnson, 1991). According to Rieppel (2011, p. 4)[†]:

> The main focus of the doctrine of Intelligent Design is what its proponents call "irreducible biological complexity." Harking back on William Paley's old "watchmaker analogy" as expounded in his notorious "Natural Theology" of 1802, the proponents of Intelligent Design highlight biological structures such as the bacterial flagellum, which is thought to reveal a mechanistic complexity so sophisticated that its evolution cannot possibly be satisfactorily explained on Darwin's theory of natural selection.[‡]

ID proponents believe in an Intelligent Designer although they officially refrain from identifying the designer. Moreover, Phy-Olsen (2011, p. xi) adds

> ID, as the movement is known, does not use religious language and does not make its appeals from the Bible. It does not seek to drive evolution out of the schools. It only asks for equal time to present its interpretation of cosmic and human origins alongside evolutionary teachings.

Several ID proponents have seized Borel's ideas on small probabilities, especially his 10^{-50} guideline for a negligible probability on a cosmic scale, in an attempt to disprove evolution. For example, in the ID literature, one can often read statements like the following (Lindsey, 2005, p. 172):

[†] The evolution-creation debate is also discussed in Pennock (2001), Ruse (2003), Forrest and Gross (2004), Skybreak (2006), Scott (2004), Ayala (2007), Seckbach and Gordon (2008), Young and Largent (2007), Lurquin and Stone (2007), Moore (2002), Manson (2003), Bowler (2007), and Young and Strode (2009).

[‡] Natural selection can be understood as the differential reproduction of different individuals due to different fitnesses of their genotypes.

The French probability expert Emile Borel said there is a single law of chance, that beyond which there is no possibility of its occurrence no matter how much time elapses. Borel calculated that number to be one chance in 10^{50}, beyond which things will never happen, even with an eternity of time (one chance in a billion is 1×10^9). To understand just how large one chance in 10^{50} is, it would take you millions of years just to count that high. If you recall that the odds of a human DNA sequence encoding itself by random chance is one chance in $10^{1,000,000,000,000}$, the evolution of man could never have happened in 4.5 billion years, or an eternity for that matter. Thus, the evolutionists' main argument, that an infinite amount of time allows for the impossible to happen, has been disproved by statistical probability studies using Borel's Law.

Thus, the ID proponents claim that evolution is as unlikely as a robot monkey actually typing the whole works of Shakespeare. However, we now show that this argument cannot be right because evolution by natural selection does not work in the completely random manner implied in the above paragraph.

Suppose a gene[†] Ge corresponds to the following sequence that consists of thirty nucleotides[‡]:

Ge: AGGGT CCTAA AATAG TATCAT AGCGT AGTCA

We assume Ge makes the organism able to successfully reproduce and transmit its genes in future generations, that is, this gene contributes to the organism's fitness. Suppose also that Ge makes the organism achieve an optimal fitness with respect to its environment. Now, the original DNA sequence could have been a variant Ge_1 of Ge with very low fitness, such as

Ge_1:   AG**AA**T CCTAA **GACAG CGTGA**T AGCGT **AACCA**

Suppose all organisms of a particular species initially had Ge_1, but mutations make some of the offspring to have a variant Ge_2 that has a higher fitness that Ge_1 but still much less than Ge:

Ge_2:   AG**AA**T CCTAA **AATAG CGTGA**T AGCGT **AACCA**

Because Ge_2 has a higher fitness that Ge_1, there will be proportionally more and more of it in future generations. That is to say, natural selection is acting on the organisms and leads to evolution. Suppose more mutations occur in the future leading to a variant, say Ge_3, which has a higher fitness that Ge_2. Because of natural selection, after some more generations there will be an overwhelming number of Ge_3 compared to Ge_2 and Ge_1. Evolution by natural selection proceeds such that even if the best variant Ge is not eventually produced the population will settle for the variant with the highest fitness. Natural selection is therefore a *cumulative* process that works by *editing* previous variants. It is not the completely random process that each time tries to shuffle the nucleotides until the best sequence (Ge) is obtained, as the ID proponents usually portray it.

[†] A gene can be thought of as a DNA segment that encodes an RNA molecule with some control function.

[‡] A nucleotide can be thought of as a subunit of DNA. There are four possible nucleotides: adenine (A), thymine (T), guanine (G), and cytosine (C).

Figure 24.1   Richard Dawkins (b. 1941).

The last argument is essentially what the eminent evolutionary biologist Richard Dawkins (b. 1941) gave in the book *The Blind Watchmaker* (Dawkins, 1987, p. 46) (Fig. 24.1). Dawkins further writes (p. 49)

> What matters is the difference between the time taken by cumulative selection, and the time which the same computer, working flat out at the same rate, would take to reach the target phrase if it were forced to use the other procedure of single-step selection: about a million million million million million years. . . . Whereas the time taken for a computer working randomly but with the constraint of cumulative selection to perform the same task is of the same order as humans ordinarily can understand, between 11 seconds and the time it takes to have lunch.

> There is a big difference, then, between cumulative selection (in which each improvement, however slight, is used as a basis for future building), and single-step selection (in which each new 'try' is a fresh one). If evolutionary progress had had to rely on single-step selection, it would never have got anywhere. If, however, there was any way in which the necessary conditions for cumulative selection could have been set up by the blind forces of nature, strange and wonderful might have been the consequences. As a matter of fact that is exactly what happened on this planet, and we ourselves are among the most recent, if not the strangest and most wonderful, of those consequences.

Thus, although it is perfectly legitimate to debate the scientific merits of the theory of evolution, attacks on it based on probabilistic arguments such as those of Lindsey (2005) and others cannot be valid.

Problem 25

Kraitchik's Neckties and Newcomb's Problem (1930, 1960)

Problem. *Paul is presented with two envelopes, and is told one contains twice the amount of money than the other. Paul picks one of the envelopes. Before he can open it, he is offered the choice to swap it for the other. Paul decides to swap his envelope based on the following reasoning:*

> Suppose the envelope I initially chose contains D dollars. This envelope contains the smaller amount with probability 1/2. Then, by swapping, I will get $2D$ dollars and make a gain of D dollars. On the other hand, the envelope could contain the larger amount, again with probability 1/2. By swapping, this time I will get only $D/2$ dollars and make a loss of $D/2$. Hence swapping leads to a net expected gain of $(1/2)(D) + (1/2)(-D/2) = D/4$, and therefore I should swap.

Briefly state why Paul's reasoning cannot be correct.

Solution. If Paul decides to swap and applies the same reasoning to the other envelope, then he should swap back to the original envelope. By applying the reasoning again and again, he should therefore keep swapping forever. Therefore, Paul's reasoning cannot be correct.

25.1 Discussion

This little problem is usually attributed to the Belgian mathematician Maurice Kraitchik (1882–1957). It appears as the puzzle "The Paradox of the Neckties" in the 1930 book *Les Mathématiques des Jeux ou Récréations Mathématiques*[†] (Kraitchik, 1930, p. 253):

> B and S each claim to possess the better necktie. They ask Z to arbitrate, the rules of the game being the following: Z would decide which tie is better, and the winner would give his tie to the loser. . . . S reasons as follows: "I know what my tie is worth. I may lose it, but I

[†] The problem also appears in the second English edition of the book (Kraitchik, 1953, p. 133).

Classic Problems of Probability, Prakash Gorroochurn.
© 2012 John Wiley & Sons, Inc. Published 2012 by John Wiley & Sons, Inc.

may also win a better one; so the game is favorable to me." But B can reason similarly, the problem being symmetrical. One wonders how the game can be favorable to both?[†]

Kraitchik acknowledges that the problem was developed in 1912 by the German mathematician Edmund Landau (1877–1938) during a course on mathematical recreations.

We know that Paul's reasoning is wrong, but the question remains: what specifically is the fallacy in Paul's argument? To see the error, assume one of the envelopes contains an amount a, and the other contains $2a$. Now, the envelope initially picked up by Paul has an amount D where

$$D = \begin{cases} a & \text{prob.} = 1/2, \\ 2a & \text{prob.} = 1/2. \end{cases}$$

If Paul had picked the lower amount (i.e., if $D = a$) then by swapping he gets $2D \,(= 2a)$, and makes a gain of a. But if he had picked the larger amount (i.e., if $D = 2a$) then by swapping he gets $D/2 \,(= a)$, and makes a loss of a. Note that both the gain and loss each amount to a. Yet in his reasoning Paul wrote the gain as D and the loss as $D/2$, thus making them have different values. This leads to the erroneous net expected gain of $(1/2)(D) + (1/2)(-D/2) = D/4$ by swapping. Had Paul realized that the profit and loss were each the same (i.e., a), he would have correctly concluded that the net expected gain is $(1/2)(a) + (1/2)(-a) = 0$. Therefore, it does not matter if he swaps or not. Thus, if D had been correctly treated as a random variable, rather than a constant, the fallacy could have been avoided.

As we mentioned before, the problem Kraitchik originally considered was slightly different. S knew the worth of his tie, while Paul did not know how much was in the envelope. Let us now outline Kraitchik's 1930 solution to the problem. Assume B's and S's ties are worth b and s dollars, respectively. Let G be S's gain, where

$$G = \begin{cases} -s & \text{if } b = 1, \\ -s & \text{if } b = 2, \\ \vdots \\ -s & \text{if } b = s - 1, \\ 0 & \text{if } b = 0, \\ b & \text{if } b = s + 1, \\ b & \text{if } b = s + 2, \\ \vdots \\ b & \text{if } b = x. \end{cases}$$

[†] This problem is in fact slightly different to the one we just considered. Paul did not know the worth of the envelope he picked, since he did not open it. On the other hand, S knows the worth of his tie.

In the last equation, Kraitchik sets an upper bound x for the worth of each tie. Now, each of the values of b is equally likely and has probability $1/x$. Therefore, the conditional expected gain of S is

$$\mathscr{E}(G|S = s) = \frac{1}{x}\left\{-s(s-1) + \sum_{b=s+1}^{x} b\right\}$$

$$= \frac{1}{x}\{-s(s-1) + (s+1) + (s+2) + \cdots + x\}$$

$$= \frac{x+1}{2} - \frac{3s^2 - s}{2x}.$$

Now s can take any of the values $1, 2, \ldots, x$, each with probability $1/x$. Therefore, the unconditional expected gain of S is

$$\mathscr{E}G = \mathscr{E}_S\mathscr{E}_G(G|S = s)$$

$$= \frac{x+1}{2} - \frac{1}{x}\sum_{s=1}^{x}\frac{3s^2 - s}{2x}$$

$$= \frac{x+1}{2} - \frac{1}{4x^2}\{x(x+1)(2x+1) - x(x+1)\}$$

$$= 0.$$

As simple as the *Two-Envelope Problem* looks, it is almost mind-boggling that it has generated such a huge literature[†] among statisticians, economists, and philosophers. Bayesians, decision theorists, and causal scientists have all jumped in the fray, and the debate is still raging. Let us now outline some of the alternative variants of the problem and resolutions that have been proposed.[‡] In 1989, Nalebuff[§] (1989) considered a variant that is closer to the problem Kraitchik actually considered:

> *Peter is presented with two envelopes, and is told one contains twice the amount of money than the other. Peter picks and opens one of the envelopes. He discovers an amount x. He is then offered the choice to swap it for the other. What should Peter do?*

Note that, following Paul in **Problem 25**, Peter also reasons that the other envelope could contain either $2x$ or $x/2$, with probability $1/2$ each, and incorrectly

---

[†] See, for example, Barbeau (2000, p. 78), Jackson et al. (1994), Castell and Batens (1994), Linzer (1994), Broome (1995), McGrew et al. (1997), Arntzenius and McCarty (1997), Scott and Scott (1997), McGrew et al. (1997), Clark and Shackel (2000), Katz and Olin (2007), and Sutton (2010).

[‡] Surprisingly, in his book *Aha! Gotcha*, the master puzzlist Martin Gardner fails to resolve the two-envelope paradox. He admits (Gardner, 1982, p. 106), "We have been unable to find a way to make this [i.e., what is wrong with the reasoning of the two players] clear in any simple manner."

[§] Nalebuff thus became the first who considered the "envelope version" of Kraitchik's problem.

concludes that swapping leads to an expected gain of $x/4$. However, Nalebuff uses a Bayesian argument to show that, once Peter discovers an amount x in his envelope, the probability that the other envelope contains either $2x$ or $x/2$ will be $1/2$ only when all values of x on $[0, \infty)$ are equally likely. Since such a probability density on x is improper and inadmissible, Nalebuff resolves the error in Peter's argument that swapping should be better.

However, a more complete resolution of the paradox is provided by Brams and Kilgour (1995) by using a similar Bayesian argument.[†] The reason a Bayesian analysis is appealing is because if the amount discovered by Peter is very large, he will be less likely to swap than if the amount is very small. This suggests placing a prior distribution on this amount and then carrying out a Bayesian analysis. Now, suppose one of the envelopes has the larger amount L and the other the smaller amount S, where $L = 2S$. Let the prior density of L be $f_L(l)$. This automatically defines a prior on S, since

$$\Pr\{L \le l\} = \Pr\{S \le l/2\}.$$

In terms of cumulative distribution functions, the above implies

$$F_L(l) = F_S(l/2),$$

which, upon differentiation, yields

$$f_L(l) = \frac{1}{2}f_S(l/2).$$

Assume that the amount in the envelope initially chosen by Peter is X with probability density function $p(x)$. Once the envelope is opened, suppose Peter observes the amount x. Then the conditional probability that Peter opened the envelope with the larger amount is given by Bayes' Theorem[‡]:

$$\Pr\{X = L | X = x\} = \frac{p(x|l)f_L(x)}{p(x)}$$

$$= \frac{p(x|l)f_L(x)}{p(x|L = l)f_L(x) + p(x|L = 1/2)f_S(x)}.$$

Now $p(x|L = l) = p(x|L = 1/2) = 1/2$, so that

$$\Pr\{X = L | X = x\} = \frac{f_L(x)}{f_L(x) + 2f_L(2x)}. \tag{25.1}$$

[†] See also Christensen and Utts (1992), which contains an error, and the follow-up paper by Blachman et al. (1996). In these papers, the *Two-Envelope Paradox* is called the *Exchange Paradox*.

[‡] See **Problem 14**.

Similarly,

$$\Pr\{X = S | X = x\} = \frac{2f_L(x)}{f_L(x) + 2f_L(2x)}. \tag{25.2}$$

Now, if $x = L$, then Peter gets $L/2 (= x/2)$ by swapping and makes a loss of $x/2$. On the other hand, if $x = S (= L/2)$, then Peter gets $L (= 2x)$ by swapping and makes a gain x. Therefore, Peter's expected gain is

$$\mathscr{E}G^* = \frac{f_L(x)}{f_L(x) + 2f_L(2x)} \times (-x/2) + \frac{2f_L(2x)}{f_L(x) + 2f_L(2x)} \times x$$

$$= \frac{x[4f_L(2x) - f_L(x)]}{2[2f_L(2x) + f_L(x)]}. \tag{25.3}$$

Equation (25.3) also shows that if there are priors such that $4f_L(2x) > f_L(x)$ (the so-called "exchange condition" for continuous distributions), then it would be better to swap. Brams and Kilgour (1995) prove that, for *any* prior $f_L(x)$, there is at least some values for x for which the exchange condition is satisfied. Some examples they give include

- **A Uniform Prior for L**: $f_L(l) = 1, 0 \le l \le 1$. The exchange condition holds for $l \le 1/2$.
- **An Exponential Prior for L**: $f_L(l) = e^{-l}, l \ge 0$. The exchange condition holds for $l \le \ln 4 \approx 1.39$.
- $f_L(l) = (1 - k)l^{-2+k}, l \ge 1, 0 < k < 1$. The exchange condition holds for all permissible values of this density (i.e., $l \ge 1$).

We next examine two further problems that are in the same spirit as Kraitchik's neckties, but each with a very different type of solution. The first one is as follows[†]:

> *You are presented with two identical envelopes, one containing an amount x, the other an amount y, where x and y are different positive integers. You pick an envelope at random that shows an amount W (where W = x or y). Devise a procedure so that you can guess with probability greater than half whether W is the smaller of the two amounts.*[‡]

A procedure can be obtained as follows. We toss a fair coin and note the number of tosses N until a head first appears. Define $Z = N + 1/2$. Then, if $Z > W$, we declare W as the smaller of the two amounts; otherwise, if $Z < W$, we declare W as the larger of the two amounts. To show that this procedure will make us correct with probability

---

[†] See also Grinstead and Snell (1997, pp. 180–181) and Winkler (2004, pp. 65, 68). The latter attributes this problem to Cover (1987, p. 158).

[‡] Of course, if one wanted the probability of being correct to be exactly half, one could make a decision by simply flipping a fair coin.

greater than half, we first note that

$$\Pr\{N = n\} = \left(\frac{1}{2}\right)^{n-1}\left(\frac{1}{2}\right) = \frac{1}{2^n}, \quad n = 1, 2, \dots$$

Therefore,

$$\begin{aligned}
\Pr\{Z < z\} &= \Pr\{N < z - 1/2\} \\
&= \Pr\{N \le z - 1\} \\
&= \sum_{n=1}^{z-1}\frac{1}{2^n} \\
&= \frac{(1/2)[1 - (1/2)^{z-1}]}{1 - 1/2} \\
&= 1 - \frac{1}{2^{z-1}}, \quad z = 1, 2, \dots
\end{aligned}$$

Since Z cannot be an integer by definition, we also have

$$\Pr\{Z > z\} = 1 - \Pr\{Z < z\} = \frac{1}{2^{z-1}}, \quad z = 1, 2, \dots$$

To fix ideas, let $x < y$. With probability half, we will pick the smaller amount and $W = x$. We will then be right if and only if $Z > x$. Similarly, with probability half, we will pick the larger amount, that is, $W = y$, and we will then be right if and only if $Z < y$. The probability that we will be right is therefore

$$\frac{1}{2}\Pr\{Z > x\} + \frac{1}{2}\Pr\{Z < y\} = \frac{1}{2}\cdot\frac{1}{2^{x-1}} + \frac{1}{2}\cdot\left(1 - \frac{1}{2^{y-1}}\right) = \frac{1}{2} + \left(\frac{1}{2^x} - \frac{1}{2^y}\right) > \frac{1}{2},$$

since $x < y$. This completes the proof. A similar calculation can show that the above procedure works if N is a variable from *any* discrete distribution with a strictly increasing distribution function and an infinite positive support (such as a Poisson distribution). On the other hand, if a variable V is chosen from any *continuous* distribution with a strictly increasing distribution function and an infinite positive support (such as an exponential distribution), then we define $Z = V$ and use the same decision rule as before.

The second problem, *Newcomb's Paradox*,[†] is a classic problem in decision theory and is as follows:

> *You are presented with two boxes, one transparent (box A) and the other opaque (box B). You may open either both boxes or only box B, and take the prize inside. A very reliable predictor has already acted in the following manner. She has put $1000*

[†] Newcomb's paradox is also discussed in Resnick (1987, p. 109), Shackel (2008), Sloman (2005, p. 89), Sainsbury (2009, Chapter 4), Lukowski (2011, pp. 7–11), Gardner (2001, Chapter 44), Campbell and Sowden (1985), Joyce (1999, p. 146), and Clark (2007, p. 142).

Figure 25.1   Robert Nozick (1938–2002).

in box A. If she thought you will open box B only, she has also placed $1,000,000 in box B. On the other hand, if she thought you will open both boxes, she has placed nothing in box B. Which box should you open?

We will now show how arguments can be made for both options. First, we make a case for opening box B only by using the principle of maximum expected utility, as follows. Consider the decision matrix for *Newcomb's Paradox* shown in Table 25.1. The matrix consists of the utilities U for each decision, conditional on whether the predictor is right.

The probability that the predictor is right is p, where p is reasonably large (say $p \approx .9$) Thus, $U = 0$ is your utility of opening box B given that the predictor predicted that you would open both boxes (i.e., given that box B is empty). On the other hand, $U = 1000$ is your utility of opening both boxes, given the predictor predicted that you would do so (i.e., given that box B is empty). The expected utility of opening box B only is

$$\mathscr{E}U_{\text{box B}} = (0)(1 - p) + (1000000)(p) = 1000000p,$$

and the expected utility of opening both boxes is

$$\mathscr{E}U_{\text{both boxes}} = (1000)(p) + (1001000)(1 - p)$$
$$= 1001000 - 1000000p.$$

Which of these two utilities is larger? Solving $\mathscr{E}U_{\text{box B}} > \mathscr{E}U_{\text{both boxes}}$, we obtain $p > 1000/1999$ ($\approx.5$). Since p is much larger than .5 (the predictor is very reliable), we see that choosing box B has the greater expected utility.

We now show that a case can also be made for opening both boxes. If the predictor has put $1,000,000 in box B, then choosing both boxes results in one winning $1,001,000 while choosing box B only results in one winning $1,000,000. Thus, choosing both boxes is better in this case. On the other hand, if the predictor has put nothing in box B, then choosing both boxes will result in one winning $1000 while choosing box B only results in one winning nothing. Thus, choosing both boxes "dominates" in both cases and seems to be always better than choosing box B only (i.e., the second row in Table 25.1 has both utilities greater than the corresponding utilities in the first row). The argument we have used here is the so-called dominance principle.

Hence, we have shown that the principles of maximum expected utility and dominance recommend opposite courses of action. Have both principles been correctly applied? The prevailing answer seems to be that the maximum utility principle has been incorrectly applied and that one should act according to the dominance principle (i.e., open both boxes). Let us explain why. First recall that the predictor already makes her decision (i.e., decides whether box B is empty or not) before one chooses to open. However, a look at Table 25.1 suggests that whether box B is empty or not depends on one's choice. Thus, the error in Table 25.1 is that of reverse causation. Using an unconditional probability α that box B is empty, the correct decision matrix is shown in Table 25.2.

This time, the expected utility of opening box B only is

$$\mathscr{E}U_{\text{box B}} = (0)(\alpha) + (1000000)(1 - \alpha) = 1000000 - 1000000\alpha.$$

The expected utility of opening both boxes is

$$\mathscr{E}U_{\text{both boxes}} = (1000)(\alpha) + (1001000)(1 - \alpha) = 1001000 - 1000000\alpha.$$

We now see that $\mathscr{E}U_{\text{both boxes}} > \mathscr{E}U_{\text{box B}}$ for all $0 < \alpha < 1$, which also agrees with the dominance principle. Thus *causal* decision theory helps us to address *Newcomb's Paradox*, resulting in a preference for opening two boxes. However, this is by no means the only word in the debate. For example, Bar-Hillel and Margalit (1972) write

> To sum up, we have tried to show that the Newcomb problem allows for just one, rather than two, 'rational' strategies. . . This strategy is to take only the covered box. It is not justified by arguing that it makes the million dollars more likely to be in that box, although

Table 25.1   Decision Matrix for Newcomb's Paradox

Decision	True state of nature	
	Box B empty	Box B not empty
Open box B only	$U=0$ (prob. $1-p$)	$U=1,000,000$ (prob. p)
Open both boxes	$U=1000$ (prob. p)	$U=1,001,000$ (prob. $1-p$)

Table 25.2   Modified Decision Matrix for Newcomb's Paradox

Decision	True state of nature	
	Box B empty (prob. α)	Box B not empty (prob. $1 - \alpha$)
Open box B only	$U = 0$	$U = 1,000,000$
Open both boxes	$U = 1000$	$U = 1,001,000$

that is the way it appears to be, but because it is inductively known to correlate remarkably with the existence of this sum in the box...

More recently, in his *Ruling Passions*, Blackburn (2001, p. 189) states

You need to become the kind of person who takes one box, even if both boxes are there (even in full view), and the second contains the bit extra which you had learned to forgo. If you are successful, then your character, adapted to a world with bizarre experimenters of this kind, is the one that a rational agent would wish for. Your action is the product a rational character would wish for. It brooks no criticism, for at the end you are the person with the big bucks. "Two boxers", misled by a false sense of rationality, have only the small bucks.

Newcomb's Paradox is named after the American theoretical physicist William A Newcomb (1927–1999), who discovered the problem in 1960. The problem was then communicated to the American political philosopher Robert Nozick (1938–2002, Fig. 25.1) by David Kruskal, a mutual friend of Newcomb and Nozick. The latter then wrote a paper on the paradox in *Essays in Honor of Carl G. Hempel* in 1969 (Nozick, 1969), and presented both solutions, but did not give a definite answer.

Problem 26

Fisher and the Lady Tasting Tea (1935)

Problem. *At a tea party, a lady claims she can discriminate by tasting a cup of tea whether the milk (M) or the tea (T) was poured first. To verify her claim, a statistician presents her in random order with eight cups of tea, and tells her there are four of each type M and T. The lady correctly identifies the four M types (and thus also correctly identifies the four T types). How can the statistician verify if her claim is true?*

Solution. Consider Table 26.1, which shows the lady's response in the tea-tasting experiment.

Conditional on the true states T and M, suppose the lady responds T with probabilities p_1 and p_2, respectively. The lady's claim is false if $p_1 = p_2$, that is, the odds ratio $\theta = \frac{p_1/(1-p_1)}{p_2/(1-p_2)} = 1$. Under the null hypothesis $H_0 : \theta = 1$, the probability of correctly answering T for each of the four true T types and not answering T for each of the four true M types is the product of the number of ways of choosing 4 T out of 4 T and the number of ways choosing 0 T out of 4 M, divided by the total number of ways of choosing 4 cups out of 8, that is,

$$p_{\text{obs}} = \frac{\binom{4}{4}\binom{4}{0}}{\binom{8}{4}} = .0143.$$

The reader will recognize that the above probability is based on a hypergeometric probability distribution.[†] There is no other table that is more extreme than the one

---

[†] The hypergeometric distribution is discussed in detail in Chapter 6 of Johnson et al. (2005) See also later in this problem.

Classic Problems of Probability, Prakash Gorroochurn.
© 2012 John Wiley & Sons, Inc. Published 2012 by John Wiley & Sons, Inc.

Table 26.1 The Lady's Response in Fisher's Tea-Tasting Experiment

		Lady's response		Sum
		T	M	
Cup's true "state"	T	4	0	4
	M	0	4	4
	Sum	4	4	8

observed, so the p-value of the test is $p = .0143$. At a level of significance level $\alpha = .05$, we therefore reject the null hypothesis that the lady's claim is false.

26.1 Discussion

The statistician in the tea-tasting experiment was none other than the foremost statistician of the twentieth century, namely Sir Ronald Aylmer Fisher (1890–1962) (Fig. 26.1).[†] In her biography of her father, Joan Fisher Box (1978, p. 134) recollects

> Already, quite soon after he [Fisher] had come to Rothamsted, his presence had transformed one commonplace tea time to an historic event. It happened one afternoon when he drew a cup of tea from the urn and offered it to the lady beside him, Dr B. Muriel Bristol, an algologist. She declined it, stating that she preferred a cup into which the milk had been poured first. "Nonsense," returned Fisher, smiling, "Surely it makes no difference." But she maintained, with emphasis, that of course it did. From just behind, a voice suggested, "Let's test her." It was William Roach, who was not long afterwards to marry Miss Bristol. Immediately, they embarked on the preliminaries of the experiment, Roach assisting with the cups and exulting that Miss Bristol divided correctly more than enough of those cups into which tea had been poured first to prove her case.[‡]

Apparently, Dr Bristol knew what she was talking about, for she identified all the eight cups correctly!

Sure enough, in the landmark 1935 book *The Design of Experiments* (Fisher, 1935), Fisher starts Chapter 2 with the mention of the experiment:

> A lady declares that by tasting a cup of tea made with milk she can discriminate whether the milk or the tea infusion was first added to the cup. We will consider the problem of designing an experiment by means of which this assertion can be tested. For this purpose let us first lay down a simple form of experiment with a view to studying its limitations and

---

[†] Fisher was also a top-notch geneticist who, together with S. Wright and J.B.S. Haldane, laid out the neo-Darwinian theory of evolution.

[‡] Kendall (1963), however, reports that Fisher denied the experiment actually took place: "The first edition of The Design of Experiments was published in 1935. It is another landmark in the development of statistical science. The experiment on tea tasting with which it opens has, I suppose, become the best-known experiment in the world, not the less remarkable because, as he once told me, Fisher never performed it."

Figure 26.1   Sir Ronald Aylmer Fisher (1890–1962). [Picture licensed under the Creative Commons Attribution-Share Alike 3.0 Unported license.]

its characteristics, both those which appear to be essential to the experimental method, when well developed, and those which are not essential but auxiliary.

In the remainder of the chapter, Fisher describes the experiment in considerable details and discusses the concepts of null hypothesis and level of significance.[†] Fisher highlights the importance of randomization as one of the key principles of experimental design. The statistical test he used and which is based on the hypergeometric distribution later came to be known as Fisher's exact test. Although in our case both margins of the 2 × 2 table were fixed, Fisher's exact test can also be used when only one or even none of the margins is fixed. The design itself is based on what is now called a single-subject randomization design. Fisher's tea-tasting experiment is

---

[†] According to David (1995, 1998), the term "null hypothesis" was first used by Fisher in his *Design of Experiments* (Fisher, 1935, p. 18), although "level of significance" had first been previously used by him in 1925 in his *Statistical Methods for Research Workers* (Fisher, 1925, p. 43). On the other hand, the term "*P* value" was first used later in W.E. Deming's *Statistical Adjustment of Data* (Deming 1943, p. 30)

Table 26.2 Hypothetical Scenario Where the Lady Identifies Only Three of the Four True T Types Correctly

		Lady's response		Sum
		T	M	
Cup's true "state"	T	3	1	4
	M	1	3	4
	Sum	4	4	8

undoubtedly "the best-known experiment of the world" (Kendall, 1963) and has been retold in countless textbooks of statistics.[†]

It is interesting to ask whether Dr Bristol was given a fair chance in the tea-tasting experiment. Would we still support her claim if she had identified only three of the four true T types correctly? Consider Table 16.2. The probability of the observed table is now

$$
p^*_{obs} = \frac{\binom{4}{3}\binom{4}{1}}{\binom{8}{4}} = .2286.
$$

There is only one 2×2 table more extreme than the one in Table 26.2, namely the table in Table 16.1. The p-value this time is therefore $.0143 + .2286 = .2429 > .05$, which means that we would now reject the lady's claim that she can discriminate between "tea first" and "milk first." This seems to suggest that the experiment was somewhat unfair because Dr Bristol needed to come up with a perfect 4 out of 4 for her claim to be statistically validated. However, a proper statistical answer to the question of whether the experiment was "fair" requires a concept that was outside of Fisher's considerations, namely that of the statistical power of the test.[‡]

To explain statistical power, let us make some brief remarks on Fisher's method of significance testing.[§] In the first step, a null hypothesis H_0 is set up. This usually represents the status-quo and the hypothesis the researcher would like to reject. Second, a test-statistic is calculated such that it represents the extent to which the

---

[†] To mention just a few examples, see Edgington (1995), Maxwell and Delaney (2004, p. 37), Schwarz (2008, p. 12), Hinkelmann and Kempthorne (2008, p. 139), Rosenbaum (2010, p. 21), Fleiss et al. (2003, p. 244), Senn (2003, p. 54), Todman and Dugard (2001, p. 32), Agresti (2002, p. 92), and Chernick and Friis (2003, p. 329). One writer, David Salsburg, has written a wonderful book on the history of statistics in the 20th century entitled *The Lady Tasting Tea: How Statistics Revolutionized Science in the Twentieth Century* (Salsburg, 2001).

[‡] That does not mean that Fisher was unaware of the issue, for in Chapter 2 of his *Design of Experiments* he discusses the sensitivity of his design.

[§] For more technical and historical details, see Chapter 14 of Spanos (2003), Chapter 15 of Cowles (2001), and Howie (2004, pp. 172–191).

sample statistic is different from the value of the parameter under the null hypothesis. The test-statistic is usually based on a random variable called a *pivotal quantity*, which is a function of both the sample and the parameter being tested for, but whose distribution is independent of the parameter under the null hypothesis. In the final step, the probability of observing a test-statistic as extreme as or more extreme than the observed test-statistic calculated (under the assumption that the null hypothesis is true). This is the *p*-value or level of significance achieved.

Note in particular that Fisher does not explicitly make use of an alternative hypothesis (H_a). However, to be able to calculate the statistical power of a test, one needs an explicit formulation of the alternative hypothesis. The power, $1 - \beta$, of the test is then the probability of rejecting the null hypothesis given that the alternative hypothesis is true, that is,

$$1 - \beta = \Pr\{\text{Reject } H_0 | H_a\}.$$

In the above β is also called the Type II error of the test. Moreover, we also need to take into account another associated quantity, called the level of significance or Type I error, of a test. This is the probability of rejecting H_0 when H_0 is in fact true, that is,

$$\alpha = \Pr\{\text{Reject } H_0 | H_0\}.$$

These two quantities, α and β, reflect two different aspects of the performance of a test statistic. The quantity α measures how likely one is to mistakenly reject a true null hypothesis, while the quantity β measures how likely one is to mistakenly fail to reject a false null hypothesis.

What is the relationship between the Type I error α and the *p*-value? Remember that α is usually a fixed quantity (such as .05 or .01) that is specified before the test is actually carried out. On the other hand, the *p*-value is calculated based on the calculated test-statistic.

Now, the *p*-value is also be defined as *the smallest value of α for which we can still reject H_0*, that is,

$$p\text{-value} \leq \alpha \Rightarrow \text{reject } H_0.$$

The notions of an explicit alternative hypothesis, and of Type I and Type II errors were some of the highlights of an alternative approach to Fisher's that the eminent statisticians Jerzy Neyman (1894–1981) (Fig. 26.2) and Egon S. Pearson (1895–1980) (Fig. 26.3) put forward in the late twenties and early thirties (Neyman and Pearson, 1928a, 1928b, 1933). Moreover, Neyman and Pearson provided the means of constructing optimal statistical tests, as opposed to the less formal pivotal method of Fisher. Finally, Neyman and Pearson advocated the use of rejection regions, rather than *p*-values, in order to make a final statistical decision. Needless to say, Neyman and Pearson's prescriptions did not flow well with Fisher and gave rise to often-heated debates between the two sides. It is also ironical that the current practice of using both null and alternative hypotheses and *p*-values is an amalgam that would not have been agreeable to either party.

Figure 26.2   Jerzy Neyman (1894–1981). [Picture licensed under the Creative Commons Attribution-Share Alike 2.0 Germany license.]

Some believe that the use of an $\alpha = .05$ (or sometimes .01) level of significance to decide whether a null hypothesis ought to be rejected is because of some deep underlying statistical theory. This is not true. Indeed, the 5% value has nothing sacrosanct in it. It stems from convention and is undoubtedly because Fisher himself advocated this number since it seemed reasonable to him. Fisher first used the 5% convention[†] in the first edition of his book *Statistical Methods for Research Workers* (1925, p. 79). He wrote

> If P is between .1 and .9 there is certainly no reason to suspect the hypothesis tested. If it is below .02 it is strongly indicated that the hypothesis fails to account for the whole of the facts. We shall not often be astray if we draw a conventional line at .05, and consider that higher values of X^2 indicate a real discrepancy.

[†] But see Cowles and Davis' very interesting article on the use of the .05 significance level before Fisher (Cowles and Davis, 1982).

Figure 26.3   Egon S. Pearson (1895–1980).

If one in twenty does not seem high enough odds, we may, if we prefer it, draw the line at one in fifty (the 2 per cent point), or one in a hundred (the 1 per cent point). Personally, the writer prefers to set a low standard of significance at the 5 per cent point, and ignore entirely all results which fail to reach this level. A scientific fact should be regarded as experimentally established only if a properly designed experiment rarely fails to give this level of significance.

Coming back to the issue as to whether Fisher's experiment was really "fair" towards Dr Bristol, we need to calculate the statistical power of the Fisher's exact test under some suitable alternative. That is, even if Dr Bristol had some reasonably good ability to discriminate T from M, how likely is it that her claim would be statistically vindicated, given the nature of the statistical test, the sample size, and the Type I error allowed? Consider Table 26.3.

Under $H_0 : \theta = 1$ (no discriminatory power), the distribution of N_{11} is hypergeometric:

$$\Pr\{N_{11} = n_{11} | H_0\} = \frac{\binom{4}{n_{11}}\binom{4}{4 - n_{11}}}{\binom{8}{4}} \quad \text{for } n_{11} = 0, 1, 2, 3, 4. \qquad (26.1)$$

Table 26.3   General Set-Up for The Lady Tea-Tasting Experiment

		Lady's response		Sum
		T	M	
Cup's true "state"	T	N_{11}	$4-N_{11}$	4
	M	$4-N_{11}$	N_{11}	4
	Sum	4	4	8

On the other hand, under $H_a : \theta \, (\neq 1)$, the distribution of N_{11} is noncentral hypergeometric:

$$\Pr\{N_{11} = n_{11}|H_a\} = \frac{\binom{4}{n_{11}}\binom{4}{4-n_{11}}\theta^{n_{11}}}{\sum_{j=0}^{4}\binom{4}{j}\binom{4}{4-j}\theta^{j}} \quad \text{for } n_{11} = 0, 1, 2, 3, 4. \quad (26.2)$$

First we set $\alpha = .01$. Before we can calculate the power, let us find the value of n_{11} such that

$$\Pr\{N_{11} \geq n_{11}|H_0\} = .01.$$

From our computation in **Problem 26,** we see that $n_{11} = 4$. We are almost ready for the power calculation now. It is only left to specify H_a. Suppose Dr Bristol's probability of responding T when the cup is truly of type T is $p_1 = .9$ and her probability of responding T when the cup is truly of type M is $p_2 = .2$. Then $H_a : \theta = \frac{p_1/(1-p_1)}{p_2/(1-p_2)} = 36.0$. Noting that the distribution of N_{11} under H_a is noncentral hypergeometric, the power of the test is, from Eq. (26.2),

$$\Pr\{N_{11} \geq 4|H_a : \theta = 36.0\} = \Pr\{N_{11} = 4|H_a : \theta = 36.0\}$$

$$= \frac{\binom{4}{4}\binom{4}{0}(36.0)^4}{\sum_{j=0}^{4}\binom{4}{j}\binom{4}{4-j}(36.0)^j}$$

$$= .679.$$

This power is much less than the 80% that we would normally require for a well-powered study. Thus our power calculations show that, based on the assumptions we have made, even if Dr Bristol in the tea-tasting experiment had excellent discrimination, the odds had been stacked against her.

Finally, let us make some comments on the origins of the hypergeometric distribution. The distribution can be traced to de Moivre's *De Mensura Sortis* (de Moivre, 1711 Hald, 1984). On p. 235, de Moivre's Problem 14[†] reads as follows:

[†] This is the same question that Huygens had previously asked in Problem 4 of the *De ratiociniis in aleae ludo* (Huygens, 1657).

Taking 12 counters, 4 white and 8 black, *A* contends with *B* that if he shall have taken out, blindfold, 7 counters 3 of them would be white; the ratio of the expectation of *A* to the expectation of *B* is sought.

De Moivre first calculates the total number of ways of choosing 7 counters out of 12 as

$$\frac{12}{1} \cdot \frac{11}{2} \cdot \frac{10}{3} \cdot \frac{9}{4} \cdot \frac{8}{5} \cdot \frac{7}{6} \cdot \frac{6}{7} = 792.$$

He then calculates the number of ways of choosing 3 white counters from 4 white counters and 4 black counters from 8 black counters as

$$4 \cdot \left(\frac{8}{1} \cdot \frac{7}{2} \cdot \frac{6}{3} \cdot \frac{5}{4} \right) = 280.$$

De Moivre also calculates the number of ways of choosing 4 white counters from 4 white counters and 3 black counters from 8 black counters:

$$1 \cdot \left(\frac{8}{1} \cdot \frac{7}{2} \cdot \frac{6}{3} \right) = 56.$$

De Moivre thus obtains the probability for *A* to obtain at least three white counters as $(280 + 56)/792 = 14/33$. We see that de Moivre has in effect used the hypergeometric distribution. In modern notation, if *X* is the number of white counters, then

$$\Pr\{X \geq 3\} = \frac{\binom{4}{3}\binom{8}{4} + \binom{4}{4}\binom{8}{3}}{\binom{12}{7}} = \frac{14}{33}.$$

On p. 236 of the *De Mensura Sortis*, de Moivre gives the formula for the general case. In general, suppose an urn contains *a* white and *b* black balls, and *n* balls are selected at random without replacement. Let *X* be the number of white balls in the sample drawn. Then *X* is said to have a hypergeometric distribution with parameters *n*, *a*, and *b*, and its probability mass function is

$$\Pr\{X = x\} = \frac{\binom{a}{x}\binom{b}{n-x}}{\binom{a+b}{n}}, \quad \text{where } \max(0, n - b) \leq x \leq \min(n, a).$$

Problem 27

Benford and the Peculiar Behavior of the First Significant Digit (1938)

Problem. *In many naturally occurring tables of numerical data, it is observed that the leading (i.e., leftmost) digit is not uniformly distributed among {1, 2, . . ., 9}, but the smaller digits appear more frequently that the larger ones. In particular, the digit one leads with a frequency close to 30% whereas the digit nine leads with a frequency close to only 4%. Explain.*

Solution. We make use of the fact that, for a given data set, changing the units of measurements should not change the distribution $p(x)$ of the first significant digit. That is, this distribution should be scale invariant:

$$p(kx) = Cp(x),$$

where k and C are constants. Integrating on both sides,

$$\int p(kx)dx = C\int p(x)dx$$

$$\frac{1}{k}\int p(kx)d(kx) = C(1)$$

$$C = \frac{1}{k}.$$

Classic Problems of Probability, Prakash Gorroochurn.
© *2012 John Wiley & Sons, Inc. Published 2012 by John Wiley & Sons, Inc.*

Therefore $kp(kx) = p(x)$. Let us now differentiate the latter w.r.t. k:

$$\frac{d}{dk}\{kp(kx)\} = 0$$

$$p(kx) \cdot 1 + k\frac{d}{dk}p(kx) = 0$$

$$p(kx) + k\frac{d}{d(kx)}p(kx) \cdot \frac{d(kx)}{dk} = 0$$

$$p(kx) + kxp'(kx) = 0.$$

Writing $u = kx$, we have $up'(u) = -p(u)$ so that

$$\int \frac{p'(u)}{p(u)}du = -\int \frac{1}{u}du \Rightarrow p(u) = \frac{1}{u}$$

is a solution. Let D be the first significant digit in base 10. Then its probability mass function is

$$Pr\{D = d\} = \frac{\int_d^{d+1} \frac{1}{u}du}{\int_1^{10} \frac{1}{u}du}$$

$$= \frac{\ln(1 + d) - \ln d}{\ln 10}$$

$$= \frac{\ln(1 + 1/d)}{\ln 10}$$

$$= \log_{10}(1 + 1/d), \quad d = 1, 2, \ldots, 9. \tag{27.1}$$

Table 27.1 shows the values of $Pr\{D = d\}$ for values of d.

Table 27.1   Probability that a First Significant Digit in Base Ten is d, as Calculated from Eq. (27.1)

d	$Pr\{D = d\}$
1	.3010
2	.1761
3	.1249
4	.0969
5	.0792
6	.0669
7	.0580
8	.0512
9	.0458

Figure 27.1   Simon Newcomb (1835–1909).

The table shows that the smaller the leading digit the more frequent it is. In particular the digit one has a theoretical frequency of 30.10% while the digit nine has a theoretical frequency of 4.58%.

27.1 Discussion

This peculiar behavior of the leading digit is known as the first-digit law or Benford's law.[†] Frank Benford (1883–1948) (Fig. 27.2) was an American physicist who rediscovered and popularized the law in 1938 (Benford, 1938). In his paper, Benford presented a table of the frequency of occurrence of first digits of 20,229 naturally occurring numbers, which included the areas of rivers, population sizes, physical constants, and so on. For each group of data presented, Benford showed a remarkable fit to the first-digit law.

[†] Benford's law is also discussed in Krishnan (2006, pp. 55–60), Havil (2008, Chapter 16), Richards (1982, Chapter 6), Tijms (2007, p. 194), Hamming (1991, p. 214), Frey (2006, p. 273), and Pickover (2005, p. 212).

Figure 27.2   Frank Benford (1883–1948).

Newcomb's law had already been observed previously. Almost 60 years prior to Benford's article, the Canadian-American mathematician and astronomer Simon Newcomb (1835–1909) (Fig. 27.1) observed in a short article in 1881 (Newcomb, 1881)[†]:

> That the ten digits do not occur with equal frequency must be evident to any one making much use of logarithmic tables, and noticing how much faster the first pages wear out than the last ones. The first significant figure is oftener 1 than any other digit, and the frequency diminishes up to 9.

He concluded

> The law of probability of the occurrence of numbers is such that all mantissas of their logarithms are equally probable.

This statement in essence implies Eq. (27.1). However, Newcomb did not present any statistical evidence for his discovery. Since the first enunciation of the law, Benford's law has been discussed by several authors, many of whom are mentioned in Raimi (1976). In particular, Pinkham (1961) showed that the $\log_{10}(1 + 1/d)$ distribution of the first digit is the only one that remains scale invariant, that is, *scale invariance implies Benford's law*, as we showed in the solution. However, no

---

[†] Another example of Stigler's law of eponymy?

Table 27.2   The First 100 Fibonacci Numbers (Right Column Courtesy of Bruce Levin)

1	121,393	20,365,011,074	3,416,454,622,906,707
1	196,418	32,951,280,099	5,527,939,700,884,757
2	317,811	53,316,291,173	8,944,394,323,791,464
3	514,229	86,267,571,272	14,472,334,024,676,221
5	832,040	139,583,862,445	23,416,728,348,467,685
8	1,346,269	225,851,433,717	37,889,062,373,143,906
13	2,178,309	365,435,296,162	61,305,790,721,611,591
21	3,524,578	591,286,729,879	99,194,853,094,755,497
34	5,702,887	956,722,026,041	160,500,643,816,367,088
55	9,227,465	1,548,008,755,920	259,695496,911,122,585
89	14,930,352	2,504,730,781,961	420,196,140,727,489,673
144	24,157,817	4,052,739,537,881	679,891,637,638,612,258
233	39,088,169	6,557,470,319,842	1,100,087,778,366,101,931
377	63,245,986	10,610,209,857,723	1,779,979,416,004,714,189
610	102,334,155	17,167,680,177,565	2,880,067,194,370,816,120
987	165,580,141	27,777,890,035,288	4,660,046,610,375,530,309
1,597	267,914,296	44,945,570,212,853	7,540,113,804,746,346,429
2,584	433,494,437	72,723,460,248,141	12,200,160,415,121,876,738
4,181	701,408,733	117,669,030,460,994	19,740,274,219,868,223,167
6,765	1,134,903,170	190,392,490,709,135	31,940,434,634,990,099,905
10,946	1,836,311,903	308,061,521,170,129	51,680,708,854,858,323,072
17,711	2,971,215,073	498,454,011,879,264	83,621,143,489,848,422,977
28,657	4,807,526,976	806,515,533,049,393	135,301,852,344,706,746,049
46,368	7,778,742,049	1,304,969,544,928,657	218,922,995,834,555,169,026
75,025	12,586,269,025	2,111,485,077,978,050	354,224,848,179,261,915,075

satisfactory statistical explanation of the first-digit phenomenon existed until Hill's 1995 paper (Hill, 1995a). In it, Hill proved that

> If distributions are selected at random (in any "unbiased" way) and random samples are then taken from each of these distributions, the significant digits of the combined sample will converge to the logarithmic (Benford) distribution.

The Benford distribution is thus the distribution of distributions. Not all distributions of numbers occurring in nature conform to Benford's law. Numbers that vary within a narrow range, such as the heights of humans or the number of airplane crashes in a given month, will not. On the other hand, measurements that vary across several orders of magnitude such as populations of countries, areas of lakes, and stock prices probably will. It has been observed that numbers that follow a "power law" usually follow the Benford distribution (Pickover, 2009, p. 274). This is the case with numbers that are in geometric progression and with the Fibonacci numbers.[†] Table 27.2 shows the first 100 Fibonacci numbers. You will note that a leading digit of one is more common

---

[†] The Fibonacci numbers are a well-known sequence of numbers whose nth term f_n is such that $f_n = f_{n-1} + f_{n-2}$, with $f_1 = 1, f_2 = 1$. For more details, see for example Posamentier and Lehmann (2007), which is an entire book on the topic.

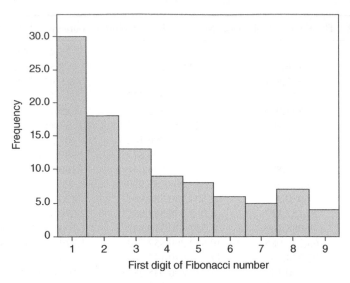

Figure 27.3   Histogram of the first digits of the Fibonacci numbers in Table 27.2.

than a two, which in turn is more common than a three, and so on. The histogram in Fig. 27.3 shows the distribution of the first digits. The frequencies are in remarkable agreement with the probabilities listed in Table 27.1.

Two other comments are in order. First, scale invariance implies base invariance, which in turn implies Benford's law (Hill, 1995a, 1995b). That is to say, if the underlying distribution of numbers is invariant to the unit of measurement used, then it is also invariant to the base system used to represent the numbers. This in turn implies the numbers must follow Benford's Law. Second, we can also obtain the probability mass functions of other significant digits, apart from the first, by using a similar argument as in our solution. Thus, if $D_b^{(i)}$ denotes the ith significant digit in base b, then the joint mass function of $D_b^{(1)}, D_b^{(2)}, \dots, D_b^{(k)}$ is

$$\Pr\left\{D_b^{(1)} = d_1, D_b^{(2)} = d_2, \dots, D_b^{(k)} = d_k\right\} = \log_b\left(1 + \frac{1}{d_1 b^{k-1} + d_2 b^{k-2} + \cdots + d_k}\right).$$
(27.2)

By using Eq. (27.2), we can obtain both conditional and marginal distributions. Suppose we want to find the marginal probability mass function of $D_{10}^{(2)}$. We have

$$\Pr\left\{D_{10}^{(1)} = d_1, D_{10}^{(2)} = d_2\right\} = \log_{10}\left(1 + \frac{1}{10d_1 + d_2}\right), \quad d_1, d_2 = 1, 2, \dots, 9,$$

implying

$$\Pr\left\{D_{10}^{(2)} = d_2\right\} = \sum_{d_1=1}^{9} \log_{10}\left(1 + \frac{1}{10d_1 + d_2}\right), \quad d_2 = 1, 2, \dots, 9.$$

Finally, we draw the reader's attention to Steven W. Smith's book *The Scientist and Engineer's Guide to Digital Signal Processing*[†] (Smith, 1999, Chapter 34). Using simulations, Smith points out that lognormal distributions[‡] with standard deviations larger than unity satisfy Benford's law quite accurately. In general, the larger the standard deviation, the more accurate the fit. Smith further adds (p. 721)

> ...the CLT [Central Limit Theorem] describes that adding many random numbers produces a normal distribution. This accounts for the normal distribution being so commonly observed in science and engineering. However, if a group of random numbers are multiplied, the result will be a normal distribution on the logarithmic scale. Accordingly, the log-normal distribution is also commonly found in Nature. This is probably the single most important reason that some distributions are found to follow Benford's law while others do not. Normal distributions are not wide enough to follow the law. On the other hand, broad lognormal distributions follow it to a very high degree.

A similar result is reported in Scott and Fasli (2001). These authors further note that real-life distributions with positive values and with medians much closer to zero than their means are likely to conform to "broad" lognormal distributions and hence satisfy Benford's law.

[†] This source was pointed to me by an anonymous reviewer.

[‡] If the random variable e^X is normally distributed then X has a lognormal distribution. The latter is discussed in detail in Chapter 14 of Johnson et al. (1995).

Problem 28

Coinciding Birthdays (1939)

Problem. *What is the least number of people that must be in a room before the probability that some share a birthday becomes at least 50%? Assume there are 365 days each year and that births are equally likely to occur on any day.*

Solution. Suppose there are n persons in the room. Then each person can have 365 possible birthdays, so that the total number of possible birthdays for the n individuals is $N = 365^n$. Let A be the event "each of the n persons has a different birthday." Then a given person has 365 possibilities for a birthday, the next person has 364 possibilities, and so on. Therefore, the total number of ways the event A can be realized is

$$n_A = 365 \times 364 \times \cdots \times (365 - n + 1).$$

Using the classical definition, the probability that all n individuals have a different birthday is

$$p_A = \frac{n_A}{N}$$

$$= \frac{365 \times 364 \times 363 \times \cdots \times (365 - n + 1)}{365^n} \tag{28.1}$$

$$= \left(\frac{365}{365}\right)\left(\frac{365 - 1}{365}\right)\left(\frac{365 - 2}{365}\right)\cdots\left(\frac{365 - (n-1)}{365}\right)$$

$$= \left(1 - \frac{1}{365}\right)\left(1 - \frac{2}{365}\right)\cdots\left(1 - \frac{n-1}{365}\right).$$

Classic Problems of Probability, Prakash Gorroochurn.
© 2012 John Wiley & Sons, Inc. Published 2012 by John Wiley & Sons, Inc.

Taking natural logarithms on both sides, we have[†]

$$\ln p_A = \ln\left(1 - \frac{1}{365}\right) + \ln\left(1 - \frac{2}{365}\right) + \cdots + \ln\left(1 - \frac{n-1}{365}\right)$$

$$\approx -\frac{1}{365} - \frac{2}{365} - \cdots - \frac{n-1}{365}$$

$$= -\frac{1 + 2 + \cdots + (n-1)}{365}$$

$$= -\frac{(n-1)n/2}{365}.$$

Thus, $p_A \approx \exp\left\{-\frac{n(n-1)}{2 \times 365}\right\}$ and the probability that at least two individuals share a birthday is

$$q_A = 1 - p_A$$

$$= 1 - \left(1 - \frac{1}{365}\right)\left(1 - \frac{2}{365}\right)\cdots\left(1 - \frac{n-1}{365}\right) \qquad (28.2)$$

$$\approx 1 - \exp\left\{-\frac{n(n-1)}{2 \times 365}\right\}.$$

We now solve $1 - \exp\left\{-\frac{n(n-1)}{2 \times 365}\right\} \geq .5$ obtaining $n \geq 23$. Since we have used the approximate formula for q_A in the inequality, we cannot be confident of this answer. Let us evaluate q_A for $n = 22$, 23, and 24 using the exact formula in Eq. (28.2). We have $q_A = .476$, .507, and .538 (to 3 d.p.), respectively. Therefore, we need at least 23 persons in the room before the probability that some share a birthday becomes at least 50%.

28.1 Discussion

The problem was first proposed by von Mises[‡] (1939). That we need only 23 people for slightly more than an even chance might appear intriguing.[§] If we apply the

---

[†] In the following we shall make use of $\ln(1 + x) \approx x$ for small x and of $1 + 2 + \cdots + m \equiv m(m + 1)/2$.

[‡] Ball and Coxeter (1987, p. 45), however, attribute it to Harold Davenport.

[§] The Birthday Problem is also discussed in Weaver (1982, p. 9), DasGupta (2010, p. 23), Gregersen (2011, p. 57), Mosteller (1987, p. 49), Terrell (1999, p. 100), Finkelstein and Levin (2001, p. 45), Hein (2002, p. 301), Sorensen (2003, p. 353), Andel (2001, p. 89), Woolfson (2008, p. 45), and Stewart (2008, p. 132).

Table 28.1   The Probability that At Least Two People have the Same Birthday When There are n Persons in a Room

Number of people in room (n)	Probability that at least two persons have the same birthday (q_A)
10	.117
20	.411
23	.507
30	.706
57	.990
366	1.00

pigeonhole principle,[†] we see that 366 persons are required for a probability of one that at least two have the same birthday. Naïve reasoning would probably suggest a number close to 183 for at least an even chance. To understand why 23 people would be sufficient, note that with this number we have $\binom{23}{2} = 253$ possible comparisons of pairwise birthdays, any of which could result in a coincidence. Table 28.1 shows values of n and q_A.

There is an easy way to misinterpret the solution of the *Birthday Problem*. Many believe it implies that, if a particular person is chosen from a group of 23, then there is slightly more than 50% chance that she will have the same birthday as at least one person from the remaining 22. This is not true. Paulos (1988, p. 36) gives a typical illustration:

> A couple of years ago, someone on the Johnny Carson show was trying to explain this. Johnny Carson didn't believe it, noted that there were about 120 people in the studio audience, and asked how many of them shared his birthday of, say, March 19. No one did, and the guest, who wasn't a mathematician, said something incomprehensible in his defense.

Our solution refers to the probability for *any* two or more people to have the same birthday, not for any *particular* person to have the same birthday as at least somebody else. This leads us to the following classic problem:

> *What is the smallest number of people so that any particular person has the same birthday as at least one other person with probability at least .5?*

Consider a group of n people. Let us calculate the probability of the event $A*$ that a particular person has a birthday in common with nobody else. Then the total number of possible birthdays is the same as before, namely $N = 365^n$. This time, though, the number of ways the event $A*$ can be realized is

---

[†] The pigeonhole principle states that if n objects are put in p boxes, and $n > p$, then at least one box receives two or more objects.

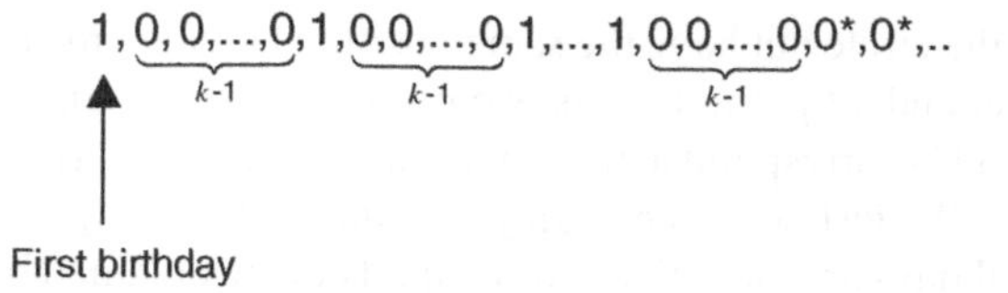

$$1, \underbrace{0, 0, \ldots, 0}_{k-1}, 1, \underbrace{0, 0, \ldots, 0}_{k-1}, 1, \ldots, 1, \underbrace{0, 0, \ldots, 0}_{k-1}, 0^*, 0^*, ..$$

First birthday

Figure 28.1   Illustration of the almost birthday problem.

$$n_{A*} = 365 \times \overbrace{364 \times 364 \times \cdots \times 364}^{n-1 \text{ terms}} = 365(364)^{n-1}.$$

Therefore, $p_{A*} = n_{A*}/N = (364/365)^{n-1}$. We set

$$1 - \left(\frac{364}{365}\right)^{n-1} \geq .5,$$

obtaining $n \geq 253.65$. Therefore, we need at least 254 people to have more than an even chance that one particular person will share the same birthday as at least somebody else. Two additional, less straightforward variants of the classical *Birthday Problem* are also worth mentioning. The first is the so-called *Almost Birthday Problem*.

What is the smallest number of people required so that at least two have a birthday within k adjacent days with probability half or more (Naus, 1968)?[†]

To solve the above, we consider a group of n people and define A_k as the event that at least two birthdays are within k adjacent days. Then we have

$$\Pr\{A_k\} = 1 - \Pr\{\bar{A}_k\} = 1 - \Pr\{\bar{A}_k | \bar{A}_1\} \Pr\{\bar{A}_1\}. \tag{28.3}$$

In the above equation, $\bar{A}_1$ is the event that no two birthdays coincide. From Eq. (28.1), we have

$$\Pr\{\bar{A}_1\} = \frac{365!}{(365 - n)!365^n}.$$

We next define the indicator variable I_j ($j = 1, 2, \ldots, 365$) on each of the days of a year such that

$$I_j = \begin{cases} 1, & \text{if there is one birthday on day } j, \\ 0, & \text{otherwise.} \end{cases}$$

Now, an arrangement such that each of n different birthdays is k days apart of the next looks as in Fig 28.1. In the above arrangement, we imagine the days of the year as arranged in a circle so that the 365th of a given year and the first day of the next year are adjacent (i.e., are within $k = 2$ adjacent days). Now, when objects are arranged in a circle, redundant permutations arise when the objects are at different

---

[†] For example, two birthdays on March 3rd and March 7th are within five adjacent days.

positions along the circle but have the same positions relative to each other. In order to avoid these redundant permutations, we consider the first birthday as fixed. The additional $(n-1)$ 1's correspond to the remaining $(n-1)$ birthdays and the starred 0's are the days after the last $(k-1)$ birthday-free days. An arrangement such that any two adjacent birthdays are more than k days apart can be obtained by considering the first 1 and the "unstarred" birthday-free days as fixed and permuting the $(n-1)$ 1's and the $(365-1)-(n-1)-n(k-1) = 365 - kn$ starred 0's. The number of such permutations is

$$\binom{n - kn + 364}{n - 1}.$$

This implies

$$\Pr\{\bar{A}_k|\bar{A}_1\} = \frac{\binom{n - kn + 364}{n - 1}}{\binom{364}{n - 1}}.$$

From Eq. (28.3) we have

$$\Pr\{A_k\} = 1 - \frac{\binom{n - kn + 364}{n - 1}}{\binom{364}{n - 1}} \cdot \frac{365!}{(365 - n)!365^n}$$

$$= 1 - \frac{(364 - kn + n)!}{(365 - kn)!365^{n-1}}. \tag{28.4}$$

We can use Eq. (28.4) to find the smallest number of people required so that at least two of them have birthdays within k adjacent days with probability .5 or more. It can also be shown that an approximation for this number is $\lceil n_k \rceil$,[†] where

$$n_k = 1.2\sqrt{\frac{365}{2k - 1}}$$

(Diaconis and Mosteller, 1989). Table 28.2 shows values of k and $\lceil n_k \rceil$.

The second less straightforward variant of the classical *Birthday Problem* is the *Multiple Birthday Problem*:

What is the smallest number of people required so that at least m of them have the same birthday with probability half or more?

The above problem can be cast as a cell occupancy question: if one randomly distributes N items over $c = 365$ cells or categories that are equally likely, what is the

---

[†] $\lceil n_k \rceil$ is the smallest integer greater or equal to n_k.

Table 28.2 Smallest Number of People Required for At Least Two of them to have Birthdays Within k Days Apart with Probability At Least 1/2

Number of days apart (k)	Smallest number of people required ($[n_k]$)
1	23
2	14
3	11
4	9
5	8
6	7
11	6

least number of items required so that at least one cell has m or more items with probability .5 or more? To answer this question, we use the method developed by Levin (1981). First, we consider a c-category multinomial distribution with parameters $(N; p_1, p_2, \ldots, p_c)$. Next let the number of items in cell j ($j = 1, 2, \ldots, c$) be $X_j \sim Po(sp_j)$,[†] where $s > 0$[‡] and the X_j's are independent. Using Bayes' Theorem,[§] we have

$$\Pr\{X_1 \le n_1, \ldots, X_c \le n_c | \sum_{j=1}^{c} X_j = N\}$$

$$= \frac{\Pr\{\sum_{j=1}^{c} X_j = N | X_1 \le n_1, \ldots, X_c \le n_c\} \Pr\{X_1 \le n_1, \ldots, X_c \le n_c\}}{\Pr\{\sum_{j=1}^{c} X_j = N\}}$$

$$= \frac{\Pr\{W = N\} \Pr\{X_1 \le n_1, \ldots, X_c \le n_c\}}{\Pr\{\sum_{j=1}^{c} X_j = N\}}. \tag{28.5}$$

Now, in Eq. (28.5) above, the left side is a multinomial distribution with parameters $(N; p_1, p_2, \ldots, p_c)$,[**] the denominator on the right is a Poisson random variable with mean s, and the random variable W is the sum of c independent truncated Poisson (TPo) random variables, that is, $W = Y_1 + Y_2 + \cdots + Y_c$, where $Y_j \sim TPo(sp_j)$ with support $0, 1, \ldots, n_j$. Denoting the cell counts in the multinomial distribution by $(N_1, N_2, \ldots, N_c)$ where $\sum_{j=1}^{c} N_j = N$, Eq. (28.5) becomes

[†] Since we will soon be using these Poisson random variables to represent the multinomial frequencies (as that of Poisson frequencies given a fixed sum).

[‡] s can theoretically be any positive real number; Levin (1981) recommends choosing $s = N$ that works satisfactorily and has an intuitive interpretation that $sp_i = Np_i$ are the expected cell counts.

[§] See **Problem 14**.

[**] This is because, in general, if $V_1, V_2, \ldots, V_c$ are independent Poisson random variables with expected values $t_1, t_2, \ldots, t_c$, respectively, then the conditional joint distribution of $V_1, V_2, \ldots, V_c$, given that $\sum_i V_i = N$, is multinomial with parameters $(N; t_1/t, t_2/t, \ldots, t_c/t)$, where $t = \sum_i t_i$. See Johnson et al. (1997, Chapter 5) for more details.

Table 28.3   Smallest Number of People (N) Required so that At Least m of them have the Same Birthday with Probability Half or More (Values Obtained from Bruce Levin)

m	N
2	23
3	88
4	187
5	313
6	460
7	623
8	798
9	985
10	1181

$$\Pr\{N_1 \leq n_1, \ldots, N_c \leq n_c\} = \frac{N!}{s^N e^{-s}} \left[\prod_{j=1}^{c} \Pr\{X_j \leq n_j\} \right] \Pr\{W = N\} \qquad (28.6)$$

Levin's method can be used for exact computation, and he also gives a so-called Edgeworth approximation for the term $\Pr\{W = N\}$ for cases where c is large and exact calculation of the convolution to get $\Pr\{W = N\}$ is arduous. Using Eq. (28.6) we are now ready to solve the *Multiple Birthday Problem*. The probability that, among N people, at least m have the same birthday is

$$1 - \Pr\{N_1 \leq m-1, \ldots, N_c \leq m-1\} = 1 - \frac{N!}{s^N e^{-s}} \left[\prod_{j=1}^{c} \Pr\{X_j \leq m-1\} \right] \Pr\{W = N\}.$$

$$(28.7)$$

Then, for a given m, one can find the smallest value of N for which the probability exceeds one-half. One can use results from preceding calculations to predict the next value of N for the next larger m. Levin has used this method to calculate the values in Table 28.3 using exact computation (personal communication).

Problem 29

Lévy and the Arc Sine Law (1939)

Problem. *A fair coin is tossed successively. Let $X_j = +1$ if the jth toss is a head, and $X_j = -1$ if the jth toss is a tail. Define the difference between the number of heads and tails after j tosses by $S_j = X_1 + X_2 + \cdots + X_j$. Show that the probability that the last time there was an equal number of heads and tails within 2n tosses occurred at the 2kth toss is $\Pr\{S_{2k} = 0\} \cdot \Pr\{S_{2n-2k} = 0\}$.*

Solution. Let $p_{2j}^{(k)} = \Pr\{S_{2j} = k\} = N_{2j}^{(k)}/2^{2j}$, where $N_{2j}^{(k)}$ is the number of ways the event $S_{2j} = k$ can be realized. We first establish an important result. We have

$$\Pr\{S_1 \neq 0, S_2 \neq 0, \ldots, S_{2j} \neq 0\} = 2\Pr\{S_1 > 0, S_2 > 0, \ldots, S_{2j} > 0\}, \qquad (29.1)$$

since the coin is fair. Using the law of total probability,

$$\Pr\{S_1 > 0, S_2 > 0, \ldots, S_{2j} > 0\} = \sum_{r=1}^{\infty} \Pr\{S_1 > 0, S_2 > 0, \ldots, S_{2j-1} > 0, S_{2j} = 2r\}.$$

Applying the Ballot Theorem (see **Problem 17**),

$$
\begin{aligned}
\Pr\{S_1 > 0, S_2 > 0, \ldots, S_{2j-1} > 0, S_{2j} = 2r\} &= \frac{N_{2j-1}^{(2r-1)} - N_{2j-1}^{(2r+1)}}{2^{2j}} \\
&= \frac{1}{2} \cdot \frac{N_{2j-1}^{(2r-1)} - N_{2j-1}^{(2r+1)}}{2^{2j-1}} \\
&= \frac{1}{2}\left\{ p_{2j-1}^{(2r-1)} - p_{2j-1}^{(2r+1)} \right\}.
\end{aligned}
$$

Classic Problems of Probability, Prakash Gorroochurn.
© 2012 John Wiley & Sons, Inc. Published 2012 by John Wiley & Sons, Inc.

Therefore, we have

$$\Pr\{S_1 > 0, S_2 > 0, \ldots, S_{2j} > 0\} = \frac{1}{2} \sum_{r=1}^{\infty} \left\{ p_{2j-1}^{(2r-1)} - p_{2j-1}^{(2r+1)} \right\}$$

$$= \frac{1}{2} \left\{ p_{2j-1}^{(1)} - p_{2j-1}^{(3)} + p_{2j-1}^{(3)} - p_{2j-1}^{(5)} + \cdots \right\}$$

$$= \frac{1}{2} p_{2j-1}^{(1)}$$

$$= \frac{1}{2} p_{2j}^{(0)}.$$

So Eq. (29.1) can be written as

$$\Pr\{S_1 \neq 0, S_2 \neq 0, \ldots, S_{2j} \neq 0\} = \Pr\{S_{2j} = 0\} = p_{2j}^{(0)}, \tag{29.2}$$

which is the important result we will soon use. We are now ready to complete the proof. Note that the probability that the last time there was an equal number of heads and tails within $2n$ tosses occurred at the $2k$th toss can be written as

$$\Pr\{S_{2k} = 0, S_{2k+1} \neq 0, \ldots, S_{2n} \neq 0\} = \Pr\{S_{2k+1} \neq 0, S_{2k+2} \neq 0, \ldots, S_{2n} \neq 0 | S_{2k} = 0\} \Pr\{S_{2k} = 0\}$$

$$= \Pr\{S_1 \neq 0, \ldots, S_{2n-2k} \neq 0\} \Pr\{S_{2k} = 0\}$$

$$= \Pr\{S_1 \neq 0, \ldots, S_{2n-2k} \neq 0\} p_{2k}^{(0)}.$$

Using the result in Eq. (29.2), we finally obtain

$$\Pr\{S_{2k} = 0, S_{2k+1} \neq 0, \ldots, S_{2n} \neq 0\} = p_{2n-2k}^{(0)} p_{2k}^{(0)},$$

as required.

29.1 Discussion

The above result can be further explored to obtain some truly intriguing and fascinating results. From **Problem 16** we know that $p_{2j}^{(0)} \sim 1/\sqrt{\pi j}$ for moderate to large j, so that

$$\Pr\{S_{2k} = 0, S_{2k+1} \neq 0, \ldots, S_{2n} \neq 0\} = p_{2n-2k}^{(0)} p_{2k}^{(0)} \approx \frac{1}{\pi \sqrt{k(n-k)}}.$$

Let T be the number of the toss where there was last an equal number of heads and tails, and let X be the fraction of the total number of tosses where this last equality occurs. Then $X = T/(2n)$ and the distribution function of X is

$$\Pr\{X \leq x\} = \Pr\{T \leq 2xn\}$$

$$= \sum_{k \leq xn} \frac{1}{\pi \sqrt{k(n-k)}}$$

$$\approx \int_0^{xn} \frac{1}{\pi \sqrt{u(n-u)}} du$$

$$= \frac{2}{\pi} \arcsin \sqrt{x}. \qquad (29.3)$$

The probability density of X is

$$f_X(x) \approx \frac{d}{dx}\left(\frac{2}{\pi}\arcsin\sqrt{x}\right) = \frac{1}{\pi\sqrt{x(1-x)}}, \quad \text{for } 0 < x < 1, \qquad (29.4)$$

when the number of the tosses $(2n)$ is large. Note that Eq. (29.4) is the beta(1/2, 1/2) distribution, which is symmetric about $x=1/2$ (see Fig. 29.2). The result in Eq. (29.3), which we shall call the *Second Arcsine Law*,[†] was first presented by the illustrious French mathematician Paul Lévy (1886–1971) for the case of Brownian motion[‡] (Lévy, 1965) (Fig. 29.1). Brownian motion can be regarded as a continuous-time analogue of a symmetric random walk. An example of the latter is the coin-tossing experiment we have considered here, where the difference between the number of heads and tails is the position of the random walk and the number of tosses is time. Roughly speaking, in the limiting case as the fluctuations of a symmetric random walk become quicker and smaller, we obtain Brownian motion.

Although it might not be apparent at first, the implications of the graph in Fig 29.2 are quite disconcerting. We see that the densities are highest near $x=0$ (start of the tosses) and near $x=1$ (end of tosses). While it is not surprising that a last equality in the number of heads and tails is very likely toward the end of the tosses, it is quite counterintuitive that such a last equality is also very likely toward the start of the tosses. Thus, if we were to toss a fair coin 1000 times, the probability that the last time there was an equal number of heads and tails happened before the 100th toss has a probability as high as approximately $(2/\pi)\arcsin\sqrt{.1} = .20$, and happened before the 500th toss has a probability of approximately .5! Thus, there is about an equal chance that in the last 500 tosses, there will be no instance where the accumulated number of heads and tails would be equal. The times at which there

---

[†] Note that arcsine here refers to the distribution function, and not to the density, of X. Also, the reader has to be careful here: what we refer to as the *Second* Arcsine Law might be called by a different number in other textbooks, since there are several arcsine laws. For more details on the arcsine distribution, see Johnson et al. (1995, Chapter 25, Sec. VII). The second arc sine law is also discussed in Feller (1968, p. 78), Lesigne (2005, p. 61), Mörters and Peres (2010, p. 136), and Durrett (2010, p. 203).

[‡] Lévy also pointed out the connection between his results for Brownian motion and corresponding ones for the coin-tossing experiment.

Figure 29.1   Paul Pierre Lévy (1886–1971). [Picture licensed under the Creative Commons Attribution-Share Alike 2.0 Germany license.]

is an equal number of heads and tails are rare. These conclusions are no doubt quite unexpected. Feller (1968, p. 79) writes

> Suppose that in a learning experiment lasting one year a child was consistently lagging except, perhaps, during the initial week. Another child was consistently ahead except, perhaps, during the last week. Would the two children be judged equal? Yet, let a group of 11 children be exposed to a similar learning experiment involving no intelligence but only chance. One among the 11 would appear as leader for all but one week, another as laggard for all but one week.

Feller's point is that common opinion would most probably judge the second student as better than the first. However, by the very nature of chance, even if the two children were equally good, one would be ahead of the other for the 51 weeks with probability approximately equal to $(2/\pi)\arcsin\sqrt{51/52} = .9$. Thus, out of 11 equally good students, we would expect approximately one to be ahead for all but the last week, as we would expect approximately one to be lagging for all but the last week.

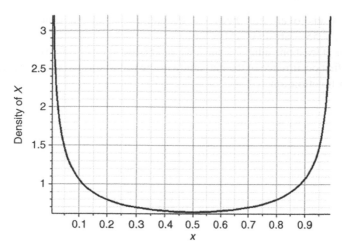

Figure 29.2   The beta(1/2, 1/2) density of x, the fraction of the total time when there was last an equal number of heads and tails.

The *First Arcsine Law*[†] was actually given by Lévy in a seminal paper (Lévy, 1939) almost 25 years prior to the second law. In it, Lévy proved a result that essentially implies that, if Y is the fraction of time that the random walk is positive (i.e., there is an excess of heads over tails), then the distribution function of Y is

$$\Pr\{Y \le y\} \approx \frac{2}{\pi}\arcsin\sqrt{y}, \quad \text{for } 0 < y < 1, \tag{29.5}$$

where the total number of steps made by the random walk is large. Equation (29.5) is the *First Arcsine Law*, and was proved for the general case of the sum of independent random variables with mean zero and variance one by Erdös and Kac (1947). The *First Arcsine Law* is as counterintuitive as the second one. For we might expect that for approximately half of the time, the random walk will be positive (i.e., there will more heads than tails) and for the other half it will be negative. That is, we would expect values of y near half to be the most likely. However, a look at Fig. 29.2 reveals the contrary to be the case. It is actually values of y close to 0 and 1 that are most likely. What this implies is intriguing but true: with high probability, the proportion of time the random walk will be positive is either very low or very high (but not close to half). Thus, if a fair coin is tossed 1000 times, the probability that there will be an excess of heads over tails for less than 10% of the time or more than 90% of the time is as high as approximately $(2/\pi)\arcsin\sqrt{.1} = .20$. On the other hand, the probability that there will be an

---

[†] The First Arcsine Law is also discussed in Feller (1968, p. 82), Lesigne (2005, p. 60), Mörters and Peres (2010, p. 136), Tijms (2007, p. 25), and Durrett (2010, p. 204).

excess of heads over tails between 45 and 55% of the time is only about $(2/\pi)(\arcsin\sqrt{.55} - \arcsin\sqrt{.45}) \approx .06$!

There is at least one other arcsine law,[†] again due to Lévy (1965). If $Z \in [0, \ 1]$ is the time of first occurrence of the maximum of the random walk, then Z has an arcsine distribution function too.

[†] See, for example, Mörters and Peres (2010, pp. 136–137).

Problem 30

Simpson's Paradox (1951)

Problem. *Consider the three events X, Y, and Z. Suppose*

$$\Pr\{X|YZ\} > \Pr\{X|\bar{Y}Z\} \quad \text{and} \quad \Pr\{X|Y\bar{Z}\} > \Pr\{X|\bar{Y}\bar{Z}\}.$$

Prove that it is still possible to have

$$\Pr\{X|Y\} < \Pr\{X|\bar{Y}\}.$$

Solution. Using the law of total probability, we have

$$\Pr\{X|Y\} = \Pr\{X|YZ\}\Pr\{Z|Y\} + \Pr\{X|Y\bar{Z}\}\Pr\{\bar{Z}|Y\}$$
$$= s\Pr\{X|YZ\} + (1-s)\Pr\{X|Y\bar{Z}\}, \tag{30.1}$$

$$\Pr\{X|\bar{Y}\} = \Pr\{X|\bar{Y}Z\}\Pr\{Z|\bar{Y}\} + \Pr\{X|\bar{Y}\bar{Z}\}\Pr\{\bar{Z}|\bar{Y}\}$$
$$= t\Pr\{X|\bar{Y}Z\} + (1-t)\Pr\{X|\bar{Y}\bar{Z}\}, \tag{30.2}$$

where $s = \Pr\{Z|Y\}$ and $t = \Pr\{Z|\bar{Y}\}$. Therefore,

$$\Pr\{X|Y\} - \Pr\{X|\bar{Y}\} = [s\Pr\{X|YZ\} - t\Pr\{X|\bar{Y}Z\}]$$
$$+ [(1-s)\Pr\{X|Y\bar{Z}\} - (1-t)\Pr\{X|\bar{Y}\bar{Z}\}] \tag{30.3}$$

We now consider the sign of Eq. (30.3). Let $t = s + \delta(-1 \leq \delta \leq 1)$, $u \equiv \Pr\{X|YZ\} - \Pr\{X|\bar{Y}Z\} \geq 0$, and $v \equiv \Pr\{X|Y\bar{Z}\} - \Pr\{X|\bar{Y}\bar{Z}\} \geq 0$. Then

$$\Pr\{X|Y\} - \Pr\{X|\bar{Y}\} = [s\Pr\{X|YZ\} - s\Pr\{X|\bar{Y}Z\} - \delta\Pr\{X|\bar{Y}Z\}]$$
$$+ [(1-s)\Pr\{X|Y\bar{Z}\} - (1-s)\Pr\{X|\bar{Y}\bar{Z}\}] + \delta\Pr\{X|\bar{Y}\bar{Z}\}$$
$$= su + (1-s)v - \delta w,$$

where $w \equiv \Pr\{X|\bar{Y}Z\} - \Pr\{X|\bar{Y}\bar{Z}\}$. Therefore, $\Pr\{X|Y\} - \Pr\{X|\bar{Y}\}$ is negative if $su + (1-s)v < \delta w$.

Classic Problems of Probability, Prakash Gorroochurn.
© 2012 John Wiley & Sons, Inc. Published 2012 by John Wiley & Sons, Inc.

30.1   Discussion

The algebra can mask the real implication of the three inequalities given in **Problem 30**. Consider the following example.[†] In a given University 1, 200 of 1000 males, and 150 of 1000 females, study economics. In another University 2, 30 of 100 males, and 1000 of 4000 females, study economics. Thus, in each university, more males than females study economics (1: 20% vs. 15%, 2: 30% vs. 25%). However, when the universities are combined, 230 of 1100 males (20.9%), and 1150 of 5000 females (23.0%), study economics. It now appears that, overall, more females than males study economics! If we define the event X that a student studies economics, the event Y that a student is male, and the event Z that a student goes to University 1, we obtain the counterintuitive set of inequalities in **Problem 30**. This is the essence of *Simpson's Paradox*[‡]: a reversal of the direction of association between two variables (gender and study economics) when a third variable[§] (university) is controlled for.

Intuitively, why does *Simpson's Paradox* occur? First note that University 2 has a higher study rate in economics for both males and females, compared to University 1 (see Table 30.1). However, out of the total 1100 males, only about 9% (i.e., 100) go to University 2. On the other hand, out of the total 5000 females, 80% (i.e., 4000) go to University 2. Thus, the university that has the higher study rate in economics, namely University 2, takes many more females than males, relatively speaking. No wonder when looking at the combined data we get the impression that a higher proportion of females than males study economics!

Simpson's Paradox shows the importance of carefully identifying the third variable(s) before an analysis involving two variables on aggregated data is carried out. In our example, when university is not controlled for, we get the wrong impression that more females than males study economics. Furthermore, from Eqs. (30.1) and (30.2), we observe that if $s = t$ then $\Pr\{X|Y\} > \Pr\{X|\bar{Y}\}$, that is, if gender is independent of university (i.e., there is no gender differences across universities), Simpson's Paradox is avoided. The tables for each university are then said to be collapsible.

An interesting alternative demonstration of *Simpson's Paradox* can be obtained through a graphical approach.[**] Consider the data in Table 30.2.

[†] Taken from Haigh (2002, p. 40), courtesy of Springer.

[‡] First named by Blyth (1972). Simpson's paradox is also discussed in Dong (1998), Rumsey (2009, p. 236), Rabinowitz (2004, p. 57), Albert et al. (2005, p. 175), Lindley (2006, p. 199), Kvam and Vidakovic (2007, p. 172), Chernick and Friis (2003, p. 239), Agresti (2007, p. 51), Christensen (1997, p. 70), and Pearl (2000, p. 174).

[§] Also sometimes called the lurking variable.

[**] See Kocik (2001), and Alsina and Nelsen (2009, pp. 33–34).

Table 30.1   Illustration of Simpson's Paradox Using University Data

| | University 1 | | | | University 2 | | | | Pooled | | | |
	E	$\bar{E}$	Total	Study rate (%)	E	$\bar{E}$	Total	Study rate (%)	E	$\bar{E}$	Total	Study rate (%)
Female	150	850	1000	15	1000	3000	4000	25	1150	3850	5000	23.0
Male	200	800	1000	20	30	70	100	30	230	370	1100	20.9

We wish to show that, given

$$\frac{a}{b} < \frac{A}{B} \quad \text{and} \quad \frac{c}{d} < \frac{C}{D}, \tag{30.4a}$$

it is still possible to have

$$\frac{a+c}{b+d} > \frac{A+C}{B+D}. \tag{30.4b}$$

We represent the proportions of the different students who study economics by vectors on a Cartesian plane such that the proportions are equal to the slopes of the corresponding lines (see Fig. 30.1). For example, the proportion of females who study economics in University 1 is a/b; we represent this proportion by the vector joining $(0, 0)$ and (b, a). Since $a/b < A/B$, the segment joining $(0, 0)$ and (b, a) has a smaller slope than the segment joining $(0, 0)$ and (B, A). Similarly for $c/d < C/D$. By addition of vectors, we see that it is possible for the slope joining $(0, 0)$ and $(b + d, a + c)$ to be larger than the slope joining $(0, 0)$ and $(B + D, A + C)$, that is, it is possible to have $(a + c)/(b + d) > (A + C)/(B + D)$.

A natural question remains: how should we combine the data from the two universities in order to obtain "correct" economics study rates for females and males? Clearly, just adding the numbers in each university is not appropriate since it gives the two proportions $(a + c)/(b + d)$ and $(A + C)/(B + D)$ for females and males, respectively (see Table 30.2). Following Tamhane and Dunlop (2000, p. 132), we calculate the adjusted proportion of females who study economics across the two

Table 30.2   General Distribution of Frequencies Across Universities and Gender

| | University 1 | | | University 2 | | | Pooled | | |
	E	$\bar{E}$	Total	E	$\bar{E}$	Total	E	$\bar{E}$	Total
Female	a	$b - a$	b	c	$d - c$	d	$a + c$	$(b + d) - (a + c)$	$b + d$
Male	A	$B - A$	B	C	$D - C$	D	$A + C$	$(B + D) - (A + C)$	$B + D$

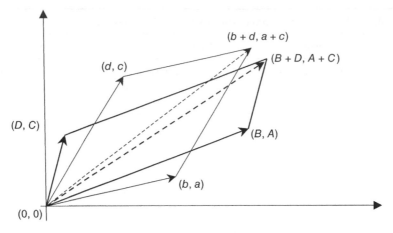

Figure 30.1   A graphical illustration of Simpson's Paradox.

universities as the sum of the weighted proportions of females in each university, where the weight is the relative size of the university:

$$\frac{a}{b}\left(\frac{b+B}{b+B+d+D}\right) + \frac{c}{d}\left(\frac{d+D}{b+B+d+D}\right) = \frac{a(b+B)/b + c(d+D)/d}{b+B+d+D}.$$

Likewise, the adjusted proportion of males who study economics across the two universities is

$$\frac{A}{B}\left(\frac{b+B}{b+B+d+D}\right) + \frac{C}{D}\left(\frac{d+D}{b+B+d+D}\right) = \frac{A(b+B)/B + C(d+D)/D}{b+B+d+D}.$$

For the data presented earlier, these formulae give 16.8% for females and 20.1% for males. We see that the directionality of the association is now preserved (i.e., a higher proportion of males than females in each university and also overall).

Simpson's Paradox is eponymously attributed to the statistician Edward H. Simpson (b. 1922) (Simpson, 1951), but a similar phenomenon was first mentioned by the eminent British statistician Karl Pearson (1857–1936) (Fig. 30.2) and his colleagues in 1899 (Pearson et al., 1899). These authors noted

> We are thus forced to the conclusion that a mixture of heterogeneous groups, each of which exhibits in itself no organic correlation, will exhibit a greater or less amount of correlation. This correlation may properly be called spurious, yet as it is almost impossible to guarantee the absolute homogeneity of any community, our results for correlation are always liable to an error, the amount of which cannot be foretold. To those who persist in looking upon all correlation as cause and effect, the fact that correlation can be produced between two quite uncorrelated characters A and B by taking an artificial mixture of two closely allied races, must come rather as a shock.

Figure 30.2   Karl Pearson (1857–1936).

Four years later, the renowned British statistician George Udny Yule (1871–1951) (Fig. 30.3), who had previously been Pearson's assistant, further delineated the problem (Yule, 1903).[†] Yule considers three attributes *A*, *B*, and *C*, such that *A* and *B* are independent of each other, given *C* or *C̄*. Yule then proves that

> . . .there will be apparent association between A and B in the universe at large unless either A or B is independent of C.

The last two examples do not really conform to the probabilities given in **Problem 30** because there is no *reversal* in the direction of the association. When an actual reversal occurs, we have a strong form of *Simpson's Paradox*. On the other hand, we have a weak form of the paradox when the three expressions hold simultaneously:

$$\Pr\{X|YZ\} = \Pr\{X|\bar{Y}Z\},$$
$$\Pr\{X|Y\bar{Z}\} = \Pr\{X|\bar{Y}\bar{Z}\},$$
$$\Pr\{X|Y\} < \Pr\{X|\bar{Y}\}.$$

(The direction of the last inequality could, of course, be reversed.)

[†] Simpson's paradox is also sometimes called Yule's paradox or Yule–Simpson's paradox.

Figure 30.3 George Udny Yule (1871–1951).

The first actual instance of the strong form of *Simpson's Paradox* was demonstrated in Cohen and Nagel's acclaimed *An Introduction to Logic and Scientific Method* (Cohen and Nagel, 1934, p. 449). These authors present tables for mortality rates from tuberculosis in Richmond and New York City in 1910 as an exercise in their book (see Table 30.3).

They then write

Notice that the death rate for Whites and that for Negroes were *lower* in Richmond than in New York, although the *total* death rate was *higher*. Are the two populations compared really *comparable*, that is, homogeneous?

Simpson's own 1951 paper was partly motivated by a set of 2 × 2 tables presented in Kendall's *Advanced Theory of Statistics* (Kendall, 1945, p. 317). In the first of these tables, Kendall displayed the frequencies for two attributes in a population, and showed that they were independent. The author then splits this 2 × 2 table into two 2 × 2 tables, one for males and one for females. Kendall then shows the two attributes are

Table 30.3 Rates from Tuberculosis in Richmond and New York City in 1910 as used by Cohen and Nagel (1934, p. 449)

	Population		Deaths		Death rate per 100,000	
	New York	Richmond	New York	Richmond	New York	Richmond
White	4,675,174	80,895	8,365	131	179	162
Colored	91,709	46,733	513	155	560	332
Total	4,766,883	127,628	8,881	286	187	226

Table 30.4 2 × 2 Tables Considered by Simpson (1951)

	Male		Female	
	Untreated	Treated	Untreated	Treated
Alive	4/52	8/52	2/52	12/52
Dead	3/52	5/52	3/52	15/52

positively associated in the male subpopulation and negatively associated in the female subpopulation. He concludes

> The apparent independence in the two together is due to the cancelling of these associations in the sub-populations.

Motivated by this example, Simpson considered the situation where two attributes are positively associated in each of two 2 × 2 tables (Simpson, 1951). He then showed that there is no net association in the aggregated 2 × 2 table (see Table. 30.4).

> This time we say that there is a positive association between treatment and survival both among males and among females; but if we combine the tables we again find that there is no association between treatment and survival in the combined population. What is the "sensible" interpretation here? The treatment can hardly be rejected as valueless to the race when it is beneficial when applied to males and to females.

We end by stressing that *Simpson's Paradox* is not a consequence of small sample size. In fact, it is a mathematical, rather than statistical, phenomenon, as can be seen from the algebraic inequalities in Eqs. (30.4a) and (30.4b).

Problem 31

Gamow, Stern, and Elevators (1958)

Problem. *A person is waiting for the elevator on the ath floor in a building and wishes to go down. The building has b (> a) floors. Assume the elevator is continually going up and down all day long. Calculate the probability that when the elevator arrives, it will be going down.*

Solution. The elevator can come down to the person from any one of the $(b - a)$ floors above or can come up from any one of the $(a - 1)$ floors below. Therefore, the probability of the elevator going down is

$$\frac{b - a}{(b - a) + (a - 1)} = \frac{b - a}{b - 1}. \tag{31.1}$$

31.1 Discussion

This intriguing little problem was adduced by the physicists George Gamow (1904–1968) (Fig. 31.1) and Marvin Stern (1935–1974), and is eponymously called the *Gamow–Stern Elevator Problem.*[†] The problem first appeared in their book *Puzzle-Math* (Gamow and Stern, 1958). As the story goes, Gamow had an office on the second floor of a seven-story building while Stern had one on the sixth floor. When Gamow wanted to visit Stern, he noticed that in about five times out of six the elevator stopping on his floor would be going downward. On the other hand, when Stern wanted to visit Gamow, in about five times out of six the elevator stopping on his floor would be going upward. This is somewhat counterintuitive because for Gamow the elevator was more likely to be coming down, yet for Stern the same elevator was more likely to be going up.

[†] The *Gamow–Stern Elevator Problem* is also discussed in Knuth (1969), Havil (2008, p. 82), Gardner (1978, p. 108), Weisstein (2002, p. 866), and Nahin (2008, p. 120).

Classic Problems of Probability, Prakash Gorroochurn.
© 2012 John Wiley & Sons, Inc. Published 2012 by John Wiley & Sons, Inc.

Figure 31.1   George Gamow (1904–1968).

By using a similar reasoning to the one in our solution, Gamow and Stern were able to explain their observations. For Gamow, $a = 2, b = 7$, so that the probability of "a down" is $(7 - 2)/(7 - 1) = 5/6$; for Stern, $a = 6, b = 7$, and the probability of "an up" is $1 - (7 - 6)/(7 - 1) = 5/6$. So far so good, but then Gamow and Stern made a slip. They reasoned that, even if the building had many elevators, the probability would still be 5/6 for the first elevator to stop on Gamow's floor to be going down. Likewise the probability for Stern would not change. Gamow and Stern actually considered the analogous situation of eastbound and westbound trains between Los Angeles and Chicago. In Chapter 3 of their book, they stated

> Since the track from here to Los Angeles is three times longer than that to Chicago, the chances are three to one that the train is to the west rather than to the east of you. And, if it is west of you, it will be going eastward the first time it passes. If there are many trains traveling between Chicago and California, as is actually the case, the situation will, of course, remain the same, and the first train passing our city after any given time is still most likely to be an eastbound one.

In fact, when there are many trains between Chicago and California, the situation does *not* remain the same. This fact was first pointed out in 1969 by Knuth (1969).

We now prove this. From Eq. (31.1), the probability of a downward elevator when there is one elevator is

$$P_1 = \frac{b-a}{b-1} = 1 - p, \tag{31.2}$$

where $p = (a-1)/(b-1)$.

Let us now consider the case of two elevators (see Fig. 31.2). We assume, without loss of generality, that $p \leq 1/2$, that is, the person G waiting is in the lower half of the building. There are two mutually exclusive ways in which the first elevator to reach G will be going down:

(i) both elevators are above G; this has probability $(b-a)^2/(b-1)^2$;

(ii) one elevator is below G and one is within $(a-1)$ floors above[†]; this has probability

$$(a-1)/(b-1) \times (a-1)/(b-1) \times 2 = 2(a-1)^2/(b-1)^2.$$

These two competing elevators have an even chance of reaching the ath floor first.
The probability for a downward elevator is therefore

$$
\begin{aligned}
P_2 &= \frac{(b-a)^2}{(b-1)^2} + 2\frac{(a-1)^2}{(b-1)^2} \cdot \frac{1}{2} \\
&= \frac{(b-a)^2 + (a-1)^2}{(b-1)^2} \\
&= \frac{[(b-1)-(a-1)]^2 + (a-1)^2}{(b-1)^2} \\
&= 1 - \frac{2(a-1)}{(b-1)} + 2\frac{(a-1)^2}{(b-1)^2} \\
&= 1 - 2p + 2p^2. \tag{31.3}
\end{aligned}
$$

Note that from Eqs. (31.2) and (31.3), P_1 and P_2 can be written as

$$P_1 = \frac{1}{2} + \frac{1}{2}(1 - 2p),$$

$$P_2 = \frac{1}{2} + \frac{1}{2}(1 - 2p)^2.$$

It therefore looks like the probability of a downward elevator when there are n elevators is

$$P_n = \frac{1}{2} + \frac{1}{2}(1 - 2p)^n, \quad n = 1, 2, 3, \ldots \tag{31.4}$$

[†] If the latter was not the case the two elevators could not compete with each other and the first elevator would always be coming up.

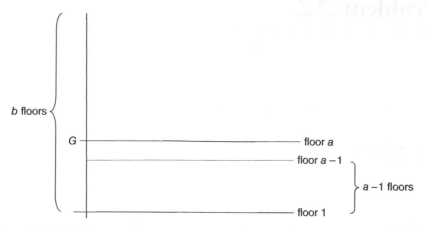

Figure 31.2   The Gamow–Stern elevator problem.

Let us prove Eq. (31.4) by induction. Clearly, the formula is correct for $n = 1$. Let us now prove that it is correct for $n + 1$ assuming it is correct for n. We have

$P_{n+1} = \Pr\{\text{first out of } (n+1)\text{ elevators is going down}\}$

$= \Pr\{\text{first out of } n \text{ elevators is going down and } (n+1)\text{th elevator is going down}\}$

$+ \dfrac{1}{2}\Pr\{\text{first out of } n \text{ elevators is going up and } (n+1)\text{th elevator is within } (a - 1)$

floors above G and going down$\}$.

The 1/2 in the above is due to the equal chance for both top and bottom elevators to reach G first (just like in the derivation of P_2 in Eq. (31.3)). Therefore,

$$P_{n+1} = P_n(1 - p) + \frac{1}{2}\{2(1 - P_n)p\}$$

$$= P_n(1 - 2p) + p$$

$$= \left\{\frac{1}{2} + \frac{1}{2}(1 - 2p)^n\right\}(1 - 2p) + p$$

$$= \frac{1}{2} + \frac{1}{2}(1 - 2p)^{n+1}.$$

Hence, we have established Eq. (31.4) by mathematical induction. Also since $|1-2p| < 1$, we have

$$\lim_{n \to \infty} P_n = \frac{1}{2}.$$

This means that, contrary to what Gamow and Stern initially believed, as there are ever more elevators, the probability of a downward elevator gets ever closer to 1/2.

Problem 32

Monty Hall, Cars, and Goats (1975)

Problem. *In a TV prize-game, there are three doors (X, Y, and Z). One of them conceals a car while the other two doors hide a goat each. A contestant chooses one of the doors, say X. The game host knows which door hides the car and opens a door, say Y, which has a goat behind it. The host then gives the contestant the options of either sticking to her original choice (i.e., X) or of switching to the other unopened door (i.e., Z). Which is the better option?*

Solution. Let C_X, C_Y, and C_Z be the events that, respectively, doors X, Y, and Z hide the car. Let H_Y be the event that the host opens door Y. Now, since the question assumes the contestant initially chooses door X and the host opens door Y, switching will be a winning strategy only if the car is behind door Z. That is, the probability of winning the car if the contestant switches is $\Pr\{C_Z \mid H_Y\}$, where by Bayes' Theorem[†]

$$
\Pr\{C_Z \mid H_Y\} = \frac{\Pr\{H_Y \mid C_Z\}\Pr\{C_Z\}}{\Pr\{H_Y \mid C_X\}\Pr\{C_X\} + \Pr\{H_Y \mid C_Y\}\Pr\{C_Y\} + \Pr\{H_Y \mid C_Z\}\Pr\{C_Z\}}
$$

$$
= \frac{(1)(1/3)}{(1/2)(1/3) + (0)(1/3) + (1)(1/3)}
$$

$$
= \frac{2}{3}.
$$

In the above calculation, we have assumed that, when the host has two possible doors to open (i.e., if the car was behind door X), she is equally likely to open any one of them. Further, since the host does not open the door that hides the car, we have $\Pr\{C_Y \mid H_Y\} = 0$, so that

$$
\Pr\{C_X \mid H_Y\} + \Pr\{C_Y \mid H_Y\} + \Pr\{C_Z \mid H_Y\} = \Pr\{C_X \mid H_Y\} + \Pr\{C_Z \mid H_Y\} = 1.
$$

[†] See **Problem 14**.

Classic Problems of Probability, Prakash Gorroochurn.
© 2012 John Wiley & Sons, Inc. Published 2012 by John Wiley & Sons, Inc.

The probability of winning the car if the contestant "sticks" is therefore

$$\Pr\{C_X \mid H_Y\} = 1 - \Pr\{C_Z \mid H_Y\} = \frac{1}{3}.$$

Thus, the contestant should always switch as she then has twice the probability of winning than if she sticks to her original choice.

32.1 Discussion

This problem, in the form presented above, was first put forward by the American biostatistician Steve Selvin in a 1975 *Letter to the Editor* to the journal *American Statistician* (Selvin, 1975a). Such has been the contentious nature of the problem that it is worth revisiting Selvin's *Letter*. In the first part of the latter, Selvin writes

It is "Let's Make a Deal" - a famous TV show starring Monty Hall.

MONTY HALL: One of the three boxes labeled A, B and C contains the keys to that new 1975 Lincoln Continental. The other two are empty. If you choose the box containing the keys, you win the car.

CONTESTANT: Gasp!

MONTY HALL: Select one of these boxes.

CONTESTANT: I'll take box B.

MONTY HALL: Now box A and box C are on the table and here is box B (contestant grips box B tightly). It is possible the car keys are in that box! I'll give you $100 for the box.

CONTESTANT: No, thank you.

MONTY HALL: How about $200?

CONTESTANT: No!

AUDIENCE: No!!

MONTY HALL: Remember that the probability of your box containing the keys to the car is 1/3 and the probability of your box being empty is 2/3. I'll give you $500.

AUDIENCE: No!!

CONTESTANT: No, I think I'll keep this box.

MONTY HALL: I'll do you a favor and open one of the remaining boxes on the table (he opens box A). It's empty! (Audience: applause). Now either box C or your box B contains the car keys. Since there are two boxes left, the probability of your box containing the keys is now 1/2. I'll give you $1000 cash for your box.

WAIT!!!!

Table 32.1   Selvin's Original Solution to the Monty-Hall Problem (Selvin, 1975a)

Keys are in box	Contestant chooses box	Monty-Hall opens bat	Contestant switches	Result
A	A	B or C	A for C or B	Loses
A	B	C	B for A	Wins
A	C	B	C for A	Wins
B	A	C	A for B	Wins
B	B	A or C	B for C or A	Loses
B	C	A	C for B	Wins
C	A	B	A for C	Wins
C	B	A	B for C	Wins
C	C	A or B	C for B or A	Loses

Is Monty right? The contestant knows that at least one of the boxes on the table is empty. He now knows that it was box A. Does this knowledge change his probability of having the box containing the keys from 1/3 to 1/2? One of the boxes on the table has to be empty. Has Monty done the contestant a favor by showing him which of the two boxes was empty? Is the probability of winning the car 1/2 or 1/3?

CONTESTANT: I'll trade you my box B for the box C on the table.

MONTY HALL: That's weird!!

HINT: The contestant knows what he is doing!

In the second part of his *Letter*, Selvin proceeds to mathematically justify the contestant's decision to switch. He presents Table 32.1 which shows that, for a given box containing the keys, switching makes the contestant win in two out of three equally likely cases.

However, only 6 months later, Selvin was back with another *Letter to the Editor*[†] (Selvin, 1975b), complaining

I have received a number of letters commenting on my "Letter to the Editor" in The American Statistician of February, 1975, entitled "A Problem in Probability." Several correspondents claim my answer is incorrect.

Selvin then provides a more technical proof that the contestant who switches doubles her chances of winning than the one who "sticks."

Let us now fast forward to 25 years later. In 1990, Craig Whitaker, a reader of *Parade Magazine* asked the very same question to columnist Marilyn vos Savant[‡]

[†] In this second Letter, the phrase "Monty Hall Problem" appears for the first time in print. The Monty Hall Problem is also discussed in Gill (2010, pp. 858–863), Clark (2007, p. 127), Rosenthal (2005, p. 216), Georgii (2008, pp. 54–56), Gardner (2001, p. 284), Krishnan (2006, p. 22), McBride (2005, p. 54), Dekking et al. (2005, p. 4), Hayes and Shubin (2004, p. 136), Benjamin and Shermer (2006, p. 75), and Kay (2006, p. 81).

[‡] In 1985, the Guinness Book of World Records gave vos Savant the world's highest IQ, with a score of 228.

Figure 32.1 Paul Erdős (1913–1996).

(b. 1946). The latter gave the same correct answer: the contestant should switch. This is when the storm really started. As a consequence of her answer, vos Savant received thousands of letters, some very acrimonious. Most responders argued that vos Savant was absolutely wrong, and that it did not matter whether the contestant switches or sticks to her original choice. The following is the argument that the proponents of the "no-switch" strategy put forward, and it does seem very convincing:

> Once the host opens a door which hides a goat, there are only two doors left. One surely hides the prize and one does not. So it would seem that it does not matter if one switches or not, since each of the two doors would apparently have the same chance (i.e. 1/2) of hiding the prize.

One can see the principle of indifference[†] at play in the above argument. Even more baffling is the fact that among those who believed this erroneous argument were many mathematicians, some of the highest caliber. Paul Erdős (1913–1996) was one of them (Fig. 32.1). Erdős was unarguably the most prolific of all mathematicians

---

[†] See pp. 135–137.

ever, having published in a whole gamut of fields including combinatorics, number theory, analysis, and probability. However, the *Monty Hall Problem* proved to be a trap for him. In his interesting biography of Paul Erdős, Hoffman (1998, p. 237) recollects the pains that mathematician Laura Vázsonyi had to go through in trying to convince Erdős on the correct solution to the problem. Hoffman writes

> "I told Erdős that the answer was to switch," said Vázsonyi, "and fully expected to move to the next subject. But Erdős, to my surprise, said, 'No, that is impossible. It should make no difference.' At this point I was sorry I brought up the problem, because it was my experience that people get excited and emotional about the answer, and I end up with an unpleasant situation. But there was no way to bow out, so I showed him the decision tree solution I used in my undergraduate Quantitative Techniques of Management course." Vázsonyi wrote out a "decision tree," not unlike the table of possible outcomes that vos Savant had written out, but this did not convince him. "It was hopeless," Vázsonyi said. "I told this to Erdős and walked away. An hour later he came back to me really irritated. 'You are not telling me why to switch,' he said. 'What is the matter with you?' I said I was sorry, but that I didn't really know why and that only the decision tree analysis convinced me. He got even more upset." Vázsonyi had seen this reaction before, in his students, but he hardly expected it from the most prolific mathematician of the twentieth century.

Let us examine the fallacy that, since there are only two doors in the final stage of the game and only one hides the car, it should not matter if we switch or not. The most intuitive argument that the probability of any one door hiding the car cannot be 1/2 was provided by vos Savant (1996) herself:

> Here's a good way to visualize what happened. Suppose there are a million doors, and you pick door number 1. Then the host, who knows what's behind the doors and will always avoid the one with the prize, opens them all except door number 777,777. You'd switch to that door pretty fast, wouldn't you?

Indeed, once the contestant makes an original choice out of d possible doors, her probability of winning is $1/d$. At the final stage, when all but one of the doors has been simultaneously opened, there are only two doors left. The initially chosen door still has a probability of $1/d$ of hiding the car while the other door now has a probability $(d-1)/d$. By switching, the contestant increases her chance of winning by a factor of $(d-1)$. If $d = 10^6$, this is an increase by a factor of 999,999.

The Monty Hall Problem can be historically traced to the *Box Problem*[†] in Bertrand's *Calculs des Probabilités* (Bertrand, 1889, p. 2). In that problem, recall the initial probability of choosing box C (containing one gold and one silver coin) is 1/3, and remains 1/3 once a silver coin is drawn from one of the boxes (the two other boxes, A and B, contain two gold and two silver coins, respectively). This is similar to the probability invariance in the Monty Hall Problem.

[†] See **Problem 18**. For an exhaustive list of the different versions of the Monty Hall Problem, see Edward Barbeau's *Mathematical Fallacies, Flaws, and Flimflam* (Barbeau, 2000, p. 86).

Figure 32.2   Martin Gardner (1914–2010). [Picture licensed under the Creative Commons Attribution-Share Alike 2.0 Germany license.]

A more recent ancestor of the Monty Hall Problem is the *Prisoner's Problem*,[†] initially posed by Gardner (1961, pp. 227–228) (Fig. 32.2):

A wonderfully confusing little problem involving three prisoners and a warden, even more difficult to state unambiguously, is now making the rounds. Three men - A, B, and C - were in separate cells under sentence of death when the governor decided to pardon one of them. He wrote their names on three slips of paper, shook the slips in a hat, drew out one of them and telephoned the warden, requesting that the name of the lucky man be kept secret for several days. Rumor of this reached prisoner A. When the warden made his morning rounds, A tried to persuade the warden to tell him who had been pardoned. The warden refused.

"Then tell me," said A, "the name of one of the others who will be executed. If B is to be pardoned, give me C's name. If C is to be pardoned, give me B's name. And if I'm to be pardoned, flip a coin to decide whether to name B or C."

"But if you see me flip the coin," replied the wary warden, "you'll know that you're the one pardoned. And if you see that I don't flip a coin, you'll know it's either you or the person I don't name."

"Then don't tell me now," said A. "Tell me tomorrow morning."

The warden, who knew nothing about probability theory, thought it over that night and decided that if he followed the procedure suggested by A, it would give A no help whatever in estimating his survival chances. So next morning he told A that B was going to be executed. After the warden left, A smiled to himself at the warden's stupidity. There were

---

[†] Not to be confused with the *Prisoner's Dilemma*, which is a Newcomb-type classic problem in decision theory (see **Problem 25**).

now only two equally probable elements in what mathematicians like to call the "sample space" of the problem. Either C would be pardoned or himself, so by all the laws of conditional probability, his chances of survival had gone up from 1/3 to 1/2. The warden did not know that A could communicate with C, in an adjacent cell, by tapping in code on a water pipe. This A proceeded to do, explaining to C exactly what he had said to the warden and what the warden had said to him. C was equally overjoyed with the news because he figured, by the same reasoning used by A, that his own survival chances had also risen to 1/2.

Did the two men reason correctly? If not, how should each have calculated his chances of being pardoned?

Of course, the two men did not reason correctly. A's probability of being pardoned remains 1/3, whilst C's probability of being pardoned is now 2/3.

Such is the charm of the *Monty Hall Problem* that Jason Rosenhouse has recently devoted a whole book to it, *The Monty Hall Problem: The Remarkable Story of Math's Most Contentious Brainteaser* (Rosenhouse, 2009). The author considers several variations of the classical Monty Hall Problem. In one of the variations[†] (p. 57),

*In a TV prize-game, there are three doors (X, Y, and Z). One of them conceals a car while the other two doors hide a goat each. A contestant chooses one of the doors, say X. The game host **does not know** which door hides the car and opens one of the other two doors. The door (say Y) happens to have a goat behind it. The host then gives the contestant the options of either sticking to her original choice (i.e., X) or of switching to the other unopened door (i.e., Z). Which is the better option?*

The key point to note in this problem is that the game host does not know which door hides the car. Thus when she picks a different door from the contestant, it is not a sure thing that it will hide a goat. This contingency therefore needs to go into the calculations. Therefore, let G_Y be the event that door Y hides a goat. Then the probability that the contestant wins if she switches is

$$
\begin{aligned}
\Pr\{C_Z \mid G_Y\} &= \frac{\Pr\{G_Y \mid C_Z\}\Pr\{C_Z\}}{\Pr\{G_Y \mid C_X\}\Pr\{C_X\} + \Pr\{G_Y \mid C_Y\}\Pr\{C_Y\} + \Pr\{G_Y \mid C_Z\}\Pr\{C_Z\}} \\
&= \frac{(1)(1/3)}{(1)(1/3) + (0)(1/3) + (1)(1/3)} \\
&= \frac{1}{2}.
\end{aligned}
$$

This time it does not matter if the contestant switches or not.

[†] See also Rosenthal (2008).

Problem 33

Parrondo's Perplexing Paradox (1996)

Problem. *Consider the following games G_1, G_2, and G_3, in each of which $1 is won if a head is obtained, otherwise $1 is lost (assuming the player starts with $0).*

G_1: A biased coin that has probability of heads .495 is tossed.

G_2: If the net gain is a multiple of three, coin A is tossed. The latter has probability of heads .095. If the net gain is not a multiple of three, coin B is tossed. The latter has probability of heads .745.

G_3: G_1 and G_2 are played in any random order.

Prove that although G_1 and G_2 each result in a net expected loss, G_3 results in a net expected gain.

Solution. For game G_1 the net expected gain is $1(.495) - $1(.505) = -$.01$.

For game G_2, let $\{X_n\}$ denote the net gain of the player after n tosses. Then, given all the prior values of the net gain, the value of $\{X_n\}$ depends only on $\{X_{n-1}\}$, that is, $\{X_n\}$ is a Markov chain.[†] Moreover, $\tilde{X}_n = X_n \bmod 3$[‡] is also a Markov chain with states $\{0, 1, 2\}$ and transition probability matrix

$$\mathbf{P}_2 = \begin{matrix} & \begin{matrix} 0 & \quad 1 & \quad 2 \end{matrix} \\ \begin{matrix} 0 \\ 1 \\ 2 \end{matrix} & \begin{bmatrix} 0 & .095 & .905 \\ .255 & 0 & .745 \\ .745 & .255 & 0 \end{bmatrix} \end{matrix}.$$

[†] A Markov chain is a random process $\{W_t, t = 1, 2, \ldots\}$ such that $\Pr\{W_{t+1} = w_{t+1} | W_t = w_t, W_{t-1} = w_{t-1}, \ldots, W_1 = w_1\} = \Pr\{W_{t+1} = w_{t+1} | W_t = w_t\}$, that is, the probability of the "future, given the present and the past, depends only on the present." Markov chains are discussed in more details in Lawler (2006), Parzen (1962), Tijms (2003), Grimmet and Stirzaker (2001), and Karlin and Taylor (1975).

[‡] That is, $\tilde{X}_n$ is the remainder when X_n is divided by 3.

Classic Problems of Probability, Prakash Gorroochurn.
© 2012 John Wiley & Sons, Inc. Published 2012 by John Wiley & Sons, Inc.

After an infinitely large number of tosses, the probabilities $\pi = [\pi_0, \pi_1, \pi_2]$ that $\{\tilde{X}_n\}$ occupies each of the states 0, 1, and 2 (i.e., the *stationary distribution* of $\tilde{X}_n$) are given by

$$\pi P_2 = \pi.$$

By solving

$$.255\pi_1 + .745\pi_2 = \pi_0,$$
$$.095\pi_0 + .255\pi_2 = \pi_1,$$
$$.905\pi_0 + .745\pi_1 = \pi_2,$$

subject to $\pi_0 + \pi_1 + \pi_2 = 1$, we obtain $\pi = [.384, .154, .462]$. The expected win for a single toss is thus $.384(.095) + .154(.745) + .462(.745) = .495$ dollars, and the expected loss is .505 dollars. Therefore game G_2 results in a net expected loss of $.01.

For game G_3, we similarly define $\{Y_n\}$ to be the net gain after n tosses, and we let $\tilde{Y}_n = Y_n \bmod 3$. Then $\{\tilde{Y}_n\}$ is a Markov chain with transition matrix

$$P_3 = \frac{1}{2}(P_1 + P_2), \tag{33.1}$$

where P_1 is the transition matrix for game G_1, that is,

$$
P_1 = \begin{array}{c} \\ 0 \\ 1 \\ 2 \end{array}
\begin{array}{ccc} 0 & 1 & 2 \\ \left[\begin{array}{ccc} 0 & .495 & .505 \\ .505 & 0 & .495 \\ .495 & .505 & 0 \end{array}\right]. \end{array}
$$

The factor 1/2 in Eq. (33.1) takes into account that, in game G_3, games G_1 and G_2 are played at random. Equation (33.1) becomes

$$
P_3 = \begin{array}{c} \\ 0 \\ 1 \\ 2 \end{array}
\begin{array}{ccc} 0 & 1 & 2 \\ \left[\begin{array}{ccc} 0 & .295 & .705 \\ .380 & 0 & .620 \\ .620 & .380 & 0 \end{array}\right]. \end{array}
$$

The stationary distribution of $\{\tilde{Y}_n\}$ is $v = [v_0, v_1, v_2]$, where

$$v P_3 = v.$$

Solving like before, we obtain $v = [.345, .254, .401]$. The expected win for a single toss is thus $.345(.295) + .254(.620) + .401(.620) = .508$ dollars, and the expected loss is .492 dollars. Therefore game G_3 results in a net expected *gain* of $.016.

33.1   Discussion

This problem is truly counterintuitive because, by randomly playing two losing games, the player comes out a winner! This is, in essence, what the Spanish physicist Juan Manuel Rodriguez Parrondo (b. 1964) (Fig. 33.2) discovered in 1996 and presented at a workshop in Torino, Italy (Parrondo, 1996).[†] The result was then first published by Harmer and Abbott in 1999 in the journal *Statistical Science* and was called *Parrondo's Paradox.*[‡] Following Parrondo's original formulation, Harmer and Abbot used the following parameters for the games:

$$\left.\begin{array}{l} G_1 : \Pr\{\text{heads}\} = .5 - \varepsilon, \ \Pr\{\text{tails}\} = .5 + \varepsilon. \\ G_2 : \text{If net gain is a multiple of } 3, \text{toss coin A}: \Pr\{\text{heads}\} = .1 - \varepsilon, \ \Pr\{\text{tails}\} = .9 + \varepsilon. \\ \quad \text{Otherwise, toss coin B}: \ \Pr\{\text{heads}\} = .75 - \varepsilon, \Pr\{\text{tails}\} = .25 + \varepsilon. \\ G_3 : \text{Randomly play } G_1 \text{ and } G_2. \end{array}\right\}$$

$$(33.2)$$

In the above, ε is assumed to be a small positive number. Using the same reasoning as the one in our solution above, the net expected gain (or value) for each of the games can be shown to be (e.g., see Epstein, 2009, pp. 74–76)

$$\mathscr{E} G_1 = -2\varepsilon,$$
$$\mathscr{E} G_2 = -1.740\varepsilon + O(\varepsilon^2),$$
$$\mathscr{E} G_3 = .0254 - 1.937\varepsilon + O(\varepsilon^2).$$

Substituting $\varepsilon = .005$, we get the same values as in our solution. Parrondo has also run simulations[§] to show how switching between two disadvantageous games can be advantageous (see Fig. 33.1).

Dworsky (2008, pp. 221–224) provides a simpler demonstration of *Parrondo's Paradox*, one that does not use Markov chains. Consider the playing board in Fig. 33.3. A player moves along black and white sections of a playing board, starting on the center. Each time two dice are thrown, and the player moves forward or backward depending on the total score. When the player reaches the rightmost black section, she wins; when she reaches the leftmost black section, she loses. In game X, if the player is in a black section, then she moves forward if the total score is 11, and backward if it is 2, 4, or 12. On the other hand, if the player is in a white section, then she moves forward if the total score is 7 or 11, and backward if it is 2, 3, or 12. If none of these scores are obtained, the player stays put. This is shown in Table 33.1(a) This table also shows the rules for game Y. Finally, in game Z, the player randomly switches between games X and Y. In Table 33.1(b), the number of ways the scores can be obtained is shown. For

---

[†] See Harmer and Abbott (2002).

[‡] Parrondo's paradox is also discussed in Epstein (2009, Chapter 4), Havil (2007, Chapter 11), Clark (2007, p. 155), Dworsky (2008, Chapter 13), Mitzenmacher and Upfal (2005, p. 177), and Darling (2004, p. 324).

[§] See http://seneca.fis.ucm.es/parr/.

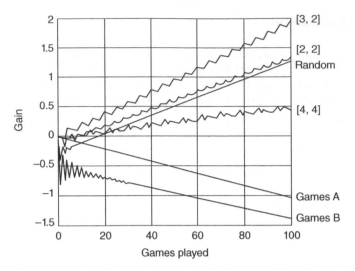

Figure 33.1   Simulations based on 5000 runs showing Parrondo's Paradox. Games A and B refer to games G_1 ad G_2 in the text. $[a, b]$ refers to G_1 and G_2 being played successively a and b times. The value of ε is taken to be .005. (Figure courtesy of Juan Parrondo.)

example, if the player is in a white section in game X, the scores 7 or 11 (first entry in Table 33.1(a)) can be obtained through (1, 6), (2, 5), (3, 4), (4, 3), (5, 2), (5, 6), (6, 1), (6, 5), a total of 8 ways (first entry in Table 33.1(b)).

Now for any one game, the probability of winning is the number of ways of moving to the right divided by the total number of ways of moving to the right or left.

Figure 33.2   Juan Parrondo (b. 1964) (Photo courtesy of Juan Parrondo.)

Figure 33.3 Playing board for Parrondo's game.

Therefore,

$$\Pr\{\text{winning game } X\} = \frac{(8)(2)}{(8)(2) + (4)(5)} = .444,$$

and

$$\Pr\{\text{winning game } Y\} = \frac{(2)(8)}{(2)(8) + (5)(4)} = .444.$$

Now, consider game Z where the player starts with any one game and randomly switches between games X and Y. Then the average number of forward moves is

$$\frac{(8+2)}{2} \cdot \frac{(2+8)}{2} = 25,$$

and the average number of backward moves is

$$\frac{(4+5)}{2} \cdot \frac{(5+4)}{2} = 20.25.$$

Hence,

$$\Pr\{\text{winning game } Z\} = \frac{25}{25 + 20.25} = .552.$$

Thus by randomly mixing two disadvantageous games, an advantageous one is obtained.

Why does *Parrondo's Paradox* actually occur? Consider the specific example given in **Problem 33**. Note that game G_1 is only slightly biased against the player.

Table 33.1 Illustration of Parrondo's Paradox

	Game X		Game Y	
	White	Black	White	Black
(a)				
Forward	7, 11	11	11	7, 11
Backward	2, 3, 12	2, 4, 12	2, 4, 12	2, 3, 12
(b)				
Forward	8	2	2	8
Backward	4	5	5	4

Concerning game G_2, if the player's net gain is a multiple of 3, then she tosses coin A that is heavily biased against her. On the other hand if the net gain is not a multiple of 3, then she tosses coin B that is less heavily biased in her favor. Because of the heavier bias of coin A, G_2 is still biased against the player. Now, when the games are mixed, the player gets to play game G_1 sometimes. Since G_1 is almost fair, the heavy bias of coin A against the player is reduced. Similarly, the bias of coin B in favor of the player becomes less. However, because coin B is tossed more often than coin A, the reduction in the bias of coin B is less than that of coin A, so that the net positive bias of B now overcomes the net negative bias of coin A. Thus, when the games are switched, the game becomes favorable to the player.

Parrondo's Paradox is regarded by many as a truly path-breaking phenomenon. For example, Epstein (2009, p. 74) states it is potentially

...the most significant advance in game-theoretic principles since the minimax process.

Pickover (2009, p. 500) adds

Science writer Sandra Blakeslee writes that Parrondo "discovered what appears to be a new law of nature that may help explain, among other things, how life arose out of a primordial soup, why President Clinton's popularity rose after he was caught in a sex scandal, and why investing in losing stocks can sometimes lead to greater capital gains." The mind-boggling paradox has applications ranging from population dynamics to the assessment of financial risk.

However, some of these claims seem to be exaggerated. After all, the winning strategy in *Parrondo's Paradox* is obtained by a rather contrived set of rules. Iyengar and Kohli (2004) admonish

One view that has caught the imagination of many people is that Parrondo's paradox suggests the possibility of making money by investing in losing stocks... Even if stock markets were exactly modeled by the types of games considered by Parrondo, the strategy to use would be the obvious one suggested by the above analysis: sell stock that has particularly poor prospects and buy stock with better prospects; this way you will always do better than if you randomize over losing stocks.

They further add

Parrondo's paradox is suggested to possibly operate in economics or social dynamics to extract benefits from ostensibly detrimental situations. Harmer and Abbott [1] suggest the example of declining birth and death rates. Each by itself has a "negative" effect on a society or an ecosystem, but declines in both together might combine with "favorable consequences" ... there is no reason, as far as we know, to believe that one or both of the rates of births and deaths have the structure of a Markov chain or that nature can somehow randomize between births and deaths, to create a realization of Parrondo's paradox.

Bibliography

Aczel, A. D., 1996 *Fermat's Last Theorem: Unlocking the Secret of an Ancient Mathematical Problem*. Delta Trade Paperbacks.

Adams, W. J., 2009 *The Life and Times of the Central Limit Theorem*, 2nd edition. American Mathematical Society.

Adrain, R., 1808 Research concerning the probabilities of the error which happen in making observations. *The Analyst Math. Museum* **1**: 93–109.

Agresti, A., 2002 *Categorical Data Analysis*, 2nd edition. Blackwell, Oxford.

Agresti, A., 2007 *An Introduction to Categorical Data Analysis*, 2nd edition. Wiley-Interscience.

Aigner, M., and G. M. Ziegler, 2003 *Proofs from THE BOOK*, 3rd edition. Springer-Verlag.

Albert, J., J. Bennett, and J. J. Cochran, 2005 *Anthology of Statistics in Sports*. SIAM.

Alsina, C., and R. B. Nelsen, 2009 *When Less is More: Visualizing Basic Inequalities*. Mathematical Association of America.

Andel, J., 2001 *Mathematics of Chance*. Wiley.

André, D., 1887 Solution directe du problème résolu par M. Bertrand. C. R. Acad. *Sci.* **105**: 436–437.

Arnauld, A., and P. Nicole, 1662 *La Logique, ou l'Art de Penser*. Chez Charles Savreux, Paris.

Arnow, B. J., 1994 On Laplace's extension of the Buffon Needle Problem. *College Math. J.* **25**: 40–43.

Arntzenius, F., and D. McCarthy, 1997 The two envelope paradox and infinite expectations. *Analysis* **57**: 42–50.

Ash, R. B., 2008 *Basic Probability Theory*. Dover, New York. (Originally published by Wiley, New York, 1970).

Ayala, F. J., 2007 *Darwin's Gift to Science and Religion*. Joseph Henry Press.

Ball, W. W. R., and H. S. M. Coxeter, 1987 *Mathematical Recreations and Essays*, 13th edition. Dover, New York.

Barbeau, E., 2000 *Mathematical Fallacies, Flaws and Flimflam*. The Mathematical Association of America.

Barbier, M. E., 1860 Note sur le problème de l'aiguille et le jeu du joint couvert. *J. Math. Pures Appl.* **5**: 272–286.

Bar-Hillel, M., and A. Margalit, 1972 Newcomb's paradox revisited. *Br. J. Phil. Sci.* **23**: 295–304.

Barnett, V., 1999 *Comparative Statistical Inference*, 3rd edition. Wiley.

Barone, J., and A. Novikoff, 1978 History of the axiomatic formulation of probability from Borel to Kolmogorov: Part I. *Arch. Hist. Exact Sci.* **18**: 123–190.

Bartoszynski, R., and M. Niewiadomska-Bugaj, 2008 *Probability and Statistical Inference*, 2nd edition. Wiley.

Bayes, T., 1764 An essay towards solving a problem in the doctrine of chances. *Phil. Trans. R. Soc. Lond.* **53**: 370–418. (Reprinted in E. S. Pearson and M. G.

Classic Problems of Probability, Prakash Gorroochurn.
© 2012 John Wiley & Sons, Inc. Published 2012 by John Wiley & Sons, Inc.

Kendall (Eds.), *Studies in the History of Statistics and Probability*, Vol. 1, London: Charles Griffin, 1970, pp. 134–53).

Beckmann, P., 1971 *A History of Pi*. St. Martin's Press, New York.

Bell, E. T., 1953 *Men of Mathematics*, Vol. 1 Penguin Books.

Bellhouse, D. R., 2002 On some recently discovered manuscripts of Thomas Bayes. *Historia Math.* **29**: 383–394.

Benford, F., 1938 The law of anomalous numbers. *Proc. Am. Phil. Soc.* **78**: 551–572.

Benjamin, A., and M. Shermer, 2006 *Secrets of Mental Math*. Three Rivers Press, New York.

Berloquin, P., 1996 *150 Challenging and Instructive Puzzles*. Barnes & Noble Books. (Originally published by Bordes, Paris, 1981).

Bernoulli, D., 1738 Specimen theoriae novae de mensura sortis. Commentarii Academiae Scientiarum Imperalis Petropolitanea V 175–192 (translated and republished as "Exposition of a new theory on the measurement of risk," *Econometrica* **22** (1954): 23–36).

Bernoulli, J., 1713 *Ars Conjectandi*. Basle.

Bernstein, P., 1996 *Against The Gods: the Remarkable Story of Risk*. Wiley, New York.

Bertrand, J., 1887 Solution d'un problème. *C. R. Acad. Sci.* **105**: 369.

Bertrand, J., 1889 *Calcul des Probabilités*. Gauthier-Villars et fils, Paris.

Bewersdorff, J., 2005 *Luck, Logic and White Lies: The Mathematics of Games*. A K Peters, Massachusetts. (English translation by David Kramer).

Billingsley, P., 1995 *Probability and Measure*, 3rd edition. Wiley.

Bingham, N. H., 2010 Finite additivity versus countable additivity. *J. Electron. d'Histoire Probab. Stat.* **6**(1): 1–35.

Blachman, N. M., R. Christensen, and J. M. Utts, 1996 Comment on 'Bayesian resolution of the exchange paradox'. *Am. Stat.* **50**: 98–99.

Blackburn, S., 2001 *Ruling Passions: A Theory of Practical Reasoning*. Oxford University Press, Oxford.

Blom, G., L. Holst, and D. Sandell, 1994 *Problems and Snapshots from the World of Probability*. Springer-Verlag.

Blyth, C. R., 1972 On Simpson's paradox and the sure-thing principle. *J. Am. Stat. Assoc.* **67**: 364–365.

Borel, E., 1909a *Éléments de la Théorie des Probabilités*. Gauthier-Villars, Paris.

Borel, E., 1909b Les probabilités dénombrables et leurs applications arithmétiques. *Rend. Circ. Mat. Palermo* **27**: 247–271.

Borel, E., 1913 La mécanique statistique et l'irréversibilité. *J. Phys.* **3**: 189–196.

Borel, E., 1920 *Le Hasard*. Librairie Félix Alcan, Paris.

Borel, E., 1962 *Probabilities and Life*. Dover, New York. (Originally published by Presses Universitaires de France, 1943).

Bowler, P. J., 2007 *Monkey Trials and Gorilla Sermons: Evolution and Christianity From Darwin to Intelligent Design*. Harvard University Press.

Box, J. F., 1978 *R. A. Fisher: The Life of a Scientist*. Wiley, New York.

Boyer, C. B., 1950 Cardan and the Pascal Triangle. *Am. Math. Monthly* **57**: 387–390.

Brams, S. J., and D. M. Kolgour, 1995 The box problem: to switch or not to switch. *Math. Magazine* **68**: 27–34.

Brémaud, P., 2009 *Initiation aux Probabilités et aux Chaînes de Markov*, 2nd edition. Springer.

Bressoud, D., 2007 *A Radical Approach to Real Analysis*, 2nd edition. The Mathematical Association of America.

Brian, E., 1996 L'Objet du doute. Les articles de d'Alembert sur l'analyse des hasards dans les quatre premiers tomes de l'*Encyclopédie*. *Recherches sur Diderot et sur l'Encyclopédie* **21**: 163–178.

Broome, J., 1995 The two envelope paradox. *Analysis* **55**: 6–11.

Buffon, G., 1777 Essai d'Arithmétique Morale. *Supplement a l'Histoire Naturelle* **4**: 46–123.

Bunnin, N., and J. Yu, 2004 *The Blackwell Dictionary of Western Philosophy*. Blackwell Publishing.

Burdzy, K., 2009 *The Search for Certainty: On the Clash of Science and Philosophy of Probability*. World Scientific Press.

Burger, E. B., and M. Starbird, 2010 *The Heart of Mathematics: An Invitation to Effective Thinking*. Wiley.

Burton, D. M., 2006 *The History of Mathematics: An Introduction*, 6th edition. McGraw-Hill.

Campbell, D. T., and D. A. Kenny, 1999 *A Primer on Regression Artifacts*. The Guilford Press.

Campbell, R., and L. E. Sowden, 1985 *Paradoxes of Rationality and Cooperation Prisoner's Dilemma and Newcomb's Paradox*. The University of British Columbia Press, Vancouver.

Cantelli, F. P., 1917 Sulla probabilità come limite della frequenza. *Atti Accad. Naz. Lincei* **26**: 39–45.

Cardano, G., 1539 *Practica arithmetice, & mensurandi singularis. In qua que preter alias cõntinentur, versa pagina demonstrabit*. Io. Antonins Castellioneus medidani imprimebat, impensis Bernardini calusci, Milan (Appears as *Practica Arithmeticae Generalis Omnium Copiosissima & Utilissima*, in the 1663 ed.).

Cardano, G., 1564 *Liber de ludo aleae*. First printed in Opera Omnia, Vol. 1, 1663 edition, pp. 262–276.

Cardano, G., 1570 *Opus novum de proportionibus numerorum, motuum, ponerum, sonorum, aliarumque rerum mensurandum*. Basel, Henricpetrina.

Cardano, G., 1935 *Ma Vie*. Paris. (translated by Jean Dayre).

Casella, G., and R. L. Berger, 2001 *Statistical Inference*, 2nd edition. Brooks, Cole.

Castell, P., and D. Batens, 1994 The two envelope paradox: the infinite case. *Analysis* **54**: 46–49.

Chamaillard, E., 1921 *Le Chevalier de Méré*. G. Clouzot, Niort.

Champernowne, D. G., 1933 The construction of decimals normal in the scale of ten. *J. London Math. Soc.* **103**: 254–260.

Charalambides, C. A., 2002 *Enumerative Combinatorics*. Chapman & Hall/CRC Press.

Chatterjee, S. K., 2003 *Statistical Thought: A Perspective and History*. Oxford University Press, Oxford.

Chaumont, L., L. Mazliak, and P. Yor, 2007 Some aspects of the probabilistic work, in *Kolmogorov's Heritage in Mathematics*, edited by E. Charpentier, A. Lesne, and K. Nikolski. Springer. (Originally published in French under the title *L'héritage de Kolomogorov en mathématique* by Berlin, 2004).

Chebyshev, P. L., 1867 Des valeurs moyennes. *J. Math. Pures Appl.* **12**: 177–184.

Chernick, M. R., and R. H. Friis, 2003 *Introductory Biostatistics for the Health Sciences: Modern Applications Including Bootstrap*. Wiley.

Chernoff, H., and L. E. Moses, 1986 *Elementary Decision Theory*. Dover, New York. (Originally published by Wiley, New York, 1959).

Christensen, R., 1997 *Log-Linear Models and Logistic Regression*, 2nd edition. Springer.

Christensen, R., and J. Utts, 1992 Bayesian resolution of the "exchange paradox". *The American Statistician* **46**: 274–276.

Chuang-Chong, C., and K. Khee-Meng, 1992 *Principles and Techniques of Combinatorics*. World Scientific Publishing, Singapore.

Chuaqui, R., 1991 *Truth, Possibility and Probability: New Logical Foundations of Probability and Statistical Inference*. North-Holland, New York.

Chung, K. L., 2001 *A Course in Probability Theory*, 3rd edition. Academic Press.

Chung, K. L., and F. AitSahlia, 2003 *Elementary Probability Theory*, 4th edition. Springer.

Clark, M., 2007 *Paradoxes from A to Z*, 2nd edition. Routledge.

Clark, M., and N. Shackel, 2000 The two-envelope paradox. *Mind* **109**: 415–442.

Cohen, M. R., and E. Nagel, 1934 *An Introduction to Logic and Scientific Method*. Harcourt Brace, New York.

Comte, A. M., 1833 *Cours de Philosophie Positive (Tome II)* Bachelier, Paris.

Coolidge, J. L., 1925 *An Introduction to Mathematical Probability*. Oxford University Press, Oxford.

Coolidge, J. L., 1990 *The Mathematics of Great Amateurs*, 2nd edition. Clarendon Press, Oxford.

Coumet, E., 1965a A propos de la ruine des joueurs: un texte de Cardan. *Math. Sci. Hum.* **11**: 19–21.

Coumet, E., 1965b Le problème des partis avant Pascal. *Arch. Intern. d'Histoire Sci.* **18**: 245–272.

Coumet, E., 1970 La théorie du hasard est-elle née par hasard? *Les Ann.* **25**: 574–598.

Coumet, E., 1981 *Huygens et la France*. Librairie Philosophique J. Vrin, Paris.

Coumet, E., 2003 Auguste Comte. Le calcul des chances, aberration radicale de l'esprit mathématique. *Math. Sci. Hum.* **162**: 9–17.

Cournot, A. A., 1843 *Exposition de la Théorie des Chances et des Probabilités*. L. Hachette, Paris.

Courtebras, B., 2008 *Mathématiser le Hasard: Une Histoire du Calcul des Probabilités*. Vuibert, Paris.

Cover, T. M., 1987 Pick the largest number, in *Open Problems in Communication and Computation*, edited by T. M. Cover and B. Gopinath. Springer, New York.

Cowles, M., and Davis, C., 1982 On the origins of the .05 level of statistical significance. *American Psychologist*, **37**(5): 553–558.

Cowles, M., 2001 *Statistics in Psychology: An Historical Perspective*, 2nd edition. Lawrence Erlbaum Associates.

Crofton, M., 1885 Probability, in *The Britannica Encyclopedia: A Dictionary of Arts, Sciences, and General Literature*, 9th edition, Vol. 19, edited by T. S. Baynes. Cambridge University Press, Cambridge.

d'Alembert, J. L. R., 1754 Croix ou pile, in *Encyclopédie ou Dictionnaire Raisonné des Sciences, des Arts et des Métiers*, Vol. 4, edited by D. Diderot and J. L. R. d'Alembert. Briasson, Paris.

d'Alembert, J. L. R., 1757 Gageure, in *Encyclopédie ou Dictionnaire Raisonné des Sciences, des Arts et des Métiers*, Vol. 7, edited by D. Diderot and J. L. R. d'Alembert. Briasson, Paris.

d'Alembert, J. L. R., 1761 *Opuscules Mathématiques*, Vol. 2 David, Paris.

d'Alembert, J. L. R., 1767 *Mélanges de Littérature, d'Histoire et de Philosophie (Tome V)* Zacharie Chatelain & Fils, Amsterdam.

Dajani, K., and C. Kraaikamp, 2002 *Ergodic Theory of Numbers*. The Mathematical Association of America.

Dale, A. I., 1988 On Bayes' theorem and the inverse Bernoulli theorem. *Historia Math.* **15**: 348–360.

Dale, A. I., 2005 Thomas Bayes, *An essay towards solving a problem in the doctrine of chances*, In *Landmark Writings in Western Mathematics 1640–1940*, edited by I. Grattan-Guinness. Elsevier.

Darling, D., 2004 *The Universal Book of Mathematics: From Abracadabra to Zeno's Paradoxes*. Wiley.

DasGupta, A., 2010 *Fundamentals of Probability: A First Course*. Springer.

DasGupta, A., 2011 *Probability for Statistics and Machine Learning: Fundamentals and Advanced Topics*. Springer.

Daston, L., 1988 *Classical Probability in the Enlightment*. Princeton University Press, Princeton.

Daston, L. J., 1979 d'Alembert critique of probability theory. *Historia Math.* **6**: 259–279.

David, F. N., 1962 *Games, Gods and Gambling: The Origins and History of Probability and Statistical Ideas from the Earliest Times to the Newtonian Era*. Charles Griffin Co. Ltd., London.

David, H. A., 1995 First (?) occurrence of common terms in mathematical statistics. *Am. Stat.* **49**: 121–133.

David, H.A., 1998 First (?) occurrence of common terms in probability and statistics - a second list, with corrections. *Am. Stat.* **52**: 36–40.

Dawkins, R., 1987 *The Blind Watchmaker*. W. W. Norton & Co., New York.

Dawkins, R., 2006 *The God Delusion*. Bantam Press.

Debnath, L., 2010 *The Legacy of Leonhard Euler: A Tricentennial Tribute*. Imperial College Press, London.

de Finetti, B., 1937 Foresignt: its logical laws, its subjective sources, in *Studies in Subjective Probability*, edited by H. E. Kyburg and H. E. Smokler. Wiley (1964).

de Finetti, B., 1972 *Probability, Induction and Statistics: The Art of Guessing*. Wiley.

de Finetti, B., 1974 *Theory of Probability: A Critical Introductory Treatment*, Vol. I. Wiley.

de Moivre, A., 1711 De Mensura Sortis, seu, de Probabilitate Eventuum in Ludis a Casu Fortuito Pendentibus. *Phil. Trans.* **27**: 213–264.

de Moivre, A., 1718 *The Doctrine of Chances, or a Method of Calculating the Probability of Events in Play*. Millar, London.

de Moivre, A., 1730 *Miscellanea analytica de seriebus et quadraturis*. Touson & Watts, London.

de Moivre, A., 1738 *The Doctrine of Chances, or a Method of Calculating the Probability of Events in Play*. 2nd edition. Millar, London.

Deep, R., 2006 *Probability and Statistics with Integrated Software Routines*. Academic Press.

DeGroot, M. H., 1984 *Probability and Statistics*. Addison-Wesley.

Dekking, F. M., C. Kraaikamp, H. P. Lopuhaä, and L. E. Meester, 2005 *A Modern Introduction to Probability and Statistics*. Springer.

Deming, W. E., 1943 *Statistical Adjustment of Data*. Wliey, New York.

de Montessus, R., 1908 *Leçons Elémentaires sur le Calcul des Probabilités*. Gauthier-Villars, Paris.

de Montmort, P. R., 1708 *Essay d'Analyse sur les Jeux de Hazard*. Quillau, Paris.

de Montmort, P. R., 1713 *Essay d'Analyse sur les Jeux de Hazard*, 2nd edition. Quillau, Paris.

Devlin, K., 2008 *The Unfinished Game: Pascal, Fermat, and the Seventeenth-Century Letter That Made the World Modern*. Basic Books, New York.

Diaconis, P., and F. Mosteller, 1989 Methods for studying coincidences. *J. Am. Stat. Assoc.* **84**: 853–861.

Diderot, D., 1875 *Pensées Philosophiques, LIX*, Vol. I, edited by J. AssÉzat. Paris, Garnier Frères.

Dodgson, C. L., 1894 *Curiosa Mathematica, Part II: Pillow Problems*, 3rd edition. Macmillan & Co., London.

Dong, J., 1998 Simpson's paradox, in *Encyclopedia of Biostatistics*, edited by P. Armitage and T. Colton. Wiley, New York.

Doob, J. L., 1953 *Stochastic Processes*. Wiley, New York.

Doob, J. L., 1996 The development of rigor in mathematical probability (1900–1950). *Am. Math. Monthly* **103**: 586–595.

Dorrie, H., 1965 *100 Great Problems of Elementary Mathematics: Their History and Solutions*. Dover, New York. (Originally published in German under the title *Triumph der Mathematik*, Physica-Verlag, Würzburg, 1958).

Droesbeke, J.-J., and P. Tassi, 1990 *Histoire de la Statisque*. Presse Universitaire de France.

Dupont, P., 1979 Un joyau dans l'histoire des sciences: Le mémoire de Thomas Bayes de 1763 *Rendiconti del Seminario Matematico dell'Università Politecnica di Torino* **37**: 105–138.

Durrett, R., 2010 *Probability: Theory and Examples*, 4th edition. Cambridge University Press, Cambridge.

Dutka, J., 1988 On the St. Petersburg paradox. *Arch. Hist. Exact Sci.* **39**: 13–39.

Dworsky, L. N., 2008 *Probably Not: Future Prediction Using Probability and Statistical Inference*. Wiley.

Earman, J., 1992 *Bayes or Bust? A Critical Examination of Bayesian Confirmation Theory*. MIT Press.

Eddington, A. S., 1929 *The Nature of the Physical World*. The Macmillan Company, New York.

Edgington, E. S., 1995 *Randomization Tests*, 3rd edition. Marcel Dekker, Inc., New York.

Edwards, A. W. F., 1982 Pascal and the problem of points. *Int. Stat. Rev.* **50**: 259–266.

Edwards, A. W. F., 1983 Pascal's problem: the 'Gambler's Ruin'. *Int. Stat. Rev.* **51**: 73–79.

Edwards, A. W. F., 1986 Is the reference to Hartley (1749) to Bayesian inference? *Am. Stat.* **40**: 109–110.

Edwards, A. W. F., 2002 *Pascal's Arithmetic Triangle: The Story of a Mathematical Idea*. John Hopkins University Press. (Originally published by Charles Griffin & Company Limited, London, 1987).

Edwards, A. W. F., 2003 Pascal's work on probability in *The Cambridge Companion to Pascal*, edited by N. Hammond. Cambridge University Press, Cambridge, pp. 40–52.

Ellis, R. L., 1844 On the foundations of the theory of probabilities. *Trans. Cambridge Phil. Soc.* **8**: 1–6.

Ellis, R. L., 1850 Remarks on an alleged proof of the "Method of Least Squares", contained in a late number of the Edinburgh Review. *Phil. Mag. [3]* **37**: 321–328. (Reprinted in The Mathematical and other Writings of Robert Leslie Ellis. Cambridge University Press, Cambridge, 1863).

Elster, J., 2003 Pascal and Decision Theory, in *The Cambridge Companion to Pascal*, edited by N. Hammond. Cambridge University Press, Cambridge, pp. 53–74.

Eperson, D. B., 1933 Lewis Carroll, mathematician. *Math. Gazette* **17**: 92–100.

Epstein, R. A., 2009 *The Theory of Gambling and Statistical Logic*, 2nd edition. Elsevier.

Erdös, P., and M. Kac, 1947 On the number of positive sums of independent random variables. *Bull. Am. Math. Soc.* **53**: 1011–1020

Erickson, G. W., and J. A. Fossa, 1998 *Dictionary of Paradox*. University Press of America.

Ethier, S. N., 2010 *The Doctrine of Chances: Probabilistic Aspects of Gambling*. Springer.

Euler, L., 1751 Calcul de la probabilité dans le jeu de rencontre. *Hist. Acad. Berl.* **(1753)**: 255–270.

Evans, F. B., 1961 Pepys, Newton, and Bernoulli trials. Reader observations on recent discussions, in the series questions and answers. *Am. Stat.* **15**: 29.

Everitt, B. S., 2006 *The Cambridge Dictionary of Statistics*, 3rd edition. Cambridge University Press, Cambridge.

Everitt, B. S., 2008 *Chance Rules: An Informal Guide to Probability Risk and Statistics*, 2nd edition. Springer.

Faber, G., 1910 Uber stetigen Funktionen. *Math. Ann.* **68**: 372–443.

Feller, W., 1935 Über den zentralen Grenzwertsatz der Wahrscheinlichkeitsrechnung. *Math. Z.* **40**: 521–559.

Feller, W., 1937 Tber das Gesetz der grossen Zahlen. *Acta Scientiarum Litterarum Univ. Szeged* **8**: 191–201.

Feller, W., 1968 *An Introduction to Probability Theory and Its Applications*, Vol. I, 3rd edition. Wiley, New York.

Fine, T. A., 1973 *Theories of Probability: An Examination of Foundations*. Academic Press, New York.

Finkelstein, M. O., and B. Levin, 2001 *Statistics for Lawyers*, 2nd edition. Springer.

Fischer, H., 2010 *A History of the Central Limit Theorem: From Classical to Modern Probability Theory*. Springer, New York.

Fisher, A., 1922 *The Mathematical Theory of Probabilities*. The Macmillan Company, New York.

Fisher, R. A., 1925 *Statistical Methods for Research Workers*. Oliver and Boyd, Edinburgh.

Fisher, R. A., 1935 *The Design of Experiments*. Oliver and Boyd, Edinburgh.

Fisher, R. A., 1956 *Statistical Methods and Scientific Inference*, 2nd edition. Oliver and Boyd, Edinburgh/London.

Fitelson, B., A. Hájek, and N. Hall, 2006 Probability, in *The Philosophy of Science: An Encyclopedia*, edited by S. Sarkar and J. Pfeifer. Taylor & Francis Group, New York.

Fleiss, J. L., B. Levin, and M. C. Paik, 2003 *Statistical Methods for Rates and Proportions*, 3rd edition. Wiley, New York.

Foata, D., and A. Fuchs, 1998 *Calculs des Probabilités*, 2nd edition. Dunod, Paris.

Forrest, B., and P. R. Gross, 2004 *Creationism's Trojan Horse: The Wedge of Intelligent Design*. Oxford University Press, Oxford.

Frey, B., 2006 *Statistics Hacks*. O'Reilly.

Galavotti, M. C., 2005 *Philosophical Introduction to Probability*. CSLI Publications, Stanford.

Galilei, G., 1620 Sopra le scoperte dei dadi. *Opere, Firenze, Barbera* **8**: 591–594.

Galton, F., 1894 A plausible paradox in chances. *Nature* **49**: 365–366.

Gamow, G., and M. Stern, 1958 *Puzzle-Math*. Viking, New York.

Gani, J., 2004 Newton, Sir Isaac, in *Encyclopedia of Statistical Sciences*, 2nd edition, edited by S. Kotz, C. B. Read, N. Balakrishnan, and B. Vidakovic. Wiley.

Gardner, M., 1959 *Hexaflexagons and Other Mathematical Diversions: The First Scientific American Book of Mathematical Puzzles & Diversions*. Simon and Schuster, New York.

Gardner, M., 1961 *The Second Scientific American Book of Mathematical Puzzles and Diversions*. Simon and Schuster, New York.

Gardner, M., 1978 *Aha! Insight*. W.H. Freeman & Co., New York.

Gardner, M., 1982 *Aha! Gotcha*. W.H. Freeman & Co., New York.

Gardner, M., 1996 *The Universe in a Handkerchief: Lewis Carroll's Mathematical Recreations, Games, Puzzles, and Word Plays*. Copernicus, New York.

Gardner, M., 2001 *The Colossal Book of Mathematics*. W.W. Norton & Company.

Gauss, C. F., 1809 *Theoria motus corporum coelestium*. Perthes et Besser, Hamburg (English Translation by C.H. Davis as *Theory of Motion of the Heavenly Bodies Moving About the Sun in Conic Sections*, Little, Brown, Boston, 1857. Reprinted in 1963, Dover, New York).

Gelman, A., J. B. Carlin, H. S. Stern, and D. Rubin, 2003 *Bayesian Data Analysis*, 2nd edition. Chapman & Hall/CRC.

Georgii, H.-O., 2008 *Stochastics: Introduction to Probability and Statistics*. Walter de Gruyter, Berlin.

Gigerenzer, G., Z. Swijtink, T. Porter, L. Daston, J. Beatty, et al. 1989 *The Empire of Chance: How Probability Changed Science and Everyday Life*. Cambridge University Press, Cambridge.

Gill, R. D., 2010 Monty Hall Problem: Solution, in *International Encyclopedia of*

Statistical Science, edited by M. Lovric. Springer.

Gillies, D. A., 1987 Was Bayes a Bayesian? *Historia Math.* **14**: 325–346.

Gillies, D. A., 2000 *Philosophical Theories of Probability*. Routledge, London.

Gliozzi, M., 1980 Cardano, Girolamo, in *Dictionary of Scientific Biography*, Vol. 3, edited by C. C. Gillispie. Charles Scribner's Sons, New York.

Gnedenko, B. V., 1978 *The Theory of Probability*. Mir Publishers (4th Printing).

Good, I. J., 1983 *Good Thinking: The Foundations of Probability and Its Applications*. University of Minnesota Press, Minneapolis.

Good, I. J., 1988 Bayes's red billiard ball is also a herring, and why Bayes withheld publication. *J. Stat. Comput. Simul.* **29**: 335–340.

Gordon, H., 1997 *Discrete Probability*. Springer-Verlag, New York.

Gorroochurn, P., 2011 Errors of Probability in Historical Context. *Am. Stat.* **65**(4): 246–254.

Gouraud, C., 1848 *Histoire du Calcul des Probabilités Depuis ses Origines Jusqu'a nos Jours*. Librairie d'Auguste Durand, Paris.

Gray, A., 1908 *Lord Kelvin: An Account of his Scientific Life and Work*. J. M. Kent, London.

Greenblatt, M. H., 1965 *Mathematical Entertainments: A Collection of Illuminating Puzzles New and Old*. Thomas H. Crowell Co., New York.

Gregersen, E., 2011 *The Britannica Guide to Statistics and Probability*. Britannica Educational Publishing.

Gridgeman, N. T., 1960 Geometric probability and the number π. *Scripta Math.* **25**: 183–195.

Grimmett, G., and D. Stirzaker, 2001 *Probability and Random Processes*, 3rd edition. Oxford University Press, Oxford.

Grinstead, C. M., and J. L. Snell, 1997 *Introduction to Probability*, 2nd edition. American Mathematical Society.

Groothuis, D., 2003 *On Pascal*. Thomson Wadsworth.

Gut, A., 2005 *Probability: A Graduate Course*. Springer.

Gyóari, E., O. H. Katora, and L. Lovász, 2008 *Horizons in Combinatorics*. Springer.

Hacking, I., 1972 The logic of Pascal's Wager. *Am. Phil. Q.* **9**: 186–192.

Hacking, I., 1980a Bayes, Thomas, in *Dictionary of Scientific Biography*, Vol. 1, edited by C. C. Gillispie. Charles Scribner's Sons, New York, pp. 531–532.

Hacking, I., 1980b Moivre, Abraham de, in *Dictionary of Scientific Biography*, Vol. 9, edited by C. C. Gillispie. Charles Scribner's Sons, New York, pp. 452–555.

Hacking, I., 1980c Montmort, Pierre Rémond de, in *Dictionary of Scientific Biography*, Vol. 9, edited by C. C. Gillispie. Charles Scribner's Sons, New York, pp. 499–500.

Hacking, I., 2006 *The Emergence of Probability*, 2nd edition. Cambridge University Press, Cambridge.

Hahn, R., 2005 *Pierre Simon Laplace, 1749–1827: A Determined Scientist*. Harvard University Press, Cambridge, MA.

Haigh, J., 2002 *Probability Models*. Springer.

Hájek, A., 2008 Probability: a philosophical overview, in *Proof and Other Dilemmas: Mathematics and Philosophy*, edited by B. Gold and R. A. Simons. The Mathematical Association of America.

Hald, A., 1984 A. de Moivre: 'De Mensura Sortis' or 'On the measurement of chance'. *Int. Stat. Rev.* **52**: 229–262.

Hald, A., 1990 *A History of Probability and Statistics and Their Applications Before 1750* Wiley, New Jersey.

Halmos, P., 1991 *Problems for Mathematicians Young and Old*. The Mathematical Association of America.

Hamming, R. W., 1991 *The Art of Probability for Scientists and Engineers*. Addison-Wesley.

Hammond, N., Ed., 2003 *The Cambridge Companion to Pascal*. Cambridge University Press, Cambridge.

Harmer, G. P., and D. Abbott, 2002 A review of Parrondo's paradox. *Fluctuation Noise Lett.* **2**: R71–R107.

Hartley, D., 1749 *Observations on Man, His Fame, His Duty, and His Expectations*. Richardson, London.

Hassett, M. J., and D. G. Stewart, 2006 *Probability for Risk Management*, 2nd edition. ACTEX Publications, Inc., Winsted, Connecticut.

Hausdorff, F., 1914 *Grundziige der Mengenlehre*. Von Leit, Leipzig.

Havil, J., 2007 *Nonplussed!: Mathematical Proof of Implausible Ideas*. Princeton University Press, New Jersey.

Havil, J., 2008 *Impossible? Surprising Solutions to Counterintuitive Conundrums*. Princeton University Press, New Jersey.

Hayes, D. F., and T. Shubin, 2004 *Mathematical Adventures for Students and Amateurs*. The Mathematical Association of America.

Hein, J. L., 2002 *Discrete Mathematics*, 2nd edition. Jones and Bartlett Publishers.

Hellman, H., 2006 *Great Feuds in Mathematics: Ten of the Liveliest Disputes Ever*. Wiley.

Henry, M., 2004 La Démonstration par Jacques Bernoulli de son Théorème, in *Histoires de Probabilités et de Statistiques*, edited by E. Barbin and J.-P. Lamarche. Ellipses, Paris.

Higgins, P. M., 1998 *Mathematics for the Curious*. Oxford University Press, Oxford.

Higgins, P. M., 2008 *Number Story: From Counting to Cryptography*. Copernicus Books.

Hill, T., 1995a A statistical derivation of the significant-digit law. *Stat. Sci.* **10**: 354–363.

Hill, T., 1995b Base-invariance implies Benford's law. *Proc. Am. Math. Soc.* **123**: 887–895.

Hinkelmann, K., and O. Kempthorne, 2008 *Design and Analysis of Experiments*, Vol. I. Wiley.

Hoffman, P., 1998 *The Man Who Loved Only Numbers: The Story of Paul Erdös and the Search for Mathematical Truth*. Hyperion, New York.

Hogben, L., 1957 *Statistical Theory: The Relationship of Probability, Credibility and Error*. W.W. Norton & Co., Inc., New York.

Howie, D., 2004 *Interpreting Probability: Controversies and Development in the Early Twentieth Century*. Cambridge University Press, Cambridge.

Howson, C., and P. Urbach, 2006 *Scientific Reasoning: The Bayesian Approach*, 3rd edition. Open Court, Illinois.

Hume, D., 1748 *An Enquiry Concerning Human Understanding*, edited by P. Millican. London (2007 edition by P. Millican, Oxford University Press, Oxford).

Humphreys, K., 2010 A history and a survey of lattice path enumeration. *J. Stat. Plann. Inference* **140**: 2237–2254

Hunter, J. A. H., and J. S. Madachy, 1975 *Mathematical Diversions*. Dover, New York (Originally published by D. Van Nostrand Company, Inc., Princeton, New Jersey, 1969).

Huygens, C., 1657 *De ratiociniis in ludo aleae*. Johannis Elsevirii, Leiden (pp. 517–534 in Frans van Schooten's *Exercitationum mathematicarum liber primus continens propositionum arithmeticarum et geometricarum centuriam*).

Huygens, C., 1920 *Oeuvres Complètes de Christiaan Huygens*, Vol. 14 Martinus Nijhoff, La Haye, pp. 1655–1666

Ihe, O. C., 2009 *Markov Process for Stochastic Modeling*. Academic Press.

Ingersoll, J., 1987 *Theory of Financial Decision Making*. Rowman and Littlefield, New Jersey.

Isaac, R., 1995 *The Pleasures of Probability*. Springer-Verlag, New York.

Iyengar, R., and R. Kohli, 2004 Why Parrondo's paradox is irrelevant for utility theory, stock buying, and the emergence of life. *Complexity* **9**: 23–27.

Jackman, S., 2009 *Bayesian Analysis for the Social Sciences*. Wiley.

Jackson, F., F. Menzies, and G. Oppy, 1994 The two envelope 'paradox'. *Analysis* **54**: 43–45.

Jacobs, K., 2010 *Stochastic Processes for Physicists: Understanding Noisy Systems*. Cambridge University Press, Cambridge.

Jaynes, E. T., 1973 The well-posed problem. *Found. Phys.* **4**: 477–492.

Jaynes, E. T., 2003 *Probability Theory: The Logic of Science*. Cambridge University Press, Cambridge.

Jeffrey, R., 2004 *Subjective Probability: The Real Thing*. Cambridge University Press, Cambridge.

Jeffreys, H., 1961 *Theory of Probability*, 3rd edition. Clarendon Press, Oxford.

Johnson, N. L., A. W. Kemp, and S. Kotz, 2005 *Univariate Discrete Distributions*, 3rd edition. Wiley.

Johnson, N. L., S. Kotz, and N. Balakrishnan, 1995 *Continuous Univariate Distributions*, Vol. II, 2nd edition. Wiley.

Johnson, N. L., S. Kotz, and N. Balakrishnan, 1997 *Discrete Multivariate Distributions*. Wiley.

Johnson, P. E., 1991 *Darwin on Trial*. Inter-Varsity Press.

Jordan, J., 2006 *Pascal's Wager: Pragmatic Arguments and Belief in God*. Oxford University Press, Oxford.

Joyce, J. M., 1999 *The Foundations of Causal Decision Theory*. Cambridge University Press, Cambridge.

Kac, M., 1959 *Statistical Independence in Probability Analysis and Number Theory*. The Mathematical Association of America.

Kac, M., and S. M. Ulam, 1968 *Mathematics and Logic*. Dover, New York (Originally published by Frederick A Praeger, New York, 1963 under the title *Mathematics and Logic: Retrospect and Prospects*).

Kaplan, M., and E. Kaplan, 2006 *Chances Are: Adventures in Probability*. Penguin Books.

Kardaun, O. J. W. F., 2005 *Classical Methods of Statistics*. Springer-Verlag, Berlin.

Karlin, S., and H. M. Taylor, 1975 *A First Course in Stochastic Processes*, 2nd edition. Academic Press.

Katz, B. D., and D. Olin, 2007 A tale of two envelopes. *Mind* **116**: 903–926.

Katz, V. J., 1998 *A History of Mathematics: An Introduction*, 2nd edition. Addison-Wesley.

Kay, S. M., 2006 *Intuitive Probability and Random Processes Using MATLAB®*. Springer.

Kelly, D. G., 1994 *Introduction to Probability*, Macmillan, New York.

Kendall, M. G., 1945 *The Advanced Theory of Statistics*, Vol. 1, *2nd revised edition*. Charles Griffin & Co. Ltd, London.

Kendall, M. G., 1963 Ronald Aylmer Fisher 1890–1962 *Biometrika* **50**: 1–15.

Kendall, M. G., and P. A. P. Moran, 1963 *Geometrical Probability*. Charles Griffin & Co. Ltd, London.

Keuzenkamp, H. A., 2004 *Probability, Econometrics and Truth: The Methodology of Econometrics*. Cambridge University Press, Cambridge.

Keynes, J. M., 1921 *A Treatise on Probability*. Macmillan & Co, London.

Khintchine, A., 1924 Ueber einen Satz der Wahrscheinlichkeitsrechnung. *Fund. Math.* **6**: 9–20.

Khintchine, A., 1929 Sur la loi des grands nombres. *C. R. Acad. Sci.* **189**: 477–479.

Khoshnevisan, D., 2006 Normal numbers are normal. *Annual Report 2006*: **15**, continued 27–31, Clay Mathematics Institute.

Khoshnevisan, D., 2007 *Probability*. American Mathematical Society.

Klain, D. A., and G.-C. Rota, 1997 *Geometric Probability*. Cambridge University Press, Cambridge.

Knuth, D. E., 1969 The Gamow–Stern elevator problem. *J. Recr. Math.* **2**: 131–137.

Kocik, J., 2001 Proofs without words: Simpson paradox. *Math. Magazine* **74**: 399.

Kolmogorov, A., 1927 Sur la loi des grands nombres. *C. R. Acad. Sci. Paris* **185**: 917–919.

Kolmogorov, A., 1929 Ueber das gesetz des iterierten logarithmus. *Math. Ann.* **101**: 126–135.

Kolmogorov, A., 1930 Sur la loi forte des grands nombres. *C. R. Acad. Sci. Paris* **191**: 910–912.

Kolmogorov, A., 1933 *Grundbegriffe der Wahrscheinlichkeitsrechnung.* Springer, Berlin.

Kolmogorov, A., 1956 *Foundations of the Theory of Probability*, 2nd edition. Chlesea, New York.

Kotz, S., C. B. Read, N. Balakrishnan, and B. Vidakovic, 2004 Normal sequences (numbers), in *Encyclopedia of Statistical Sciences*, 2nd edition, edited by S. Kotz, C. B. Read, N. Balakrishnan, and B. Vidakovic. Wiley.

Kraitchik, M., 1930 *La Mathématique des Jeux ou Récréations Mathématiques.* Imprimerie Stevens Frères, Bruxelles.

Kraitchik, M., 1953 *Mathematical Recreations*, 2nd revised edition. Dover, New York (Originally published by W. W. Norton and Company, Inc., 1942).

Krishnan, V., 2006 *Probability and Random Processes*. Wiley-Interscience.

Kvam, P. H., and B. Vidakovic, 2007 *Nonparametric Statistics with Applications to Science and Engineering*. Wiley.

Kyburg Jr., H. E., 1983 *Epistemology and Inference*. University of Minnesota Press, Minneapolis.

Lange, K., 2010 *Applied Probability*, 2nd edition. Springer.

Laplace, P.-S., 1774a Mémoire de la probabilité des causes par les evénements. Mémoire

de l'Académie Royale des Sciences de Paris (savants étrangers) **Tome VI**: 621–656.

Laplace, P.-S., 1774b Mémoire sur les suites récurro-récurrentes et sur leurs usages dans la théorie des hasards. Mémoire de l'Académie Royale des Sciences de Paris (savants étrangers) **Tome VI**: 353–371.

Laplace, P.-S. 1776 Recherches sur l'intégration des équations différentielles aux différences finies and sur leur usage dans la théorie des hasards. Mémoire de l'Académie Royale des Sciences de Paris (Savants Etranger) **7**.

Laplace, P.-S., 1781 Sur les probabilités. Histoire de l'Académie Royale des Sciences, année 1778 **Tome VI**: 227–323.

Laplace, P.-S., 1810a Mémoire sur les approximations des formules qui sont fonctions de très grands nombres et sur leur application aux probabilités. Mémoire de l'Académie Royale des Sciences de Paris 353–415 (Reprinted in Oeuvres Complete de Laplace XII, pp. 301–345).

Laplace, P.-S., 1810b Supplément au Mémoire sur les approximations des formules qui sont fonctions de très grands nombres et sur leur application aux probabilités. Mémoire de l'Académie Royale des Sciences de Paris 559–565 (Reprinted in Oeuvres Complete de Laplace XII, pp. 349-353).

Laplace, P.-S., 1812 *Théorie Analytique des Probabilités.* Mme Ve Courcier, Paris.

Laplace, P.-S., 1814a *Essai Philosophique sur les Probabilités.* Courcier, Paris (6th edition, 1840, translated by F.W. Truscott, and F. L. Emory as *A Philosophical Essay on Probabilities*, 1902. Reprinted 1951 by Dover, New York).

Laplace, P.-S., 1814b *Théorie Analytique des Probabilités*, 2nd edition. Mme Ve Courcier, Paris.

Laplace, P.-S., 1820 *Théorie Analytique des Probabilités*, 3rd edition. Mme Ve Courcier, Paris.

Lawler, G. F., 2006 *Introduction to Stochastic Processes*, 2nd edition. Chapman & Hall/CRC.

Lazzarini, M., 1902 Un' applicazione del cacolo della probabilità alla ricerca sperimentale di un valore approssimato di p. *Periodico di Matematica* **4**: 140.

Legendre, A. M., 1805 *Nouvelles Méthodes Pour la Détermination des Orbites des Comètes*. Courcier, Paris.

Lehman, R. S., 1955 On confirmation and rational betting. *J. Symbolic Logic* **20**: 251–262.

Leibniz, G. W., 1768 *Opera Omnia*. Geneva.

Leibniz, G. W., 1896 *New Essays Concerning Human Understanding*. The Macmillan Company, New York. (Original work written in 1704 and published in 1765).

Leibniz, G. W., 1969 *Théodicée*. Garnier-Flammarion, Paris (Original work published in 1710).

Le Roux, J., 1906 Calcul des Probabilités, in *Encyclopédie des Sciences Mathématiques Pures et Appliquées*, Tome I, Vol. 4, edited by J. Molk. Gauthier-Villars, Paris, pp. 1–46.

Lesigne, E., 2005 *Heads or Tails: An Introduction to Limit Theorems in Probability*. American Mathematical Society (Originally published as *Pile ou Face: Une Introduction aux Théorèmes Limites du Calcul des Probabilités*, Ellipses, 2001).

Levin, B., 1981 A representation for multinomial cumulative distribution functions. *Ann. Stat.* **9**: 1123–1126

Lévy, P., 1925 *Calculs des Probabilités*. Gauthier Villars, Paris.

Lévy, P., 1939 Sur certains processus stochastiques homogènes. *Compos. Math.* **7**: 283–339.

Lévy, P., 1965 *Processus Stochastiques et Mouvement Brownien*, 2nd edition. Gauthier-Villars, Paris.

Liagre, J. B. J., 1879 *Calculs des Probabilités et Théorie des Erreurs*, 2nd edition. Librairie Polytechnique, Paris.

Lindeberg, J. W., 1922 Eine neue Herleitung des Exponentialgesetzes in der Wahrscheinlichkeitsrechnung. *Math. Z.* **15**: 211–225.

Lindley, D. V., 2006 *Understanding Uncertainty*. Wiley.

Lindsey, D. M., 2005 *The Beast In Sheep's Clothing: Exposing the Lies of Godless Human Science*. Pelican Publishing.

Linzer, E., 1994 The two envelope paradox. *Am. Math. Monthly* **101**: 417–419.

Loehr, N., 2004 Note in André's reflection principle. *Discrete Math.* **280**: 233–236.

Lukowski, P., 2011 *Paradoxes*. Springer, New York.

Lurquin, P. F., and L. Stone, 2007 *Evolution and Religious Creation Myths: How Scientists Respond*. Oxford University Press, Oxford.

Lyapunov, A. M., 1901 Nouvelle forme du théorème sur la limite de probabilité. *Mémoires de l'Académie Impériale des Sciences de St. Petersburg* **12**: 1–24.

Lyon, A., 2010 Philosophy of probability, in *Philosophies of the Sciences: A Guide*, edited by F. Allhoff. Wiley-Blackwell.

Macmahon, P. A., 1915 *Combinatory Analysis*, Vol. I. Cambridge University Press, Cambridge.

Maher, P., 1993 *Betting on Theories*. Cambridge University Press, Cambridge.

Mahoney, M. S., 1994 *The Mathematical Career of Pierre de Fermat, 1601–1665* 2nd edition. Princeton University Press.

Maistrov, L. E., 1974 *Probability Theory: A Historical Sketch*. Academic Press, New York.

Manson, N. A., 2003 *God and Design: The Teleological Argument and Modern Science*. Routledge.

Markov, A. A., 1906 Extension of the law of large numbers to magnitudes dependent on one another. *Bull. Soc. Phys. Math. Kazan.* **15**: 135–156.

Marques de Sá, J. P., 2007 *Chance The Life of Games & the Game of Life.* Springer-Verlag.

Maxwell, J. C., 1873 *A Treatise on Electricity and Magnetism.* Clarendon Press, Oxford.

Maxwell, S. E., and H. D. Delaney, 2004 *Designing Experiments and Analyzing Data: A Model Comparison Perspective,* 2nd edition. Lawrence Erlbaum, Mahwah, NJ.

McBride, G. B., 2005 *Using Statistical Methods for Water Quality Management.* Wiley-Interscience.

McGrew, T. J., D. Shier, and H. S. Silverstein, 1997 The two-envelope paradox resolved. *Analysis* **57**: 28–33.

Meester, R., 2008 *A Natural Introduction to Probability Theory,* 2nd edition. Birkhäuser, Verlag.

Mellor, D. H., 2005 *Probability: A Philosophical Introduction.* Taylor & Francis Group, New York.

Menger, K., 1934 Das unsicherheitsmoment in der wertlehre. *Zeitschrift für Nationalöcokonomie* **51**: 459–485.

Merriman, M., 1877 A list of writings relating to the method of least squares, with historical and critical notes. *Trans. Conn. Acad. Arts Sci.* **4**: 151–232.

Meusnier, N., 2004 Le Problème des Partis Avant Pacioli, in *Histoires de Probabilités et de Statistiques,* edited by E. Barbin and J.-P. Lamarche. Ellipses, Paris.

Meusnier, N., 2006 Nicolas, neveu exemplaire. *J. Electron. d'Histoire Probab. Stat.* **2**(1).

Mitzenmacher, M., and E. Upfal, 2005 *Probability and Computing: Randomized Algorithms and Probabilistic Analysis.* Cambridge University Press, Cambridge.

Montucla, J. F., 1802 *Histoire des Mathématiques (Tome III)* Henri Agasse, Paris.

Moore, J. A., 2002 *From Genesis to Genetics: The Case of Evolution and Creationism.* University of California Press.

Mörters, P., and Y. Peres, 2010 *Brownian Motion.* Cambridge University Press, Cambridge.

Mosteller, F., 1987 *Fifty Challenging Problems in Probability with Solutions.* Dover, New York. (Originally published by Addison-Wesley, MA, 1965).

Nahin, P. J., 2000 *Dueling Idiots and Other Probability Puzzlers.* Princeton University Press, NJ.

Nahin, P. J., 2008 *Digital Dice: Computer Solutions to Practical Probability Problems.* Princeton University Press, NJ.

Nalebuff, B., 1989 The other person's envelope is always greener. *J. Econ. Perspect.* **3**: 171–181.

Naus, J. I., 1968 An extension of the birthday problem. *Am. Stat.* **22**: 27–29.

Newcomb, S., 1881 Note on the frequency of use of the different digits in natural numbers. *Am. J. Math.* **4**: 39–40.

Newton, I., 1665 Annotations from Wallis, manuscript of 1665, in *The Mathematical Papers of Isaac Newton 1,* Cambridge University Press.

Neyman, J., and E. S. Pearson, 1928a On the use and interpretation of certain test criteria for purposes of statistical inference. *Part I. Biometrika* **20A**: 175–240.

Neyman, J., and E. S. Pearson, 1928b On the use and interpretation of certain test criteria for purposes of statistical inference. *Part II. Biometrika* **20A**: 263–294.

Neyman, J., and E. S. Pearson, 1933 The testing of statistical hypotheses in relation to probabilities *a priori. Proc. Cambridge Phil. Soc.* **29**: 492–510.

Nickerson, R. S., 2004 *Cognition and Chance: The Psychology of Probabilistic Reasoning.* Laurence Erlbaum Associates.

Niven, I., 1956 *Irrational Numbers.* The Mathematical Association of America.

Northrop, E., 1944 *Riddles in Math: A Book of Paradoxes.* D. Van Nostrand, New York.

Nozick, R., 1969 Newcomb's Problem and Two Principles of Choice, in *Essays in Honor of Carl G. Hempel*, edited by N. Rescher. Reidel, Dordrecht, pp. 107–133.

O'Beirne T. H., 1965 *Puzzles and Paradoxes*. Oxford University Press, Oxford.

Olofsson, P., 2007 *Probabilities: The Little Numbers That Rule Our Lives*. Wiley.

Oppy, G., 2006 *Arguing About Gods*. Cambridge University Press, Cambridge.

Ore, O., 1953 *Cardano, the Gambling Scholar*. Princeton University Press.

Ore, O., 1960 Pascal and the invention of probability theory. *Am. Math. Monthly* **67**: 409–419.

Pacioli, L., 1494 *Summa de arithmetica, geometrica, proportioni, et proportionalita*. Paganino de Paganini, Venezia.

Palmer, M., 2001 *The Question of God: An Introduction and Sourcebook*. Routledge.

Paolella, M. S., 2006 *Fundamental Probability: A Computational Approach*. Wiley.

Paolella, M. S., 2007 *Intermediate Probability: A Computational Approach*. Wiley.

Papoullis, A., 1991 *Probability, Random Variables, and Stochastic Processes*, 3rd edition. McGraw-Hill, Inc.

Parmigiani, G., and L. Y. T. Inoue, 2009 *Decision Theory: Principles and Approaches*. Wiley-Interscience.

Parrondo, J. M. R., 1996 How to cheat a bad mathematician, in *EEC HC&M Network on Complexity and Chaos* (#ERBCHRX-CT940546), ISI, Torino, Italy (unpublished).

Parzen, E., 1962 *Stochastic Processes*. SIAM.

Pascal, B., 1665 *Traité du Triangle Arithmétique, avec Quelques Autres Petits Traités sur la Même Matière*. Desprez, Paris (English translation of first part in Smith (1929), pp. 67–79).

Pascal, B., 1858 *Oeuvres Complètes de Blaise Pascal (Tome Second)* Librairie de L. Hachette et Cie, Paris.

Pascal, P., 1670 *Pensées*. Chez Guillaume Desprez, Paris.

Paty, M., 1988 d'Alembert et les probabilités, in *Sciences à l'Epoque de la Révolution Française. Recherches Historiques*, edited by R. Rashed. Wiley.

Paulos, J. A., 1988 *Innumeracy: Mathematical Illiteracy and its Consequences*. Hill and Wang, New York.

Pearl, J., 2000 *Causality: Models, Reasoning, and Inference*. Cambridge University Press, Cambridge.

Pearson, E. S.Ed., 1978 *The History of Statistics in the 17th and 18th Centuries, Against the Changing Background of Intellectual, Scientific and Religious Thought. Lectures by Karl Pearson Given at University College London During Academic Sessions 1921–1933* Griffin, London.

Pearson, K., 1900 *The Grammar of Science*, 2nd edition. Adam and Charles Black, London.

Pearson, K., 1924 Historical note on the origin of the normal curve of errors. *Biometrika* **16**: 402–404.

Pearson, K., A. Lee, and L. Bramley-Moore, 1899 Mathematical contributions to the theory of evolution: VI – genetic (reproductive) selection: inheritance of fertility in man, and of fecundity in thoroughbred racehorses. *Phil. Trans. R. Soc. Lond. A* **192**: 257–330.

Pedoe, D., 1958 *The Gentle Art of Mathematics*. Macmillan, New York.

Pennock, R. T., 2001 *Intelligent Design Creationism and its Critics*. MIT.

Pepys, S., 1866 *Diary and Correspondence of Samuel Pepys*, Vol. IV. J.P. Lippincott & Co., Philadelphia.

Petkovic, M. S., 2009 *Famous Puzzles of Great Mathematicians*. American Mathematical Society.

Phy-Olsen, A., 2011 *Evolution, Creationism and, Intelligent Design*. Greenwood.

Pichard, J.-F., 2004 Les Probabilités au Tournant du XVIIIe Siècle, in *Autour de la*

Modélisation des Probabilités, edited by M. Henry. Presses Universitaires de France, Paris.

Pickover, C. A., 2005 *A Passion for Mathematics*. Wiley.

Pickover, C. A., 2009 *The Math Book*. Sterling Publishing Co. Inc., New York.

Pillai, S. S., 1940 On normal numbers. *Proc. Indian Acad. Sci. Sect. A*, **12**: 179–194.

Pinkham, R., 1961 On the distribution of the first significant digits. *Ann. Math. Stat.* **32**: 1223–1230

Pitman, J., 1993 *Probability*. Springer.

Plackett, R. L., 1972 The discovery of the method of least squares. *Biometrika* **59**: 239–251.

Poincaré, H., 1912 *Calcul des Probabilités*, 2nd edition. Gauthier-Villars, Paris.

Poisson, S. D., 1837 *Recherches sur la Probabilité des Jugements en Matières Criminelles et Matière Civile*. Bachelier, Paris.

Pólya, G., 1920 Über den zentralen grenzwertsatz der wahrscheinlichkeitsrechnung und das momentenproblem. *Math. Z.* **8**: 171–181.

Popper, K. R., 1957 Probability magic or knowledge out of ignorance. *Dialectica* **11**: 354–374.

Porter, T. M., 1986 *The Rise of Statistical Thinking, 1820–1900* Princeton University Press, Princeton, NJ.

Posamentier, A. S., and I. Lehmann, 2004 *π: A Biography of the World's Most Mysterious Number*. Prometheus Books.

Posamentier, A. S., and I. Lehmann, 2007 *The Fabulous Fibonacci Numbers*. Prometheus Books.

Proschan, M. A., and B. Presnell, 1998 Expect the unexpected from conditional expectation. *Am. Stat.* **52**: 248–252.

Pruss, A. R., 2006 *The Principle of Sufficient Reason: A Reassessment*. Cambridge University Press, Cambridge.

Rabinowitz, L., 2004 *Elementary Probability with Applications*. A K Peters.

Raimi, R., 1976 The first digit problem. *Am. Math. Monthly* **83**: 521–538.

Ramsey, P. F., 1926 Truth and possibility, in *Studies in Subjective Probability*, edited by H. E. Kyburg and H. E. Smokler. Wiley (1964).

Renault, M., 2008 Lost (and found) in translation: André's actual method and its application to the generalized ballot problem. *Am. Math. Monthly* **115**: 358–363.

Rényi, A., 1972 *Letters on Probability*. Akadémiai Kiadó, Budapest.

Rényi, A., 2007 *Foundations of Probability*. Dover, New York. (Originally by Holden-Day, California, 1970).

Resnick, M. D., 1987 *Choices: An Introduction to Decision Theory*. University of Minnesota Press.

Richards, S. P., 1982 *A Number for Your Thoughts*. S. P. Richards.

Rieppel, O., 2011 *Evolutionary Theory and the Creation Controversy*. Springer.

Riordan, J., 1958 *An Introduction to Combinatorial Analysis*. Wiley, New York.

Rosenbaum, P. L., 2010 *Design of Observational Studies*. Springer.

Rosenhouse, J., 2009 *The Monty Hall Problem: The Remarkable Story of Math's Most Contentious Brainteaser*. Oxford University Press, Oxford.

Rosenthal, J. S., 2005 *Struck by Lightning: The Curious World of Probabilities*. Harper Collins Edition.

Rosenthal, J. S., 2006 *A First Look at Rigorous Probability Theory*, 2nd edition. World Scientific.

Rosenthal, J. S., 2008 Monty Hall, Monty Fall, Monty Crawl. *Math Horizons* **16**: 5–7.

Ross, S., 1997 *A First Course in Probability*, 5th edition. Prentice Hall, New Jersey.

Ross, S., and E. Pekoz, 2006 *A Second Course in Probability*. *Probabilitybookstore. com.*

Rumsey, D., 2009 *Statistics II for Dummies*. Wiley.

Ruse, M., 2003 *Darwin and Design: Does Evolution Have a Purpose?* Cambridge University Press, Cambridge.

Sainsbury, R. M., 2009 *Paradoxes*, 3rd edition. Cambridge University Press, Cambridge.

Salsburg, D., 2001 *The Lady Tasting Tea: How Statistics Revolutionized Science in the Twentieth Century*. W. H. Freeman & Co., New York.

Samueli, J. J., and J. C. Boudenot, 2009 *Une Histoires des Probabilités des Origines à 1900* Ellipses, Paris.

Samuelson, P. A., 1977 St. Petersburg Paradoxes: defanged, dissected, and historically described. *J. Econ. Lit.* **15**: 24–55.

Sarkar, S., and J. Pfeifer, 2006 *The Philosophy of Science: An Encyclopedia*. Routledge.

Savage, L. J., 1972 *The Foundations of Statistics*, 2nd revised edition. Dover, New York. (Revised and enlarged edition of the 1954 edition published by Wiley).

Scardovi, I., 2004 Cardano, Gerolamo, in *Encyclopedia of Statistical Sciences*, 2nd edition, edited by S. Kotz, C. B. Read, N. Balakrishnan,and B. Vidakovic. Wiley.

Schay, G., 2007 *Introduction to Probability with Statistical Applications*. Birkhauser, Boston.

Schell, E. D., 1960 Samuel Pepys, Isaac Newton, and probability. *Am. Stat.* **14**: 27–30.

Schneider, I., 2005a Abraham de Moivre, *The Doctrine of Chances* (1718, 1738, 1756), in *Landmark Writings in Western Mathematics 1640–1940*, edited by I. Grattan-Guinness. Elsevier.

Schneider, I., 2005b Jakob Bernoulli, *Ars Conjectandi* (1713), in *Landmark Writings in Western Mathematics 1640–1940*, edited by I. Grattan-Guinness. Elsevier.

Schwarz, W., 2008 *40 Puzzles and Problems in Probability and Mathematical Statistics*. Springer.

Schwarzlander, H., 2011 *Probability Concepts and Theory for Engineers*. Wiley, New York.

Scott, A. D., and M. Scott, 1997 What's in the two envelope paradox? *Analysis* **57**: 34–41.

Scott, E. C., 2004 *Evolution vs. Creationism: An Introduction*. Greenwood Press.

Scott, P. D., and M. Fasli, 2001 Benford's law: an empirical investigation and a novel explanation. CSM Technical Report 349, Department of Computer Science, University of Essex.

Seckbach, J., and R. Gordon, 2008 *Divine Action and Natural Selection: Science Faith and Evolution*. World Scientific.

Selvin, S., 1975a Letter to the Editor: A problem in probability. *Am. Stat.* **29**: 67.

Selvin, S., 1975b Letter to the Editor: On the Monty Hall Problem. *Am. Stat.* **29**: 134.

Seneta, E., 1984 Lewis Carroll as a probabilist and mathematician. *Math. Sci.* **9**: 79–94.

Seneta, E., 1993 Lewis Carroll's "Pillow Problem": On the 1993 centenary. *Stat. Sci.* **8**: 180–186.

Senn, S., 2003 *Dicing with Death: Chance, Risk and Health*. Cambridge University Press, Cambridge.

Shackel, N., 2007 Betrand's paradox and the principle of indifference. *Phil. Sci.* **74**: 150–175.

Shackel, N., 2008 Paradoxes in probability theory, in *Handbook of Probability: Theory and Applications*, edited by T. Rudas. Sage Publications.

Shafer, G., 2004 St. Petersburg paradox, in *Encyclopedia of Statistical Sciences*, 2nd edition, edited by S. Kotz, C. B. Read, N. Balakrishnan,and B. Vidakovic. Wiley.

Shafer, G., and V. Vovk, 2001 *Probability and Finance: It's Only a Game*. Wiley.

Shafer, G., and V. Vovk, 2006 The sources of Kolmogorov's *Grundbegriffe*. *Stat. Sci.* **21**: 70–98.

Sheynin, O., 1971 Newton and the classical theory of probability. *Arch. Hist. Exact Sci.* **7**: 217–243.

Sheynin, O., 2005 *Ars Conjectandi*. Translation of Chapter 4 of Bernoulli's Ars Conjectandi (1713) into English.

Sheynin, O. B., 1978 S D Poisson's work on probability. *Arch. Hist. Exact Sci.* **18**: 245–300.

Shiryaev, A. N., 1995 *Probability*, 2nd edition. Springer.

Shoesmith, E., 1986 Huygen's solution to the gambler's ruin problem. *Historia Math.* **13**: 157–164.

Simpson, E. H., 1951 The interpretation of interaction in contingency tables. *J. R. Stat. Soc. B* **13**: 238–241.

Singpurwalla, N. D., 2006 *Reliability and Risk: A Bayesian Perspective*. Wiley.

Skybreak, A., 2006 *The Science of Evolution and the Myth of Creationism: Knowing What's Real and Why It Matters*. Insight Press, Chicago.

Sloman, S., 2005 *Causal Models: How People Think about the World and its Alternatives*. Oxford University Press, Oxford.

Smith, D. E., 1929 *A Source Book in Mathematics*. McGraw-Hill Book Company, Inc., New York.

Smith, S. W., 1999 *The Scientist and Engineer's Guide to Digital Signal Processing*, 2nd edition. California Technical Publishing, San Diego, California.

Solomon, H., 1978 *Geometric Probability*. SIAM, Philadelphia.

Sorensen, R., 2003 *A Brief History of the Paradox: Philosophy and the Labyrinths of the Mind*. Oxford University Press, Oxford.

Spanos, A., 1986 *Statistical Foundations of Econometric Modelling*. Cambridge University Press, Cambridge.

Spanos, A., 2003 *Probability Theory and Statistical Inference: Econometric Modeling with Observational Data*. Cambridge University Press, Cambridge.

Stapleton, J. H., 2008 *Models for Probability and Statistical Inference: Theory and Applications*. Wiley.

Stewart, I., 2008 *Professor Stewart's Cabinet of Mathematical Curiosities*. Basic Books.

Stigler, S. M., 1980 Stigler's law of eponymy. *Trans. N. Y. Acad. Sci.* **39**: 147–158.

Stigler, S. M., 1983 Who discovered Bayes' theorem? *Am. Stat.* **37**: 290–296.

Stigler, S. M., 1986 *The History of Statistics: The Measurement of Uncertainty Before 1900* Harvard University Press, Cambridge, MA.

Stigler, S. M., 1999 *Statistics on the Table: The History of Statistical Concepts and Methods*. Harvard University Press.

Stigler, S. M., 2006 Isaac Newton as a probabilist. *Stat. Sci.* **21**: 400–403.

Stirling, J., 1730 *Methodus Differentialis*. G. Strahan, London.

Stirzaker, D., 2003 *Elementary Probability*, 2nd edition. Cambridge University Press, Cambridge.

Sutton, P., 2010 The epoch of incredulity: a response to Katz and Olin's 'A tale of two envelopes'. *Mind* **119**: 159–169.

Székely, G. J., 1986 *Paradoxes in Probability Theory and Mathematical Statistics*. Kluwer Academic Publishers.

Tabak, J., 2004 *Probability and Statistics: The Science of Uncertainty*. Facts on File, Inc.

Tamhane, A. C., and D. D. Dunlop, 2000 *Statistics and Data Analysis: From Elementary to Intermediate*. Prentice-Hall, NJ.

Tanton, J., 2005 *Encyclopedia of Mathematics*. Facts on File, Inc.

Terrell, G. R., 1999 *Mathematical Statistics: A Unified Introduction*. Springer.

Thompson, B., 2007 *The Nature of Statistical Evidence*. Springer.

Tijms, H., 2007 *Understanding Probability: Chance Rules in Everyday Life*, 2nd edition. Cambridge University Press, Cambridge.

Tijms, H. C., 2003 *A First Course in Stochastic Models*. Wiley.

Todhunter, I., 1865 *A History of the Mathematical Theory of Probability From the Time of Pascal to That of Laplace*. Macmillan, London (Reprinted by Chelsea, New York, 1949, 1965).

Todman, J. B., and P. Dugard, 2001 *Single-Case and Small-n Experimental Designs: A Practical Guide to Randomization Tests*. Laurence Erlbaum Associates.

Tulloch, P., 1878 *Pascal*. William Blackwood and Sons, London.

Uspensky, J. V., 1937 *Introduction to Mathematical Probability*. McGraw-Hill, New York.

Vakhania, N., 2009 On a probability problem of Lewis Carroll. *Bull. Georg. Natl. Acad. Sci.* **3**: 8–11.

van Fraassen, B. C., 1989 *Laws and Symmetry*. Oxford University Press.

van Tenac, C., 1847 *Album des Jeux de Hasard et de Combinaisons*. Gustave Havard, Paris.

Venn, J., 1866 *The Logic of Chance*. Macmillan, London.

von Kries, J., 1886 *Die Principien der Wahrscheinlichkeitsrechnung*. J.C.B. Mohr, Tübingen.

von Mises, R., 1928 *Wahrscheinlichkeit, Statistik und Wahrheit*. Springer, Vienna.

von Mises, R., 1939 Ueber Aufteilungs und Besetzungs-Wahrscheinlichkeiten. *Revue de la Faculté des Sciences de l'Université d'Istanbul* **4**: 145–163.

von Mises, R., 1981 *Probability, Statistics, and Truth*, 2nd revised English edition. Dover, New York. (Originally published by George Allen & Unwin Ltd., London, 1957).

von Plato, J., 2005 A. N. Kolmogorov, *Grundbegriffe der Wahrscheinlichkeitsrechnung* (1933), in *Landmark Writings in Western Mathematics 1640–1940*, edited by I. Grattan-Guinness. Elsevier.

vos Savant, M., 1996 *The Power of Logical Thinking*. St. Martin's Press.

Weaver, W., 1956 Lewis Carroll: mathematician. *Sci. Am.* **194**: 116–128.

Weaver, W., 1982 *Lady Luck: The Theory of Probability*. Dover, New York (Originally published by Anchor Books, Doubleday & Company, Inc., Garden City, New York, 1963).

Weisstein, E., 2002 *CRC Concise Encyclopedia of Mathematics*, 2nd edition. CRC Press.

Whittaker, E. T., and G. Robinson, 1924 *The Calculus of Observations: A Treatise on Numerical Mathematics*. Balckie & Son, London.

Whitworth, W. A., 1878 Arrangements of *m* things of one sort and *n* things of another sort under certain of priority. *Messenger Math.* **8**: 105–114.

Whitworth, W. A., 1901 *Choice and Chance*, 5th edition. Hafner, New York.

Wiles, A., 1995 Modular elliptic curves and Fermat's last theorem. *Ann. Math.* **141**: 443–551.

Williams, L., 2005 Cardano and the gambler's habitus. *Stud. Hist. Phil. Sci.* **36**: 23–41.

Winkler, P., 2004 *Mathematical Puzzles: A Connoisseur's Collection*. A K Peters, MA.

Woolfson, M. M., 2008 *Everyday Probability and Statistics: Health, Elections, Gambling and War*. Imperial College Press.

Yaglom, A. M., and I. M. Yaglom, 1987 *Challenging Mathematical Problems with Elementary Solutions Vol. I (Combinatorial Analysis and Probability Theory)* Dover, New York (Originally published by Holden-Day, Inc., San Francisco, 1964).

Yandell, B. H., 2002 *The Honors Class: Hilbert's Problems and Their Solvers*. A K Peters, Massachusetts.

Young, C. C., and M. A. Largent, 2007 *Evolution and Creationism: A Documentary and Reference Guide*. Greenwood Press.

Young, M., and P. K. Strode, 2009 *Why Evolution Works (and Creationism Fails)* Rutgers University Press.

Yule, G. U., 1903 Notes on the theory of association of attributes in statistics. *Biometrika* **2**: 121–134.

Zabell, S. L., 2005 *Symmetry and Its Discontents*. Cambridge University Press, Cambridge.

Photo Credits

- Figure 1.1: Wikimedia Commons (Public Domain), http://commons.wikimedia. org/wiki/File:Gerolamo_Cardano.jpg
- Figure 2.1: Wikimedia Commons (Public Domain), http://commons.wikimedia. org/wiki/File:Justus_Sustermans_-_Portrait_of_Galileo_Galilei,_1636.jpg
- Figure 4.1: Wikimedia Commons (Public Domain), http://commons.wikimedia. org/wiki/File:Blaise_pascal.jpg
- Figure 4.2: Wikimedia Commons (Public Domain), http://commons.wikimedia. org/wiki/File:Pierre_de_Fermat.jpg
- Figure 5.1: Wikimedia Commons (Public Domain), http://commons.wikimedia. org/wiki/File:Christiaan-huygens3.jpg
- Figure 6.3: Wikimedia Commons (Public Domain), http://commons.wikimedia. org/wiki/File:GodfreyKneller-IsaacNewton-1689.jpg
- Figure 7.4: ©Bernhard Berchtold
- Figure 7.5: Wikimedia Commons (Public Domain), http://commons.wikimedia. org/wiki/File:Todhunter_Isaac.jpg
- Figure 8.1: Wikimedia Commons (Public Domain), http://commons.wikimedia. org/wiki/File:Bernoulli_family_tree.png
- Figure 8.2: Wikimedia Commons (Public Domain), http://commons.wikimedia. org/wiki/File:Jakob_Bernoulli.jpg
- Figure 8.5: Wikimedia Commons (Public Domain), http://commons.wikimedia. org/wiki/File:Simeon_Poisson.jpg
- Figure 8.6: Wikimedia Commons (Public Domain), http://commons.wikimedia. org/wiki/File:Chebyshev.jpg
- Figure 8.7: Wikimedia Commons (Public Domain), http://commons.wikimedia. org/wiki/File:Andrei_Markov.jpg
- Figure 8.8: ©Trueknowledge.com
- Figure 8.9: ©Trueknowledge.com

Classic Problems of Probability, Prakash Gorroochurn.
© 2012 John Wiley & Sons, Inc. Published 2012 by John Wiley & Sons, Inc.

- Figure 8.10: Wikimedia Commons (Public Domain), http://en.wikipedia.org/wiki/File:Francesco_Paolo_Cantelli.jpg
- Figure 10.3: Wikimedia Commons (Public Domain), http://commons.wikimedia.org/wiki/File:Abraham_de_moivre.jpg
- Figure 10.6: Wikimedia Commons (Public Domain), http://commons.wikimedia.org/wiki/File:Carl_Friedrich_Gauss.jpg
- Figure 10.8: Wikimedia Commons (Public Domain), http://commons.wikimedia.org/wiki/File:Aleksandr_Lyapunov.jpg
- Figure 11.2: Wikimedia Commons (Public Domain), http://commons.wikimedia.org/wiki/File:Daniel_Bernoulli_001.jpg
- Figure 11.3: Public Domain, http://en.wikipedia.org/wiki/File:WilliamFeller.jpg
- Figure 12.1: Wikimedia Commons (Public Domain), http://commons.wikimedia.org/wiki/File:Maurice_Quentin_de_La_Tour_-_Jean_Le_Rond_d%27Alambert_-_WGA12353.jpg
- Figure 14.2: Wikimedia Commons (Public Domain), http://commons.wikimedia.org/wiki/File:David_Hume.jpg
- Figure 14.3: Wikimedia Commons (Public Domain), http://commons.wikimedia.org/wiki/File:Thomas_Bayes.gif
- Figure 14.5: Wikimedia Commons (Public Domain), http://commons.wikimedia.org/wiki/File:Pierre-Simon_Laplace.jpg
- Figure 14.8: Wikimedia Commons (Public Domain), http://commons.wikimedia.org/wiki/File:John_Venn.jpg
- Figure 14.9: Wikimedia Commons (Public Domain), http://commons.wikimedia.org/wiki/File:Richard_von_Mises.jpeg
- Figure 14.10: Wikimedia Commons (Public Domain), http://commons.wikimedia.org/wiki/File:Bruno_De_Finetti.jpg
- Figure 14.11: Wikimedia Commons (Public Domain), http://commons.wikimedia.org/wiki/File:John_Maynard_Keynes.jpg
- Figure 15.1: Wikimedia Commons (Public Domain), http://commons.wikimedia.org/wiki/File:Gottfried_Wilhelm_von_Leibniz.jpg
- Figure16.3: Wikimedia Commons (Public Domain), http://commons.wikimedia.org/wiki/File:Georges-Louis_Leclerc,_Comte_de_Buffon.jpg
- Figure 17.1: Wikimedia Commons (Public Domain), http://commons.wikimedia.org/wiki/File:Bertrand.jpg
- Figure 19.1a: Wikimedia Commons (Public Domain), http://commons.wikimedia.org/wiki/File:Bertrand1-figure_with_letters.png
- Figure 19.1b: Wikimedia Commons (Public Domain), http://commons.wikimedia.org/wiki/File:Bertrand2-figure.png
- Figure 19.1c: Wikimedia Commons (Public Domain), http://commons.wikimedia.org/wiki/File:Bertrand3-figure.png

Index

Note: A page number followed by an "f" denotes a footnote reference.

Printed and bound by CPI Group (UK) Ltd, Croydon, CR0 4YY

27/10/2024

14580345-0001